The Mathematical Resilience Book

This book is about mathematical resilience: what it is, why it is important, how learners can develop it and how teachers can teach in ways that help learners become mathematically resilient. Teaching for mathematical resilience helps mitigate mathematics anxiety, which is often hidden from view as poor behaviour, avoidance and underachievement and can have long-term implications especially for people's employment opportunities and career progression.

Written by a team of expert contributors that specialise in teaching every age of learner from primary school through to adult, the book shows that everyone can progress in mathematics if they have Mathematical Resilience and sets out practical strategies to support learners in developing this important attribute. Chapters cover:

- How to measure mathematics anxiety and resilience

- Tools for building mathematical resilience

- Building a resilient mathematical learning environment

- The power of coaching

- Working with groups, individuals and support staff

- Supporting parents and carers.

The world faces huge challenges that involve mathematical thinking, including climate change and economic injustice. This valuable text shows teachers how they can help learners to engage positively with mathematical learning and reach their full potential.

Sue Johnston-Wilder is an associate professor of Maths Education at the University of Warwick and a Coach for Mathematical Resilience. Her background is secondary maths teaching, teacher development, mentoring and coaching. She has specialised in addressing maths anxiety and has co-developed the Mathematical Resilience framework. She is a co-founder of the International Mathematical Resilience Network.

Clare Lee is a senior lecturer at The Open University teaching at master's and doctoral levels. Previously she taught mathematics at secondary schools for over 20 years. She has published many books and papers focusing on teaching mathematics effectively and allowing learners access to this important, exciting and useful subject.

"This book advocates building mathematical resilience in learners, both adult and younger, so that they will adapt to challenge, persevere and thrive. It argues that mathematical resilience underpins lifelong learning in a changing society. It provides inspiration and practical advice for teachers to transform the learning environment but it also aims to provoke change more widely in policy and attitudes towards mathematics."

Dr Jonathan Cole, *School of Mechanical & Aerospace Engineering, Queen's University Belfast*

"For far too long, the emotional side of learning mathematics has been neglected. I enthusiastically embrace the thorough exploration of mathematics resilience in this comprehensive book. It is skillfully written by a vibrant community of mathematics educators, and expertly edited, making it an invaluable resource for mathematics teachers at all levels, ranging from young learners to university students."

Dr Julie Crowley, *Munster Technological University*

"Anxiety around mathematics is a serious problem in our education system which has the effect of disenfranchising a huge proportion of the population. Lack of confidence in their own mathematical ability prevents many adults from making informed choices about their finances or health care and from progressing in their chosen career. Parents who suffer from maths anxiety may pass their fears on to their children embedding feelings of hopelessness and disempowerment. This book is essential reading for educators, parents and anyone who believes that numeracy is as essential as literacy in enabling young people to take a full and active role in society."

Lindsay Nicholson, MBE, *University of Warwick*

"Brilliant and essential. Everyone needs to read this book as a next-generation act of resilience-building. Its panoramic exploration decodes every imaginable aspect of the mathematics teaching and learning environment, within and beyond the classroom. Its optimism is reinforced by the definitive guide."

Dr Yuqian (Linda) Wang, *Durham University*

"As someone who works in mathematics support in higher education, I encounter students with significant levels of mathematics anxiety on an almost daily basis. There is no other subject in the school curriculum that engenders so much anxiety. When did you ever encounter someone who broke out into a cold sweat at the thought of having to read some history? Or some biology? Having reasonable fluency with mathematical concepts is essential for being an informed member of 21st century society – and yet many are prevented from having such fluency because of their mathematical anxiety, not because they cannot actually "do the maths". This book which brings together the current state of knowledge, not just relating to mathematics anxiety and resilience research but also, and more importantly, good practice in how to support those who are mathematically anxious and to develop mathematical resilience addresses a very important need. The present reality is that too many people have mathematics anxiety and it is essential that the education system recognises this and that appropriate teaching is put in place. However, we need to go beyond that – and this book does this – to ask why do so many people develop mathematics anxiety and how can we stop this happening in the first place. Prevention is better than cure and the final chapters of this book address how mathematics education might be reformed to reduce the number of people who develop mathematics anxiety in the first place. This book meets a significant need."

Professor Duncan Lawson, MBE, *Coventry University*

"I am delighted to see this new book addressing maths resilience. Maths anxiety is now recognised worldwide with Sue and Clare having to take much of the credit for this level of awareness.

The book has chapters written by authors /experts from around the world, from Australia to Argentina, from the USA to Kenya, some 11 countries in all, each bringing new perspectives and insights to this important educational issue.

The focus on resilience sets a positive tone, though, of course, anxiety is addressed, too. Although the academic focus is on maths, there is much that will be pertinent to other subjects and areas of study, such as the sciences. Education is not something that is offered up to the ether. It is for people, and we must factor them in. I feel that is stating the obvious, but the fact that contributions to this book are worldwide supports the suggestion that we educators need to address the issue.

I like that the chapters cover the age range from children to adults. Sadly, it is unlikely that learners outgrow their problems. Often it is quite the reverse.

So, I welcome this book and I am sure it will have a positive influence on many lives, including the parents who often feel helpless in the face of the problems of motivating their children. I hope they, too, will fell that this is an important and useful book.

I congratulate Sue and Clare for collecting together such an inspiring and international collection of expert authors. This book will make a significant contribution to a pervasive problem in education."

Steve Chinn, *Visiting Professor at the University of Derby*

"Comprehensive and evidence-based, this brand-new resource on resilience in maths intertwines critical topics like maths anxiety and growth mindset, offering a holistic perspective on nurturing resilience among learners. The approach not only enlightens but also equips educators, parents and teachers with many suggestions to transform their teaching methods. If you care at all about how young people learn maths, make sure this book is on your reading list."

Dr Sipho Morrison, *University of London*

The Mathematical Resilience Book

How Everyone Can Progress in Mathematics

Edited by Sue Johnston-Wilder and Clare Lee

LONDON AND NEW YORK

Designed cover image: © Getty Images

First published 2024
by Routledge
4 Park Square, Milton Park, Abingdon, Oxon OX14 4RN

and by Routledge
605 Third Avenue, New York, NY 10158

Routledge is an imprint of the Taylor & Francis Group, an informa business

© 2024 selection and editorial matter, Sue Johnston-Wilder and Clare Lee; individual chapters, the contributors

The right of Sue Johnston-Wilder and Clare Lee to be identified as the authors of the editorial material, and of the authors for their individual chapters, has been asserted in accordance with sections 77 and 78 of the Copyright, Designs and Patents Act 1988.

All rights reserved. No part of this book may be reprinted or reproduced or utilised in any form or by any electronic, mechanical, or other means, now known or hereafter invented, including photocopying and recording, or in any information storage or retrieval system, without permission in writing from the publishers.

Trademark notice: Product or corporate names may be trademarks or registered trademarks, and are used only for identification and explanation without intent to infringe.

British Library Cataloguing-in-Publication Data
A catalogue record for this book is available from the British Library

Library of Congress Cataloguing-in-Publication Data
Names: Johnston-Wilder, Sue, editor. | Lee, Clare S., editor.
Title: The mathematical resilience book : how everyone can progress in mathematics / edited by Sue Johnston-Wilder and Clare Lee.
Description: Milton Park, Abingdon, Oxon ; New York : Routledge, 2024. | Includes bibliographical references and index.
Identifiers: LCCN 2023059104 (print) | LCCN 2023059105 (ebook) | ISBN 9781032368924 (hbk) | ISBN 9781032368948 (pbk) | ISBN 9781003334354 (ebk)
Subjects: LCSH: Mathematics--Study and teaching. | Mathematics--Study and Teaching--Psychological aspects. | Math anxiety. | Learning, Psychology of.
Classification: LCC QA11.2 .M27735 2024 (print) | LCC QA11.2 (ebook) | DDC 510.71--dc23/eng/20240223
LC record available at https://lccn.loc.gov/2023059104
LC ebook record available at https://lccn.loc.gov/2023059105

ISBN: 978-1-032-36892-4 (hbk)
ISBN: 978-1-032-36894-8 (pbk)
ISBN: 978-1-003-33435-4 (ebk)

DOI: 10.4324/9781003334354

Typeset in Melior
by MPS Limited, Dehradun

We dedicate this book to everyone working to build mathematical resilience and to everyone who experiences mathematics anxiety and is yet to develop mathematical resilience. We also dedicate the work to the memory of Professor Audrey Msimanga, with love and respect.

Contents

List of figures — xi
List of tables — xiii
List of contributors — xiv
Foreword — xix
Preface — xxi
Acknowledgements — xxii
List of acronyms and abbreviations — xxiii

Introduction — 1
CLARE LEE AND SUE JOHNSTON-WILDER

Part 1: Mathematical resilience

1 Mathematical resilience — 9
CLARE LEE AND SUE JOHNSTON-WILDER

2 Recognising and measuring mathematics anxiety and resilience — 25
THOMAS E. HUNT AND DOMINIC PETRONZI

3 Tools for building mathematical resilience — 40
SUE JOHNSTON-WILDER AND CLARE LEE

Part 2: Teaching for mathematical resilience

4 Building a resilient mathematical learning environment — 61
ROBERT WARD-PENNY AND JOHN THOMAS

5 Understanding the power of a coaching approach — 75
DEBBIE INGLIS AND SUE JOHNSTON-WILDER

6 Working with groups of learners — 92
ROBERT WARD-PENNY

7 Helping individual learners to make mathematics manageable   103
   JANET KILPATRICK BAKER

8 Working with groups of support staff   114
   GEORGIE FORD AND SUE JOHNSTON-WILDER

9 Teacher-led mathematical resilience research   127
   BEN SINCLAIR, TELMA SILVEIRA PARA, MASHA APOSTOLIDU AND AÏCHA HADJI-SONNI

Part 3: Working within the wider community

10 Communicating ideas about mathematical resilience to parents   149
   ROSEMARY RUSSELL AND DONNA WRIGHT

11 Working in Further Education with adult learners   166
   HOLLY HESHMATI AND JOHN MORGAN

12 Mathematical resilience for lifelong learning   180
   CLARE LEE

Part 4: International considerations

13 Mathematics anxiety as a global problem   193
   INTERNATIONAL AUTHORS

14 Mathematical resilience global developments   220
   INTERNATIONAL AUTHORS

Part 5: Looking forward

15 Mathematical resilience – What needs to change?   249
   CLARE LEE AND SUE JOHNSTON-WILDER

16 Resilience-building problem-solving tasks – The future   261
   GAYE WILLIAMS

17 Continuing to work for mathematical resilience   278
   SUE JOHNSTON-WILDER AND CLARE LEE

Index   291

Figures

1.1	The four factors of mathematical resilience	10
1.2	Promoting a growth mindset	17
1.3	Understanding the value of mathematics	18
1.4	The factor of struggle	19
1.5	The factor of support	20
2.1	The distribution of total MAS-UK scores (Hunt et al., 2011)	28
2.2	The first 5 items on the Children's Mathematics Anxiety Scale UK (Petronzi et al., 2018)	29
3.1	The Growth Zone Model	42
3.2	Enhanced Growth Zone Model (Para and Johnston-Wilder, 2023)	44
3.3	Hand Model of the Brain (Siegel, 2010)	45
3.4	The Ladder Model, showing accessible and inaccessible maths (Johnston-Wilder et al., 2020)	49
4.1	Two one-dot shapes	66
4.2	A one-dot triangle and a two-dot rectangle	66
4.3	A three-strand approach for developing departmental practice	68
4.4	Teacher empowerment and development cycle	70
6.1	Cards from a page of *Mathematical Team Games* (Lucas, 2014)	98
6.2	A 'Diffy' with four steps	99
7.1	The Growth Zone Model	103
7.2	The three As Model: Autonomy, Awareness and Action	104
7.3	Ada learns to climb. Copyright: R. Baker-Frampton	110
7.4	The Ladder Model (Johnston-Wilder et al., 2020)	111
7.5	Jack's Growth Zone Model	112
9.1	Methodology schema – Participatory Action Research Cycle (Source: Pará and Johnston-Wilder, 2023)	129
9.2	The Toolkit (Source: Johnston-Wilder et al., 2020)	133
9.3	A summary of the four key themes and subthemes	134
9.4	The mathematics resilience card folded (left) and unfolded (right)	137
9.5	The mathematics resilience poster, tool 1 to 3 (from left to right)	139

9.6	The monster metaphor (Source: Maths ACTive, 2023)	142
10.1	The Growth Zone Model	151
10.2	My Growth Zone Chart, created by Key Stage 3 pupil	153
10.3	I am a good mathematician poster	155
10.4	Moving out of the "not ready yet" zone poster	156
10.5	Moving out of the comfort zone poster	157
10.6	Keeping in the growth zone poster	158
10.7	Learning ladder poster	159
11.1	MAS scores for October 2019	170
11.2	Learners' feelings before and after the first intervention	172
13.1	Mathematics anxiety parameters (adapted from Carmo and Henklain, 2022)	195
13.2	South African performance in mathematics and science (Mullis et al., 2020)	206
13.3	Students' distribution by level of performance in Mathematics since 2013 (Ministerio de Education, 2022)	208
14.1	The Growth Zone Model in action	230
14.2	Schematic diagram showing how children can be assisted to navigate mathematics growth zone through mathematical story picture books	232
14.3	"I managed to do divisions. I am happy. Fran."	234
14.4	The 5 pillars	235
14.5	Learners' illustrations of their feelings when learning mathematics	236
14.6	Ana using the Growth Zone Model	237
14.7	My brain is like a muscle. With effort and dedication I can achieve whatever I propose to myself	238
14.8	With many, many rungs we help children to see that mathematics is more accessible than they think	238
14.9	The key to Mathematical Resilience	239
16.1	Display on the board, recording number of boxes with each number of blue Smarties	266

Tables

2.1	Suggested self-report mathematics anxiety scales for different learners	27
2.2	Example items from the Mathematical Resilience Scale (Kooken et al., 2016)	33
8.1	Prompts for support familiarisation phase	122
9.1	The phases of the Participatory Action Research Cycle (Pará and Johnston-Wilder, 2023)	129
9.2	The mathematical resilience grid	138

Contributors

Folake Adelabu is senior lecturer at Walter Sisulu University, South Africa. Her research and publications focus on the assessment of student teachers, teachers mathematical content knowledge, and the use of technology. She is presently focusing on reducing mathematics anxiety in high school students, and technical vocational education students' perspectives on mathematics instruction.

Karina Lumena de Freitas Alves is a PhD student in Psychology at the Federal University of São Carlos (UFSCar), Brazil. She works with individual pedagogical tutoring of adults/children with Maths learning difficulties, and she is an invited teacher at the Brazilian Institute for Educator Training (IBFE).

Masha Apostolidu is a mathematics teacher in Further Education at Lewisham College, UK. Her background is in international not-for-profit management and is currently working on her MSc Psychology dissertation with the University of Derby. Her research interests include psychological aspects of performance in maths and the role of neurocognitive processes in perception of cognition and behaviour.

Janet Kilpatrick Baker is a senior lecturer at Arden University, UK, where she is developing teaching effectiveness and learner engagement. Her career spans teaching, consulting, lecturing, research and leadership from primary to Higher Education. She researches learner motivation and impact of anxiety, particularly ways learners can learn to manage emotions and learn mathematics successfully.

Sakyiwaa Boateng is senior lecturer at Walter Sisulu University, South Africa. Passionate about science teacher professional development, she is looking at breaking down the disjunction between science/culture and meaning. Her current research is on Mathematics and Science anxiety, and interventions to build resilience among pre-service science teachers and high school science learners.

João dos Santos Carmo is a professor in the Department of Psychology at UFSCar, Brazil. Advisory Professor at the Graduate Program in Psychology at UFSCar. Researcher at the National Institute of Science and Technology on Behaviour,

Cognition and Teaching (INCT-ECCE). Coordinates the Laboratory of Applied Studies to Learning and Cognition (LEAAC), Psychology Department at UFSCar.

Tawanda Chinengundu is a lecturer at the University of Pretoria, South Africa. Over 25 years, he has taught Engineering Graphics and Design Technology at schools and tertiary institutions in Zimbabwe and South Africa. He has published on technical and vocational education. His research interests are: instructional design; integration of ICT; anxiety and resilience.

Allison Dillard is a former adjunct math professor at Irvine Valley College, USA, host of the hit Allison Loves Math Podcast, and author of multiple Amazon bestselling books, including the Crush Math Experiments, Crush Hypothesis Testing, and the Love Math Journal.

Silvia Renata Figiacone is director of the Clinical Neuropsychologist Degree in the Instituto Universitario del Hospital Italiano de Buenos Aires, Argentina. Also, Director of NeuroEduca, an interdisciplinary team working with children who have specific learning disorders, she has worked with Inés Zerboni and Brian Butterworth on dyscalculia in Argentina and Latin America.

Georgie Ford has worked within NHS acute settings and has lectured and examined in mental health and Psychology. Georgie's specialist interest is aligning curricula with health. Georgie developed the popular emotional recovery framework for education post covid. Georgie achieved the Gold Carnegie Award in mental health.

Aïcha Hadji-Sonni graduated in mathematics, computer sciences, education and coaching. She is a doctoral student in the Department of Education Studies at the University of Warwick, UK. She was a teacher of mathematics in disadvantaged areas in France and her research interest is the affective domain of mathematics education.

Holly Heshmati is an associate professor at the Centre for Teacher Education-University of Warwick, UK. As part of her work with student teachers, Holly supports adult learners with building and developing mathematical resilience to improve motivation, ensure engagement and improvement of outcome in mathematics.

Thomas E. Hunt is an associate professor in Psychology at the University of Derby, UK, where he leads the Mathematics Anxiety Research Group. His research focuses on how to measure maths anxiety, understanding how it develops, how it relates to a range of emotional, cognitive, and behavioural factors, and how to reduce it.

Debbie Inglis is a former primary school leader and mathematics consultant, who specialises in coaching and training leaders and their teams to develop tools and strategies to increase resilience and confidence, create solution-focused teams, and avoid burnout. Her training style is calm and supportive, creating a psychologically safe space as paramount.

Sue Johnston-Wilder is an associate professor of Maths Education and a Coach for Mathematical Resilience. Her background is in secondary maths teaching, teacher-development, mentoring and coaching. Over the last 15 years, she specialised in addressing maths anxiety and has co-developed the Mathematical Resilience framework. She is co-founder of the International Mathematical Resilience Network.

Royda Kampamba is a lecturer in science education at The Copper-belt University in Zambia. Her research interests include: chemistry education, mainstreaming ESD, innovative teaching, Indigenous Knowledge. Currently, she is working on anxiety and resilience in STEM education among undergraduates and pre-service chemistry teachers.

Clare Lee works as a senior lecturer teaching at the master's and doctoral levels. Prior to working in universities, she taught mathematics at secondary schools for over 20 years. She has published many books and papers focusing on teaching mathematics effectively and allowing learners access to this important, exciting and useful subject.

Njaru Mbogo Harrison has been a preschool and primary school teacher for two decades. Currently fine-tuning a PhD in mathematics pedagogical simulation at the University of Nairobi, Kenya, he has secured a second PhD studentship in building sustainable mathematical resilience among grade three children at Gonville and Caius College, Cambridge.

John Morgan is a lecturer in Engineering, Computing and STEM with the Open University, UK. For many years, he has witnessed students struggle with maths anxiety. Some have avoided careers that involve maths as a result. John is carrying out PhD research to help adult learners overcome their anxiety and improve their resilience.

Brighton Mudadigwa is a lecturer in chemistry and physics education in Pretoria. Brighton's research is in teaching strategies that foster conceptual understanding and meaningful learning using pedagogical link-making in STEM subjects. Brighton leads a group of researchers investigating STEM anxieties in preservice teachers and high school students in South Africa and Zambia.

Telma Silveira Pará is a researcher and lecturer at FAETEC – Technical School Support Foundation in Rio de Janeiro working in the research fields of Mathematics Education, Methods and Teaching Techniques, STEM/STEAM Education, topics in Neuroscience and Psychology of Learning. Founding member of the Brazilian Antenna of Mathematics at Fluminense Federal University (www.antenabrasil.uff.br).

Dominic Petronzi is a senior lecturer in Psychology at the University of Derby, UK. He is a member of the Mathematics Anxiety Research Group. His research focuses on understanding the development of maths anxiety, how it relates to educational experiences/career trajectories, and approaches to reduce it.

Japcy Margarita Quiceno has a PhD in Clinical and Health Psychology with knowledge in Positive Psychology. She has served as full professor at several Colombian universities, and she boasts over 90 scientific publications, and a book focused on resilience intervention. Expertise in designing and coordinating programmes that promote psycho-emotional health and resilience.

Rosemary Russell expertise is parental engagement with mathematics. She has had several books and articles published on this subject. Rosemary is currently focusing on helping parents develop and nurture mathematical resilience in their children, through her author talks. Her PhD was from the University of Bristol, UK.

Maria Ryan lectures in Business at Mary Immaculate College, School of Post-Primary Education, Ireland. Her research explores mathematics anxiety among adults using mathematics life story. Maria delivers mathematics anxiety awareness training to pre-service and in-service teachers across education sectors. Maria is co-chair of the Mathematics Resilience Network - Ireland Branch.

Ben Sinclair taught secondary mathematics before beginning his PhD in Mathematics Education at the University of Warwick, UK. His project focuses on embedding principles from Acceptance and Commitment Training into classrooms to address mathematics anxiety and promote mathematical resilience. His academic interests include mathematics task design, and instruction and mastery.

John Thomas is an experienced teacher of mathematics, currently head of the department in a selective school in Southeast England and an EdD student. He has worked in a range of educational establishments including non-selective schools and colleges of further education, informing his commitment to building resilience and confidence in students.

Stefano Vinaccia is a psychologist, with a PhD in clinical and health psychology. Colombian Psychology Award winner in 1998 and scientific research award in psychology in 2018. He has published 190 articles, 16 chapters, and four books. Currently, he is an Emeritus Researcher at the Universidad del SINU, Monteria, Colombia.

Robert Ward-Penny is a lecturer in Mathematics Education at The Open University, UK. He has previously worked as a secondary mathematics teacher in the Midlands, London, and the US, and has contributed to several textbook series and books for teachers. His current interests include the philosophy of mathematics education and cross-curricular working.

Gaye Williams an honorary senior fellow at the International Centre for Classroom Research (ICCR Hub) at The University of Melbourne, Australia, has studied intellectual and affective qualities of activity in mathematics lessons. She identified resilience as a crucial characteristic of students who problem solve creatively, and that resilient teachers can build this student characteristic.

Donna Wright is a senior education improvement adviser for Solihull Local Authority with a responsibility for mathematics in schools. She is also co-director of Bright Pi Education Consultancy. With a passion for mathematics teaching, Donna aims to support schools to improve the teaching and learning of mathematics, working alongside parents.

Nomzamo Xaba is Life Sciences lecturer at University of Zululand, South Africa. She is finalising her PhD "Investigating meaning making, and opportunities that biology lecturers create during instruction" at the University of the Witwatersrand. She is passionate about effective instructional practices. Her research interest areas include: anxiety and resilience; language issues; integration of ICT.

Abdulvahap Yorgun has worked as counsellor in high schools, and guidance and research centres and lectures Trauma Counseling. He currently works at Bayraklı Guidance and Research Centre, Turkey, coordinating school psychological services in the district. He has focused on mathematics anxiety since 2021, composing many papers and articles on this subject.

Foreword

Professor Margaret Brown
King's College London and Maths Anxiety Trust

I am delighted to have been asked to provide the Foreword for this important book aimed at building mathematical resilience in learners of all ages.

After working as a mathematics teacher, first at primary and then at secondary level, then a teacher trainer, I moved into research in mathematics education in the 1970s. My research started off with a focus on assessing progress in mathematics learning, which led into studying the mathematics curriculum, and then into ways of teaching and teachers' beliefs.

Rather late in the day I realised that actually an essential step towards learning was the willingness of the learners to engage. Moreover, the more closely specified the standards and curriculum coverage demanded of schools, teachers and learners were becoming, the greater the problem of learner attitudes, and specifically of mathematics anxiety. Accountability was leading to performativity; increased national pressure on schools, teachers and students to do well in examinations was leading to a system which values teacher demonstration, learner imitation and curriculum coverage, and works against engagement, inspiration, excitement, curiosity, and against the depth, quality and applicability of learning.

As noted in other parts of this book, this effect is not limited to one or even a small number of countries, and everywhere the increasing importance of mathematics for economic and technical development is appreciated. This means that parents and carers, as well as teachers, are also putting pressure on their children to do well at mathematics, in order to make sure they have the necessary skills and qualifications to succeed in well-paid jobs.

Although for all these reasons it appears to be increasingly more prevalent, the existence of mathematics anxiety has been acknowledged for many decades, starting in the 1950s in the United States and the UK and elsewhere by the 1970s. There have, certainly in every recent generation and probably for much longer, been people who report that they experienced their mathematics lessons as a form of juvenile torture, and who can still recall occasions of classroom shame, humiliation and panic. Some have rationalised their fears by dismissing mathematics as irrelevant and those who are good at it as 'nerds'. Indeed, the British Prime Minister, Rishi Sunak, has recently described the UK as having an 'anti-maths mindset', but international surveys like the OECD PISA studies show that the UK is not alone in this, although some countries seem to be more successful in nurturing a more positive stance.

Thankfully a number of people over the years have tried to tackle the problem of negative attitudes to mathematics. Some, like the Maths Anxiety Trust, founded by Shirley Conran, of which I am currently president, have focused on awareness-raising, campaigning and publicising research and resources. Others have energetically researched possible solutions on a broad level, via changes in teaching styles, curriculum and professional development. Yet others have centred on initiatives aimed at supporting individual students.

This book brings together probably the most important branch of this work, that concerned with strengthening students' mathematical resilience, a cause that has long been championed successfully by the editors of this book, Sue Johnston-Wilder and Clare Lee, and their colleagues, partners and associates in the UK and internationally. I first had the privilege and joy of working with Sue as a colleague researcher in the 1980s, and Clare a little later. I have been delighted to keep in touch with them personally and with their academic trajectory since then. I believe that this work on mathematical resilience has already helped significantly many students of different ages, and their teachers, to overcome anxiety and negative attitudes. However, there is still huge scope to spread the word to a wider audience, as the need is great and ever-growing.

I am sure that this book will contribute to the increase of that influence, and will improve the lives of many.

Margaret Brown
November, 2023

Preface

We have been writing about the disabling effects of mathematics anxiety and how to work in ways that minimise the chances of anxiety developing, and mitigate its effects where anxiety is already present, for 15 years. We have termed this way of facilitating mathematical learning *developing mathematical resilience*. Mathematical Resilience is a positive stance towards mathematical learning, where the learner knows that struggle will be required but also knows that they will both need and find appropriate support. The resilient learner knows that they can grow their mathematical understanding with effort from them and support from others.

The work to develop the challenge that developing mathematical resilience in learners presents is now snowballing. It is time for a book which can bring the principles, ideas and tools to a wider audience. Fortunately, Routledge and the reviewers agreed, for which we are very appreciative.

The International Mathematical Resilience Network (MRN) launched in July 2022 and is developing as a place to share the effective ideas and deep, researched and evidenced principles explained in this book around the world. Many members of the MRN have contributed to this book. We look forward to working on further publications with this wonderful group.

Acknowledgements

Our first acknowledgement must go to those teachers and learners who worked with us when we started seeking evidence that the ideas in this book were important and workable. Our grateful thanks go to them and the many others who work to learn mathematics themselves and those who work to enable others to do so with resilience and to manage anxiety. They have contributed to our thinking in big ways and small ways – and we acknowledge that with thanks.

We give special acknowledgement to Shirley Conran, OBE, and Hon Fellow UCL, for her stalwart work raising awareness of mathematics anxiety, setting up the Maths Anxiety Trust and encouraging so many of us in this important work. At the age of 91, Shirley is taking a sabbatical from MAT to write another book. She is a powerful role model for us all – thank you Shirley.

We gratefully acknowledge the encouragement and contributions that we have received from the authors in this book. Without the encouragement of this group, we would not have started and without their contributions, the breadth and depth of ideas presented in this book would not have been generated.

Especial thanks must go to Ben Sinclair for convening the conference to bring the authors together from around the world and to Robert Ward-Penny for his unstinting creative support in the editing of the chapters in this book.

List of acronyms and abbreviations

AMSP   Advanced Mathematics Support Programme
FE   Further Education
GCSE   General Certificate of Education (an external examination taken at age 16 across much of the UK).
HE   Higher Education
HEI   Higher Education Institutions
ICT   Information and Communication Technology
MA   Mathematics Anxiety
MR   Mathematical Resilience
MRN   Mathematical Resilience Network
MAT   Maths Anxiety Trust

Introduction

Clare Lee and Sue Johnston-Wilder

This book is about mathematical resilience: what it is, why it matters, and the challenge it presents. Mathematical resilience is a term for a collection of attitudes and behaviours which together enable people to learn mathematics in a psychologically healthy way. Mathematically resilient learners have the self-efficacy, self-safeguarding skills, motivation and perseverance needed to overcome difficulties, learn mathematics, and go on to flourish as users of mathematics in society. Teachers (and others) can support the growth of mathematical resilience. We do not talk about mathematically resilient teaching, rather teaching for mathematical resilience; that is making pedagogical choices that develop a positive learning environment which protects everyone's well-being and gives everyone the best chance of succeeding as healthy learners of mathematics.

Introducing mathematical resilience

In psychological terms, resilience is a characteristic generated by the interaction of a person and their context (Herrman et al., 2011). Resilience helps people withstand and recover from difficulties. Resilience can help learners deal with struggle and take a purposeful view of the reality of the situation they face (Coutu, 2002) rather than slipping into anxiety, avoidance or anger to protect themselves. Although there are some innate factors and advantages which mean some people might be more likely to develop resilience than others, everyone's resilience can be fostered in a nurturing environment.

The idea of mathematical resilience as a specific construct grew out of a recognition of the especial value of positive attitudes and behaviours in learning mathematics. When learning mathematics, learners often need to be able to deal with challenging work, demanding goals, and stressful situations. Our experience and research have repeatedly established that, when learners work in a mathematical resilience supportive environment, they develop habits which enable them to be willing to approach learning with motivation (Ryan and Deci, 2018) and self-efficacy (Bandura, 1997). All learners can develop more mathematical resilience and a supportive learning environment is one way that helps them do so. Parents and others outside schools and colleges can also work in ways that promote mathematical resilience.

A mathematically resilient learner accepts that challenge is important, and that making mistakes is part of the learning process. They feel that they have some control over what they are doing and know how they might access support if needed. Mathematical resilience brings with it the positivity and perseverance which people require to learn mathematics, and then go on to use mathematics in their lives.

Why is mathematical resilience important?

There are many reasons why mathematical resilience deserves specific attention. The most immediate area of concern is schools and colleges, where there is a widespread history of disaffection and underachievement (Nardi and Steward, 2003; Boaler, 2016) in mathematics. There is a desperate need for all children and young people to be enabled to willingly engage in learning mathematics, but some common classroom approaches instead expose learners to adverse experiences and construct barriers to progress.

The importance of mathematical resilience is not limited to mathematics learning environments. Mathematics is everywhere in modern life, from everyday numeracy skills to understanding statistics on the news. Mathematically resilient individuals have access to mathematical thinking, to model and understand the world. In addition, mathematical qualifications often act as gatekeepers to further education and employment, so learners who have developed mathematical resilience are better supported to make decisions about and access their personal goals. There is also evidence of a monetary benefit associated with continuing to study mathematics to a higher level (Smith, 2018): for instance, people who take A-level mathematics earn an average of 11% more over their lifetime (AMSP, 2019) than those who do not. Working with learners in ways that nurture their mathematical resilience may also be of benefit to wider society, as more students will be prepared to study subjects with high mathematical content, and more people will be equipped to model complex ideas and find the solutions needed to cope with for example, climate change.

Another reason why mathematical resilience deserves specific attention is that mathematical learning poses some unique cognitive and affective challenges. These features of mathematics, when taught in a way that does not recognise the issues, can lead children and adults to feel apprehensive and anxious about using mathematics, which is detrimental to their well-being. Where anxiety is the result of the way mathematics is taught, this often leads to avoidance and worse. Psychologists term these reactions mathematics anxiety. Mathematics anxiety interferes with learning at every level. It has been detected in children in primary classroom and by the time a child is in secondary school, correctly identifying and addressing that anxiety can be very difficult. Being even slightly mathematics anxious will usually mean that a person will avoid anything to do with mathematics or numbers, in order to avoid the uncomfortable feelings that numbers evoke. This means that, when considering an apprenticeship, they choose the courses least likely to expose them to studying or using mathematics (Johnston-Wilder et al., 2014). Not enough engineers and nurses are trained because mathematics is a gatekeeper subject for these professions, as it is for many others, and those with moderate or severe mathematics anxiety cannot face the perceived risk to their well-being to study sufficiently to succeed in the required mathematics examinations (Brown et al., 2008).

Against the background of these concerns, we devised the construct of mathematical resilience. Where those that teach mathematics, or otherwise support those that are learning mathematics, work in ways known to develop mathematical resilience, they help everyone take a positive stance towards the mathematics they encounter in their lives, school careers and beyond.

The challenge of mathematical resilience

The chapters in this book contain a challenge to rebalance planning and actions as teachers in order to promote a psychologically healthy mathematics learning environment. You will not find a tick list or a to-do list in this book, as creating a learning environment that nurtures mathematical resilience is not about doing this and not that. It is, in common with all effective teaching, about thinking through the actions that constitute teaching mathematics with the well-being of the learner in mind. Helping learners to become more mathematically resilient requires the application of positive psychology to teaching mathematics. Commonly accepted ways of teaching mathematics have resulted in the widespread problems of avoidance and defensive behaviour which are prevalent in mathematics classrooms around the world (Pisa, 2012) and have led to mathematics anxiety for many. Consequently, it is also clear that the way mathematics is taught must change. Teaching in ways that help learners develop mathematical resilience allows learners to engage with mathematics with understanding and to overcome the obstacles that mathematics itself can present, whilst managing any feeling of anxiety which the nature of the subject may cause. Where significant anxiety has already developed, developing mathematical resilience will enable learners to overcome the barriers that anxiety can cause and re-engage with mathematical learning. Teaching is fundamentally about relationships and teaching for mathematical resilience ensures that those relationships are nurturing, respectful and warm. Where teachers work with learners in ways that engender trust, not anxiety, learners want to learn and make progress and the teacher is trusted to suggest ways that learning can happen.

In this book, we regard all those who are working alongside those who are learning mathematics as teachers. Hence the 'teacher' addressed in this book may be a professional working in a classroom, a neighbour trying to help, a parent, a learning advisor or coach in further education or someone providing learning materials online for individuals to access. An individual may also be their own teacher, recognising their own difficulties when approaching mathematics, using the tools advocated in this book to help reshape their own stance towards mathematics.

Currently, ideas about mathematical resilience are being taken up by academics and teachers worldwide. The ideas speak to those concerned with helping others learn mathematics in Brazil, Australia, France, South Africa and Ireland, to name just a few, as well as the four UK nations. It seems to be the under-explored myths and customs that surround the teaching of mathematics that are generating avoidance and anxiety in learners in these countries. Teachers are teaching as society expects them to, and how they learned themselves, and thus harm for many is perpetuated. This book is about promoting ways that mathematics can be learned without generating significant anxiety, harm and the need to self-safeguard (Benson, 2000). Teachers who already teach in this way exist in sufficient numbers for us to know both that it is possible and that it enables learners to succeed in examinations and in their lives. However, it is not yet part of the training routinely offered to student teachers, nor is it understood or promoted by OFSTED, the English body that inspects schools, although they do seem to understand the power of the resilient practices promoted in this book.

It seems to us teaching that promotes mathematical resilience is a requirement not an option. The rest of the book provides ideas about how the teaching of mathematics to learners of all ages can promote the well-being of learners as they engage with the difficulties and stumbling blocks that mathematics itself presents. Teaching for mathematical resilience in the ways recommended in this book should prevent the development of mathematics anxiety but such anxiety may already be an issue in learners of any age. Teaching in ways that allow the development of mathematical resilience acknowledges that anxiety may already be present and offers ways that learners can overcome the barrier which mathematics anxiety may present and continue to learn. Mathematical resilience gives people the skills and the psychological safety to think mathematically, and that skill is vital in our ever-changing world.

The rest of the book

The book is divided into three parts. In every section, the focus is on helping anyone who is working with learners of mathematics support those learners to develop mathematical resilience. Each chapter presents ideas ready to take up and use, but also makes clear the evidence on which these ideas are based.

The first section explains the fundamentals behind the concept of mathematical resilience in Chapter 1 and how mathematics anxiety and its effects can be measured in Chapter 2. The basic tools that can be used to help learners understand and develop their own resilience are presented in Chapter 3.

The second section looks at how teaching can be undertaken in ways that will develop mathematical resilience. The chapters focus on different situations and differing age groups. Chapter 4 looks at how a learning environment that fosters mathematical resilience in learners can be established, focusing mainly on school age learners. Chapter 5 looks at how the power of coaching can be used in schools and beyond to help learners overcome the barriers mathematical learning presents. Chapter 6 considers how working collaboratively can be used to help learners develop mathematical resilience whilst Chapter 7 focuses on the individual and their learning needs. Chapter 8 looks at how including support staff and other adults in schools and colleges can make sure that developing mathematical resilience is part of a whole school endeavour. Chapter 9 considers how teachers can use the principles of action research and design-based research to embed the principles of action research into their practice.

The final section looks beyond schools and classrooms. It starts by considering how parents can be helped to understand their role in supporting their children in developing mathematical resilience. Chapter 11 focuses specifically on developing mathematical resilience in those learning mathematics as adults whilst Chapter 12 looks at the characteristics that learners will need to continue their learning life-long. Chapters 13 and 14 then look globally at how understanding the need for mathematical resilience is gaining traction worldwide and at how other countries and cultures are acting to enable their learners to learn mathematics without anxiety and fear. Chapter 15 looks at what needs to change if mathematical resilience is to become an established part of the educational system, Chapter 16 looks at developments in Australia and how learners there are becoming optimistic problem solvers. Chapter 17 looks forward, summing up the messages in this book and asking the reader to continue what has been started and is needed by so many.

References

AMSP (2019). *Why it pays to study maths at A level.* https://amsp.org.uk/news/why-it-pays-to-study-maths, accessed 24 November 2022.

Bandura, A. (1997). *Self-efficacy: The exercise of control.* New York, NY: Freeman.

Benson, H. (2000). *The relaxation response.* New York, NY: HarperCollins Publishers.

Boaler, J. (2016). *Mathematical mindsets: Unleashing students' potential through creative math, inspiring messages and innovative teaching.* Hoboken, NJ: Jossey-Bass.

Brown, M., Brown, P., and Bibby, T. (2008). "I would rather die": Reasons given by 16-year-olds for not continuing their study of mathematics. *Research in Mathematics Education*, 10(1), 3–18. 10.1080/14794800801915814

Coutu, D. (2002). *How resilience works.* Harvard Business Review, May 2002.

Herrman, H., Stewart, D. E., Diaz-Granados, N., Berger E. L., Jackson B., and Yuen, T. (2011). What is resilience? *The Canadian Journal of Psychiatry*, 56(5), 258–265. 10.1177/070674371105600504

Johnston-Wilder, S., Brindley, J., and Dent, P. (2014). *A survey of mathematics anxiety and mathematical resilience among existing apprentices.* London: Gatsby Charitable Foundation.

Nardi, E., and Steward, S. (2003). Is mathematics T.I.R.E.D? A profile of quiet disaffection in the secondary mathematics classroom. *British Educational Research Journal*, 29(3), 345–367. http://www.jstor.org/stable/1502257

PISA (2012). *Results: Ready to learn (volume III): Students' engagement, drive and self-beliefs.* Paris: OECD Publishing; 2013. 10.1787/9789264201170-en

Ryan, R., and Deci, E. (2018). *Self-determination theory: Basic psychological needs in motivation, development, and wellness.* New York, NY: Guilford Press.

Smith, A. (2018). *Review of post-16 mathematics.* London: DfE.

PART I
Mathematical resilience

Mathematical resilience
Clare Lee and Sue Johnston-Wilder

Introduction

Mathematical resilience enables learners to have the positivity and perseverance needed to learn mathematics and to continue to learn the mathematics required for individuals and society to flourish in a changing world. Mathematical resilience refers to both the process of becoming positioned to address any difficulties that arise when learning mathematics and the outcome of successfully adapting to difficulties and challenges. It is having the flexibility and ability to adjust to the emotional struggles that may present themselves. For example, someone may feel awkward when they find they need more time and support to understand ideas than others who find those ideas more straightforward. There are also external demands made by mathematics, which of its nature requires hard work to comprehend, and resilience is needed here as well. Mathematical resilience is not something that someone has inherently, it is nurtured in an environment where the difficulties presented by mathematics itself, the way the subject may have previously been taught and the way that society views mathematics, are recognised and addressed.

> At primary school I did not understand the work in class, every lesson was completing questions in a textbook. I used to copy the person next to me as I was frightened to ask the teacher for help. Then in secondary school the maths got harder, and I got stressed in each lesson. I had good knowledge of times tables but when asked in front of the class, my memory would go blank, and I would get stressed, and I could not think of the answer quickly enough.

Psychological safety and well-being are ideas which should be at the heart of all teaching and learning. An environment that allows learners to develop mathematical resilience is a place of safety, inclusion and nurture as learners progress in their mathematical thinking. We see it as essential that all teachers of mathematics, to learners of any age, see themselves as someone whose job is to nurture everyone's mathematical learning and to do that, they must ensure learners' psychological safety and well-being. Thus, it is important that teachers understand and recognise the detrimental impact of mathematics anxiety to learners' well-being. Where anxiety has already been established by being exposed to adverse prior experiences, or common teaching approaches, such as those encouraging an over-reliance on memorised routines and procedures, learners may self-safeguard by avoiding engaging with mathematics, and can become either excessively passive or aggressive if challenged. Anxiety, how it is caused and the impact it has, will be discussed later; here how to establish an environment that supports well-being will be the focus.

An environment that develops mathematical resilience

Our research has led us to understand that the many aspects of an environment that allows learners to develop mathematical resilience can be divided into four inter-related factors. The four factors are as follows: growth mindset, struggle, value and support (Figure 1.1).

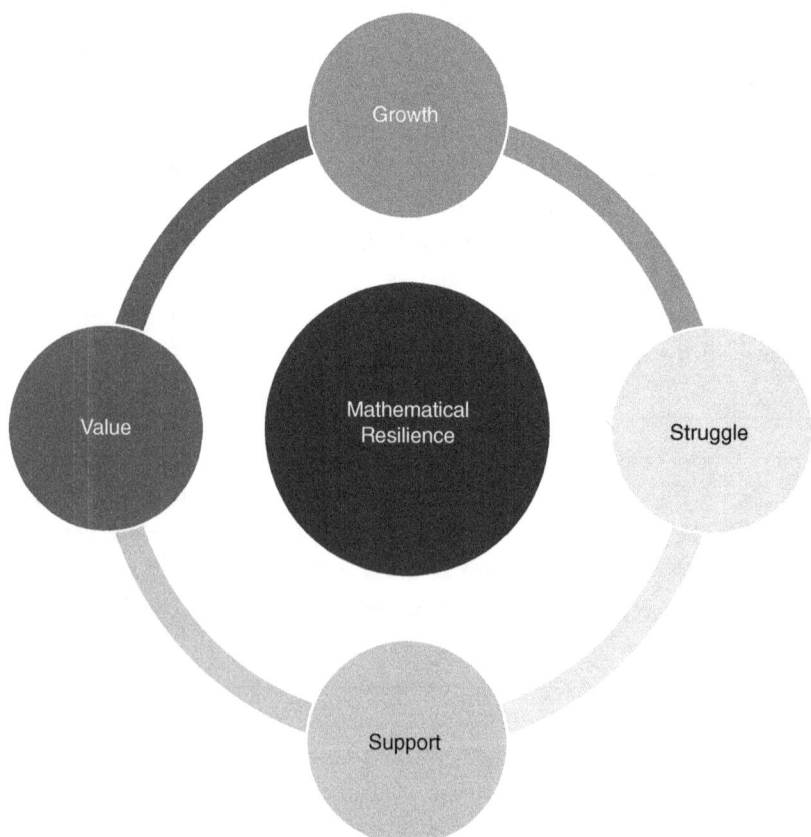

Figure 1.1 The four factors of mathematical resilience.

Growth Mindset: a clear commitment to and practical outworking of the belief that everyone can grow their mathematical understanding from where they are and can achieve an understanding of mathematics at whatever level they wish, over time.

Struggle: an understanding that learning and comprehending mathematics can be challenging and requires struggle, and that everyone, at whatever level they are working with mathematics, can meet barriers to overcome. Accepting challenge is important as well as understanding that making mistakes is part of the process. Learners are encouraged to know that they have the right to ask the questions they need to ask as they struggle, with support and collaboration, towards their learning goals.

Value: there are two aspects to this factor, an environment that develops mathematical resilience emphasises the value of mathematics within society and also that the learner is valued within the learning community. The environment helps learners understand that mathematics is of value because it is everywhere and in every decision. Such an environment shows the learners the personal value of mathematics to them in the satisfaction of a problem solved or an understanding reached, and how the learning of mathematics aligns with their own values. It also demonstrates the value of the learner in the learning environment, making them feel included and an appreciated part of the learning community.

Support: every member of an environment that develops mathematical resilience knows they have a right to the kind of support that will help them progress, and that their views and ideas are listened to. Commonly, help will be found in an ethos of respect and collaboration and the offering of support to one another, but it will also involve access to books and to digital technology or just being allowed time to think. The right of each learner to feel they are both supported by, and able to offer support to, others in their learning endeavours are vital parts of building mathematical resilience.

These four factors interact with each other to promote a welcoming, respectful and inclusive environment, where the learners form an effective learning relationship with the teacher, with each other and with mathematics, and where their mathematical well-being can be established. If one factor is missing or there is too much of one and not enough of another, then the environment will not be as effective. For example, if there is too much support and not enough struggle, or challenge, the learners will become too dependent on the teacher and will not find their own way to engage with mathematics or indeed believe that they can use and control mathematical ideas. Too much talk about how everyone can grow and progress with not enough support will prevent the learners believing that they can grow their understanding and may cause anxiety.

Specific measures that can be taken to enable learners to approach mathematics with resilience are detailed in Chapter 3 and in every other chapter in this book.

Here the background behind the practical measures which are needed in the classroom will be presented and discussed.

The importance of a growth mindset

The theory of a growth mindset (Dweck, 2000) is one of the vital parts of the kind of learning environment that allows learners to develop mathematical resilience. Where teachers fully understand and put into practice the growth theory of learning as explained by Dweck (2000), their learners will thrive. Where teachers both believe, and act, as though everyone is able to progress with their mathematical learning, given effort from the learner and access to appropriate support, learners will themselves start to believe that they can understand and progress with mathematics.

Yeager and Dweck (2012) showed both the difference that having a growth mindset makes and that persuading learners to adopt this theory for themselves is not hard. They concluded that learners who are taught and believe that they can develop and grow their intellectual abilities in mathematics tend to show higher achievement rates than those who believe that their intellectual abilities are innate and therefore fixed. Those who hold a growth theory of learning believe it is worth persevering when they come across difficulties and therefore are better positioned to succeed than those who believe their intellectual abilities in mathematics are innate. Those who hold a fixed theory of learning tend to regard finding difficulties as proof that they may have reached the limits of their understanding rather than as a prompt to seek better support or to think things through again.

It is not difficult to change mindsets. Yeager et al. (2016) showed that using an intervention that teaches the idea that people's academic and intellectual abilities can be developed through actions such as making more effort, changing strategies, and asking for help, can make a difference to the way that learners approach their learning. Conveying information about neuroplasticity is also helpful, for example saying, "the brain is like a muscle—it gets stronger (and smarter) when you exercise it." However, it is not enough to define growth mindset and illustrate it with a metaphor. Enabling learners to adopt a growth theory of learning must also involve recommending concrete actions that can be taken, such as exercising the brain by working on challenging material that requires sustained thinking, and offering and supporting learners to work with this kind of material in the learning environment.

> When I started university, I still had maths lessons. The teacher there taught me about the connections in maths and then my confidence grew, and things started to make sense. I still have mind blanks if put on the spot or under pressure, but I have a lot more confidence. I also have to check my answers with a calculator if I am doing maths for someone else, as I stress out about it being wrong.

Furthermore, in order for learners to benefit from being introduced to the ideas of a growth mindset, all the adults that the learners interact with must be persuaded to adopt, or at least act as if they adopt, the theory that everyone can grow their mathematical learning from where they are. The fixed mindset is prevalent in society; it is why most schools consider they must use setting, especially in mathematics. Many people believe that some learners just cannot learn mathematics and therefore it is reasonable to put them in a low set and teach them a restricted curriculum. Whether a learner can understand a given concept depends on what prior knowledge and experiences they have. Each learner will have had different experiences, both inside and outside of school. They may have missed some lessons which prevents them recalling the concepts on which a new idea rests. They may also have experienced teaching that has caused them to develop mathematics anxiety which will be detrimental to their learning. It is unsurprising that interventions trying to establish a growth mindset sometimes do not have the effect that might be expected (Yeager and Dweck, 2020). It is not the messages about growth mindset that are at fault, it is that those messages are not acted on fully and consistently by schools and society.

Teaching is a difficult enterprise but the more seriously each teacher takes the psychological safety and well-being of their learners and acts on the belief that everyone can grow their understanding, the more each person will learn, with consequences that will be beneficial to the individual and society.

Relationships build mathematical resilience

A further part of ensuring the well-being of learners is developing an encouraging and inclusive ethos in learning relationships. Hattie and Yates (2013) were clear that relationships matter when it comes to successful learning. When teachers want to develop effective relationships that nurture learning they show learners that they matter, consistently challenge learners to grow, offer support, allow power to be shared by offering mutual respect and consistently expand learners' possibilities (Sethi and Scales, 2020). Research conducted by McKay and Macomber (2023) found that where warm and nurturing relationships are established, learners are willing to work hard and they show confidence in positive outcomes accruing from their hard work. Focusing on nurturing relationships has been shown by Split et al. (2012) to be a key factor in providing teaching that allows at-risk children to overcome adversity. They demonstrated that nurturing relationships resulted in higher attainment and that where there was conflict within a teacher-child relationship the learner would attain less.

> We were put into sets at high school, I was in the low set – the teacher was sharp and frightening, and I felt 'bullied' into learning. When I was asked to answer a question, I felt pressured to reply but I didn't know the answer. The teacher seemed frustrated and said that he had ALREADY told me HOW to do it.

Schools which allow and encourage their teachers to form trusting and nurturing relationships with their learners encourage high attainment. All too often the pressure put on schools from government policies results in teachers being under pressure and not having their own well-being considered. Where teachers are employed in a context which results in hostile attitudes within the schools and colleges in which they teach, Ali et al. (2019) found that that hostility is often reflected in teachers' relationships with their learners. Teachers put under pressure may resort to rigid and conflict-inducing relationships to protect their own well-being. Sava (2002) sees rigidity and conflict inducement as psychological maltreatment. Using fear and intimidation to instil discipline is an example of such maltreatment and such negative control measures interfere with learning most of the time. Sava (2002) considers these acts of teacher maltreatment will lead to negative teacher-learner interactions and to learners considering their teachers as the cause of their school problems. Unfortunately, such antipathy to learning is unlikely to be short-lived. Hyman and Snook (1999) discovered a variety of symptoms in learners that seemed to develop from conflict-inducing relationships, including neurotic traits, behaviour extremes, withdrawal or avoidance, and anxiety. There were indications that such traits may last the child's entire life.

> A teacher I knew told me about teaching boys in a bottom set in Year 9. When she first taught them, whenever she would ask them a question (even one she knew they'd done), their immediate response was "Don't know" and they would stare at her with 'deer in the headlights' looks. As she has built a relationship with them, this has evolved to less panicked looks and they will now attempt to answer questions with her help.

Schools that support and encourage their teachers, rather than use the accountability system to pressurise them, both allow their teachers to protect their own well-being and allow the whole school to become a nurturing learning environment, allowing their learners to progress. The outside culture and educational system do not have to affect the learning environment, but often they do.

Some pressure can be motivating and productive, but too much pressure can be problematic causing distress. Some learners will arrive at school under too much pressure from their parents, and teachers sometimes must act to counteract this pressure. Parents can negatively affect mathematical learning in several ways. Parents have been known to insist that their children remember the procedures they were taught at school. This may get in the way of the thinking, connecting and reasoning that are required for the understanding that a learner may seek. The learner may then go onto develop anxiety, preventing the positive relationship with mathematics that allows for optimal progress. Parents may offer rote responses such as when you multiply by ten just add a zero, which can cause confusion when the child is learning to work with decimals. Perhaps even more damaging, parents may inadvertently excuse their child from trying by insisting on the fixed mindset view that "I could never do maths, you are just like me."

Learners arrive in any mathematics learning environment with a many different experiences and backstories. We would like them all to arrive curious about mathematical ideas, and ready and willing to learn more. However, given the myths and misunderstandings about mathematics and its learning in society, and the importance placed on succeeding with mathematics qualifications, it is currently more likely that many will arrive anxious and concerned about possibly experiencing threats to their well-being. If they feel this way, then they are unlikely to feel safe enough or motivated to learn.

Meeting learners needs

Establishing safe, trusting and nurturing learning relationships is part of attending to the psychological safety and well-being of learners. Ryan and Deci (2018) explain that such learning relationships allow learners to develop autonomous motivation by meeting their basic psychological needs for competence, relatedness and autonomy. These ideas form Ryan and Deci's (2018) self-determination theory. The three basic psychological needs of competence, relatedness and autonomy must be met before learners can safely engage with learning. The factors of an environment that builds mathematical resilience also meet Ryan and Deci's (2018) basic psychological needs as required for self-determination or so that a learner develops the autonomous motivation that is necessary to thrive in society. Thwarting or not meeting any one of these psychological needs is detrimental to a learner's well-being to the extent that they may become anxious, experience stress and be less motivated to engage with mathematical learning. To explain more:

Autonomy: This is the learner's need to feel in control of what they are doing, that what they are doing is of value to them and will be of value in the community in which they interact. Learners must therefore feel that they understand what is being asked of them and that they are making the decision to engage in the actions required as they can trust the teacher is asking them to act to their benefit. Where they experience autonomy, they feel that they want to, or are motivated to, engage in the learning experiences offered as these experiences align with their values.

Competence: This is the learner's need to feel that they are making an effective contribution to the environment and that they can exercise, expand, and express the capacities they have. When individuals know that they are developing their skills and understanding, their need for competence is met and they will feel successful and proficient. Thus, the ideas of assessment for learning (Black et al., 2003) are vital to meeting the need for competence, and competitive games where only a few succeed are not. Also, part of meeting the need for competence is encouraging learners to support one another. This works both ways: the learner who requires support develops their proficiency and experiences success; the learner who offers support has their skills and understanding validated.

Relatedness: This is the need to feel valued, connected, and have a sense of belonging. An environment that meets the need for relatedness is responsive, sensitive and caring and asks that all members within it form good learning relationships with one another. Learners are valued within this environment and allowed the autonomy to fit in and understand that their ideas, questions and solutions are always met with interest and respect.

An environment that builds mathematical resilience meets the basic psychological needs of its members and promotes their motivation and well-being. This is true for the learners and the teachers. Where teachers are expected to teach in certain ways, for example, telling the learners that they must remember the steps in a given process and testing them to see if they have, they are thwarting the learners' need for autonomy, competence and relatedness. They are thwarting learners' autonomy, since the learners are rarely asked how they would like to learn and if they would prefer to work in other ways. They are thwarting the learners' feelings of competence, as competence is achieved through understanding and being in control of ideas, not through remembering a sequence of poorly understood steps. They may also be thwarting learners need for relatedness by not allowing the learners to support one another. They are therefore possibly contributing to distress and anxiety. Most teachers join the profession to help young people succeed. To be placed in a position where they must thwart their learners' psychological needs, and cause stress and anxiety, means the teachers' needs are also not being met and that will affect their own well-being.

The environment also needs to be a brave space. If the psychological safety and well-being of all the actors within the environment and their mathematical learning is to be supported, then all those actors must experience respect and have a voice which is listened to. Those learners who come to the environment with established mathematics anxiety will take some time and intervention to convince them that they will not be laughed at or humiliated if they speak up or ask a question. They will have kept quiet to protect their psychological safety and well-being for some time. They will need to be brave to speak out and accept the support they need. A brave space is where everyone is encouraged to talk through the issues they have experienced, the mistakes that they have made and either talk about or invite suggestions about how to move forward will help. A brave space deliberately includes everyone, acknowledges the courage it takes to speak out and emphasises the supportive connections that can be built by doing so. It builds competence through relatedness and offers autonomy.

An environment which builds mathematical resilience

As considered earlier, there are four factors to constructing an environment that builds mathematical resilience and meets the learners' psychological needs. These are growth mindset, value, struggle, and support (Figure 1.1).

The learning environment must convey the growth mindset (Dweck, 2000) (Figure 1.2). The ethos in the environment will expect, and importantly support, everyone to succeed in progressing their mathematical learning. It will use carefully chosen and challenging activities and tasks where the learners' efforts will result in progress and success. The learners will know that the fixed mindset is prevalent but that it is inaccurate. If they make mistakes or find problems which take some time and thought to solve, those with a fixed mindset will fear this indicates their inability to engage with mathematical ideas. Those with fixed mindsets avoid challenge as they live in constant fear of their limited intelligence (which they believe to be

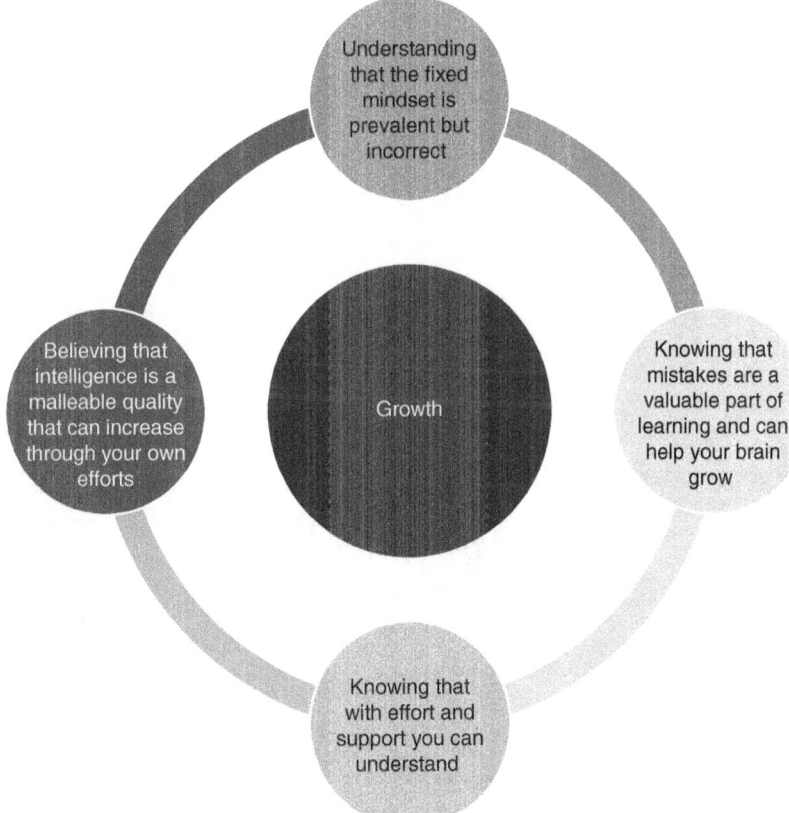

Figure 1.2 Promoting a growth mindset.

unchangeable) being exposed and that they will be found to be 'stupid'. Rows of sums all the same which can be simply and quickly answered are comforting as they confirm that their fixed level of intelligence has not been reached, yet. However, without accepting challenge, with support, mathematical learning does not grow. Teachers developing mathematical resilience counteract the fixed mindset message by clearly stating that everyone can progress from where they are with support and encouragement and that capability can be increased with effort. They will also help learners to understand that mistakes are a valuable part of learning as they help brains grow understanding and show that they are engaging in the kind of work that will progress their learning.

The environment that develops mathematical resilience in learners will also help learners understand the value of mathematics to them and that they are of value in that environment. The learning offered will exemplify the mathematical thinking that is used extensively within society (see for example Pozzi, Noss and Hoyles, 1998; Roberts, 2002). The teacher will be aware that school mathematics can be likened to a chameleon (Lee and Johnston-Wilder, 2013) hiding in plain sight in the real world, and thus will make clear where and how mathematics can be found and thereby demonstrate its value to the learner (Figure 1.3).

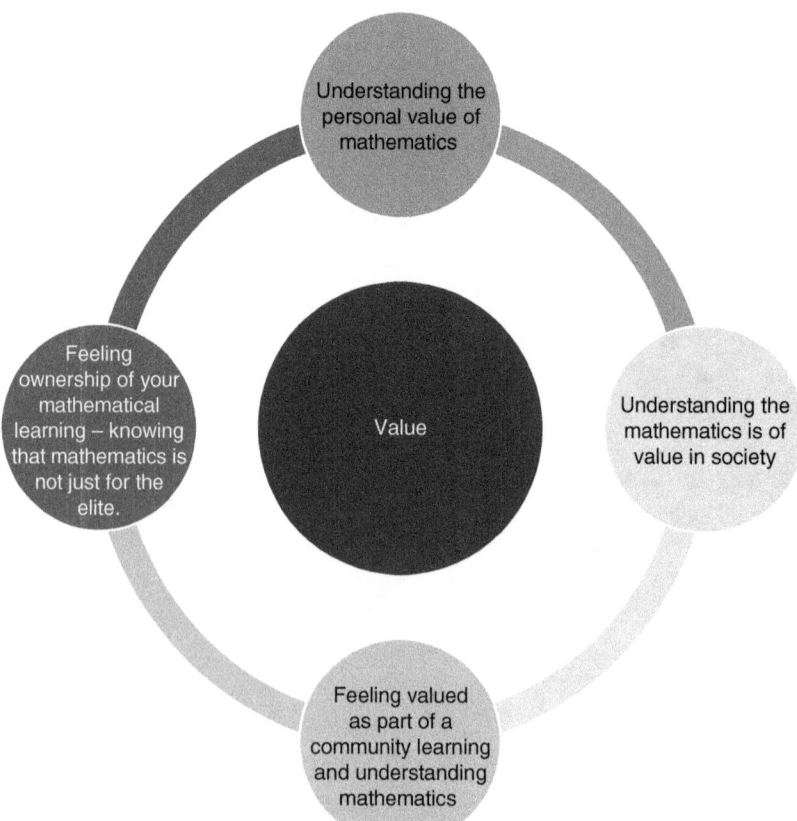

Figure 1.3 Understanding the value of mathematics.

The idea of value also encompasses ensuring each learner feels included and valued in a community of practice where mathematical understanding is sought and where mathematical ideas and concepts are used. Working collaboratively on challenging tasks (see Swan, 2006 for examples as well as chapter 6 in this book) is likely to reinforce learners' feelings of ownership of the ideas and concepts that make up mathematics, further provoking the idea that mathematics is of personal value to them.

It became clear to us that learners rarely discuss how to work at learning mathematics as they learn (Lee and Johnston-Wilder, 2013). Within the mathematically resilient environment, teachers will help the learners understand how to work successfully at learning mathematics. They will make plain that learning mathematics requires courage, effort, perseverance and struggle. Alongside this, they will emphasise that no-one is expected to struggle alone, and that support in ways that the learner needs will always be available (Figure 1.4).

Path smoothing (Wigley, 1992), which means providing hints and tips or examples that can easily be followed, can be a teachers' default position when a learner is struggling. Stigler and Hiebert (1999) found that teachers felt they were not doing their job properly if their learners struggled. However, taking this

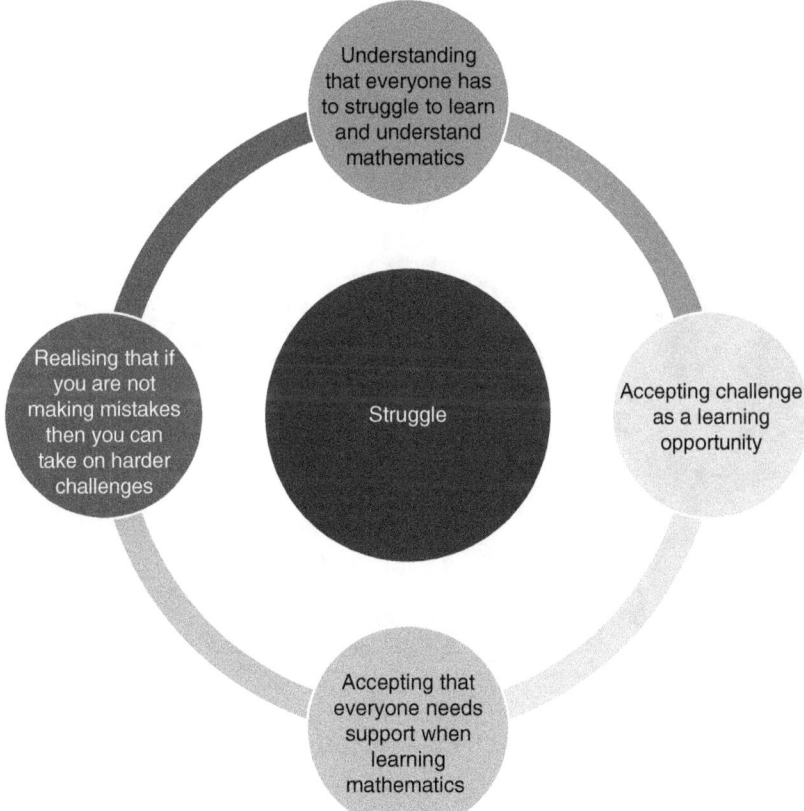

Figure 1.4 The factor of struggle.

approach results in learners feeling they 'can't do it' on their own when faced with a mathematical question or task that cannot be quickly answered. Getting stuck (Mason, 2015) and struggling towards a solution, with rough working, making mistakes and taking time, are intrinsic parts of learning and using mathematics, but learners seem to need to have this made explicit (Lee and Johnston, 2013) and possibly modelled by their teacher. Becoming a resilient learner involves coming to know both that everyone needs support to overcome the barriers that learning mathematics can present to anyone and how to seek the support appropriate for them when stuck and before the struggle becomes unproductive or overwhelming (Figure 1.5).

Struggle without appropriate support leads to counterproductive anxiety. What constitutes appropriate support differs with the individual and the situation. Too often learners feel their only option is to wait for the teacher to explain the ideas again, which often means a lengthy and sometimes fruitless wait. Teaching for mathematical resilience means helping learners know that they have the right to be supported. Learners are helped to know that everyone needs support, where that support may be found and that deciding to access support by, for example, talking

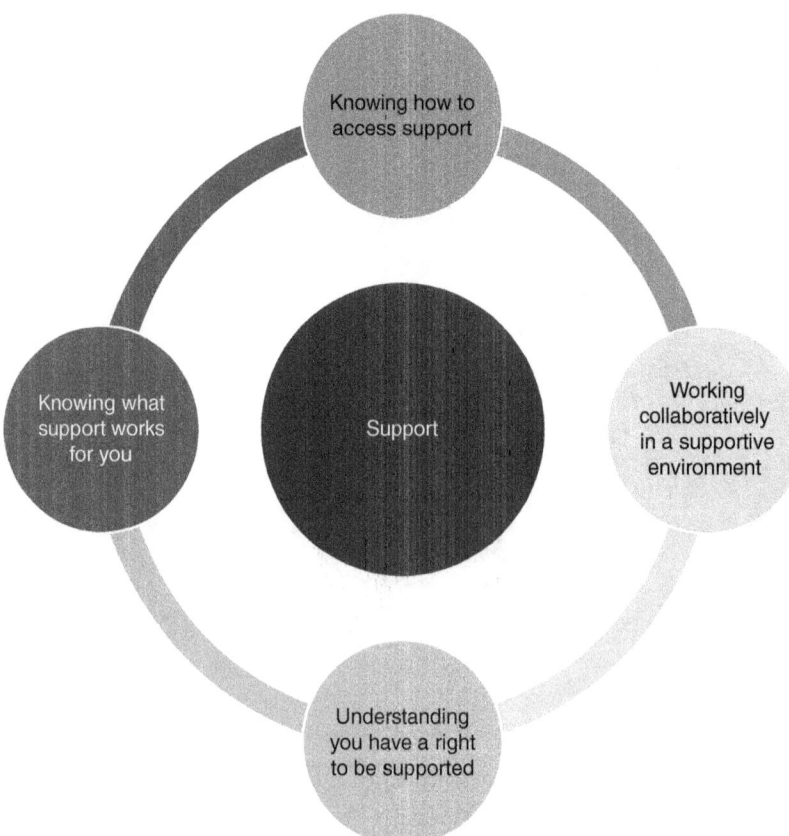

Figure 1.5 The factor of support.

through the ideas with a friend, looking in a book or online for help or just sitting quietly mulling over what to do, is important in progressing with mathematics.

A supportive environment is often, but not always, a collaborative environment. Where learners work together on tasks that require involvement from all collaborators, they support one another's thinking and reasoning. We have seen over and over again how collaboration allows inter-mental thinking to develop between individuals, which results in the intra-mental development of higher mental functions within individuals (Vygotsky, 1978). Collaboration can be very supportive but sometimes individuals prefer to quietly think or look up ideas in books or on the internet. Getting the support needed is important and spending time with learners, helping them think through how best they can be supported is time well spent.

Mathematics anxiety

Throughout this chapter, we have discussed how common teaching practices often result in mathematics anxiety. Here we will go into a little more detail about what mathematics anxiety is and how it may develop. Chapter 2 will offer more depth in this discussion. Mathematics anxiety is detrimental to a person's career prospects and also damages their day-to-day life. If someone has mathematics anxiety, they will avoid thinking about anything mathematical because it causes them to experience uncomfortable and sometimes traumatic feelings. If a person avoids thinking about mathematical ideas, then even when politicians present accurate graphs and honestly describe the certainty or uncertainty of a particular prediction being offered, they are incapable of listening effectively. Poor choices can result from not be able to listen to the mathematics; a misunderstanding of probabilities associated with risk brings more fear and more anxiety in its wake. It seems that mathematics anxiety can be the result of the pedagogical choices that many teachers feel compelled to make in UK schools and schools worldwide (Ashcraft and Krause, 2007; Smith, 2008).

> When I was at primary school, if I didn't understand something in maths there was an attitude towards me like, 'I've told you how to do it once'. This made me incredibly anxious, and that anxiety grew into a fear of the subject because I knew myself that I didn't know how to do the maths and that became embarrassing. I was embarrassed because I didn't want the other children to know that I couldn't do it. I didn't trust that the teachers were going to help me. The teacher would shout so everyone in the class knew who didn't understand the work.

The worldwide assessment of levels of mathematics anxiety conducted by The Organisation for Economic Co-operation and Development (OECD, 2014) confirmed the prevalence of mathematics anxiety globally. They found 59% of 15–16-year-olds in their global survey could be considered to be mathematics anxious and 31% were very nervous about mathematics lessons. Mathematics anxiety has a statistically significant effect on performance (Barroso et al., 2021); Luttenberger et al. (2018) confirm the outcomes of having mathematics anxiety are not only lower performance in mathematics-related situations, but there are also long-term effects which lead to sub-optimal learning in other subjects and lead learners to choose courses and vocations that are least likely to expose them to mathematics.

There are several factors known to develop mathematics anxiety. Dowker et al. (2016) and Luttenberger et al. (2018) confirm the factors which increase mathematics anxiety include:

- teachers' and parents' attitudes toward learners' ability in mathematics. Such attitudes are based in the prevalent idea that some people can do mathematics, but others cannot, described by Dweck (2000) as having an entity or fixed mindset. Holding the opposite, incremental, or growth mindset is vital for effective learning of mathematics and helps learners build resilience.

- societal stereotypes, for example, attitudes towards females' mathematical abilities, can be a factor in the development of mathematics anxiety. A further factor here is "elitism" (Nardi and Steward, 2003), a harmful but prevalent myth that leads to people seeing mathematics as the province of an elite few that most people cannot aspire to join. The anxiety here may stem from being asked to work on mathematics when holding the belief that they cannot possibly be expected to engage with these ideas.

- having low self-efficacy or motivation (Ryan and Deci, 2018) in mathematics, which intensifies mathematics anxiety and which many common pedagogic practices such as "quick quizzes" and competitive "games" confirm in many learners (Ashcraft and Krause, 2007).

There are measures which can be taken to reduce the risk of development of mathematics anxiety, or to mitigate its effect once developed, by educational institutions, teachers, learners or parents. These measures are the subject of the chapters in this book. You will find a specific focus in each chapter on teaching for mathematics resilience to a variety of age groups as well as how schools can help parents and carers build resilience in young people.

Conclusion

Helping learners develop mathematical resilience is important if sufficient people are to go onto study mathematics or other subjects that require mathematical thinking at higher levels. Such study is needed to provide the expertise to offer solutions to the complex problems caused by pandemics, climate change and problems we cannot yet foresee. Teaching in ways that will enable learners to become mathematical resilient will increase their mathematically well-being and decrease ill-being, thereby reducing the stress and anxiety in society. Increasing learners' ability to manage the anxieties that learning mathematics can and does cause, to overcome barriers to their learning and to reach an understanding of the complex, connected system that is termed mathematics is the aim of this book. Read on to find discussions of the impact and prevalence of mathematics anxiety and ideas and tools that will help you work with your learners, whatever age they are, in ways that mean they grow their mathematical resilience.

References

Ali, M., Ashraf, B., and Shuai, C. (2019). Teachers' conflict-inducing attitudes and their repercussions on students' psychological health and learning outcomes. *International Journal of Environmental Research on Public Health*, 16(14), 2534.

Ashcraft, M., and Krause, J. (2007). Working memory, math performance, and math anxiety. *Psychonomic Bulletin & Review*, 14, 243–248. 10.3758/BF03194059

Barroso, C., Ganley, C. M., McGraw, A. L., Geer, E. A., Hart, S. A., and Daucourt, M. C. (2021). A meta-analysis of the relation between math anxiety and math achievement. *Psychological Bulletin*, 147(2), 134–168. 10.1037/bul0000307

Black, P., Harrison, C., Lee, C., Marshall, B., and William, D. (2003). *Assessment for learning- putting it into practice.* Maidenhead, U.K.: Open University Press.

Dowker, A., Sakar, A., and Looi, C. (2016). Mathematics anxiety: What have we learned in 60 years? *Frontiers in Psychology.* https://doi.org/10.3389/fpsyg.2016.00508

Dweck, C. (2000). *Self-theories: Their role in motivation, personality, and development.* Philadelphia, PA: Psychology Press.

Hattie, J., and Yates, G. (2013). *Visible learning and the science of how we learn.* Abingdon UK: Routledge.

Hyman, I., and Snook, P. (1999). *Dangerous schools. What we can do about the physical and emotional abuse of our children.* San Francisco: Jossey-Bass Publishers.

Lee, C., and Johnston-Wilder, S. (2013). Learning mathematics—Letting the pupils have their say. *Educational Studies in Mathematics*, 83, 163–180. 10.1007/s10649-012-9445-3

Luttenberger, S., Wimmer, S., and Paechter, M. (2018). Spotlight on math anxiety. *Psychological Research and Behaviour Management*, 2018(11), 311–322. 10.2147/PRBM.S141421

Mason, J. (2015). On being stuck on a mathematical problem: What does it mean to have something come-to-mind? *Lumat: International Journal of Math, Science and Technology Education*, 3, 101–121. 10.31129/lumat.v3i1.1054

McKay, C., and Macomber, G. (2023). The importance of relationships in education: Reflections of current educators. *Journal of Education*, 203(4), 751–758. 10.1177/00220574211057044

Nardi E., and Steward, S. (2003). Is Mathematics T.I.R.E.D? A profile of quiet disaffection in the secondary mathematics classroom. *British Educational Research Journal*, 29(3), 345–367. http://www.jstor.org/stable/1502257

OECD. (2014). PISA 2012 results in focus organisation for economic co-operation and development. https://www.oecd.org/pisa/keyfindings/pisa-2012-results-overview.pdf

Pozzi, S., Noss, R., and Hoyles, C. (1998). Tools in practice, mathematics in use. *Educational Studies in Mathematics*, 36, 105–122. 10.1023/A:1003216218471

Roberts, G. (2002). *SET for success. The supply of people with science, technology, engineering and mathematics skills.* London: HM Treasury.

Ryan, R., and Deci, E. (2018). *Self-determination theory: Basic psychological needs in motivation, development, and wellness.* Guilford Press.

Sava, F. (2002). Causes and effects of teacher conflict-inducing attitudes towards pupils. *Teaching and Teacher Education*, 18, 1007–1021.

Sethi, J., and Scales, P. C. (2020). Developmental relationships and school success: How teachers, parents, and friends affect educational outcomes and what actions students say matter most. *Contemporary Educational Psychology*, 63, 101904.

Smith, A. (2008). *Making mathematics count.* London: The Stationary Office.

Split, J. L., Hughes, J. N., Wu, J.-Y., and Kwok, O.-M. (2012). Dynamics of teacher-student relationships. *Child Development*, 83(4): 1180–1195. 10.1111/j.1467-8624.2012.01761.x

Stigler, J. W., and Hiebert, J. (1999). *The teaching gap: Best ideas from the world's teachers for improving in the classroom.* New York: The Free Press.

Swan, M. (2006). *Collaborative learning in mathematics: A challenge to our beliefs and practices.* Leicester, UK: NIACE

Wigley, A. (1992). Models for teaching mathematics. *Mathematics Teaching*, 141, 4–7.

Vygotsky, L. S. (1978). *Mind in society: The development of higher psychological processes.* Cambridge, MA: Harvard University Press.

Yeager, D., and Dweck C. (2012). Mindsets that promote resilience: When students believe that personal characteristics can be developed. *Educational Psychologist*, 47(4), 302–314. 10.1080/00461520.2012.722805

Yeager, D., and Dweck, C. (2020). What can be learned from growth mindset controversies? *The American Psychologist*, 75(9), 1269–1284. 10.1037/amp0000794

Yeager, D., Romero, C., Paunesku, D., Hulleman, C., Schneider, B., Hinojosa, C., Lee, H., O'Brien, J., Flint, K., Roberts, A., Trott, J., Greene, D., Walton, G. M., and Dweck, C. (2016). Using design thinking to improve psychological interventions: The case of the growth mindset during the transition to high school. *Journal of Educational Psychology*, 108(3), 374–391. 10.1037/edu0000098

Recognising and measuring mathematics anxiety and resilience

Thomas E. Hunt and Dominic Petronzi

Introduction

Mathematics anxiety is central to the discussion of psychological safety and well-being within mathematics education. It is a core issue in addressing mathematical resilience and is prevalent across many groups of learners. For example, in a study of over 300 adult learners, Hunt and Maloney (2022) found that mathematics anxiety was generally higher among those with lower mathematical resilience. More recently, Hunt and colleagues (paper in preparation) studied over 1,600 14–16-year-olds from 7 different countries and again found a significant negative relationship between these variables. After many years of research, it is clear to us that mathematics anxiety is one of the main psychological barriers for mathematics learning, performance, and general engagement with mathematics. It has been suggested that children start school with intrinsically positive attitudes towards mathematics (Nicolaidou and Philippou, 2003), but research has indicated that mathematics anxiety can develop in children as young as four years of age (e.g., Petronzi et al., 2017) with attitudes formed because of classroom-based negativity (Fraser and Honeyford, 2013) such as experiencing failure and negative peer evaluation.

Given the prevalence and potentially long-term impact of mathematics anxiety, the development of suitable measures is important, particularly for those who may experience challenges associated with dyscalculia or mathematics related learning difficulties. Our understanding of the general cognitive implications of experiencing mathematics anxiety, for example, limited working memory leading to an affective drop in performance (Ashcraft and Moore, 2009), emotional implications (Luo et al., 2009; Young et al., 2012) and avoidance via the 'no-attempt' error (Chinn, 2012) further emphasises the need for early identification of children at risk. Quantitative research into mathematics anxiety almost always adopts some form of self-report rating scale of mathematics anxiety. Within classrooms especially, such scales provide researchers with a way of establishing the extent of mathematics anxiety in the learners, which means relationships can be tested with other psychological variables, but also behaviours too, such as mathematics performance. Scales also mean that differences can be tested between groups and over time, e.g., assessing the effectiveness of an intervention for reducing mathematics anxiety. Also, a perhaps under-recognised use of scales is as part of the initial phase of intervention work or action research. In some cases, working with such scales might be the first time that a learner has properly considered concepts such as mathematics anxiety and resilience and scales can act as an aid to reflect on these. Here we discuss mathematics anxiety rating scales alongside other ways in which mathematics anxiety might be recognised and measured. As we do this, we will also consider how mathematical resilience can be, and has been, measured.

Self-report scales for mathematics anxiety

Dreger and Aiken (1957) are often cited as being the first to empirically study mathematics anxiety, beginning with what they termed "number anxiety". They did this by including an additional three mathematics-specific items to an existing anxiety scale. The wording of two of these items suggests that Dreger and Aiken conceptualised numerical anxiety in terms of nervousness and "freezing up" in relation to "doing" arithmetic and "seeing" mathematics problems. The distinction between doing and seeing mathematics is an important one. Whilst some people find actual engagement with a mathematics task, "doing", to be anxiety provoking, for others simple exposure to mathematical content, "seeing", or even thinking about it, can induce anxiety. If teachers understand how mathematics anxiety might be engendered or increased in individual learners, these threats to well-being could be avoided or the threat mitigated.

It was not until 1972 that the first purpose-built scale was designed to measure mathematics anxiety: The Mathematics Anxiety Rating Scale (MARS; Richardson and Suinn, 1972). Since its publication, many further mathematics anxiety scales have been suggested, based on MARS, including more concise versions and variations of it. A second family of self-report mathematics anxiety measures for adults is based on the Abbreviated Mathematics Anxiety Scale, or AMAS (Hopko et al., 2003). A typical format for mathematics anxiety scales is to ask the learner to respond on a scale of 1 to 5 to show how anxious they would feel during an event specified by a series of items, such as "Thinking about an upcoming math test one day before". One of the reasons for the popularity of the AMAS is its short length (9 items), although it has proven to be a valid and reliable measure. However, it usually requires some modification for use outside of the American education system.

The Mathematics Anxiety Scale UK (MAS-UK, Hunt et al., 2011) is a 23-item self-report measure that has been used extensively since its publication. It is designed with a UK audience in mind and was specifically created for adults and older adolescents. It covers a wider range of mathematics situations compared to other scales and is closely related to the American Shortened Mathematics Anxiety Rating Scale (Alexander and Martray, 1989). A 10-item version of the MAS-UK has also been piloted for secondary school pupils and is available from the authors of this chapter on request. The Mathematics Anxiety Scale (Betz, 1978) is another popular scale which has been modified according to specific groups of learners (see Table 2.1).

Table 2.1 Suggested self-report mathematics anxiety scales for different learners

Scale	Suggested suitability/group of learners	No. Items (specific to mathematics anxiety)
Children's Mathematics Anxiety Scale UK (Petronzi et al., 2018)	Lower primary	19
Child Math Anxiety Questionnaire (Ramirez et al., 2013, developed from MARS-EC, Suinn et al., 1988)	Lower-to-mid primary	8
Scale for Early Mathematics Anxiety (Wu et al., 2012)	Mid primary	20
Mathematics Attitude and Anxiety Questionnaire (Dowker et al., 2012)	Mid-to-late primary	7
Children's Anxiety in Math Scale (Jameson, 2013)	Primary	16
Modified Abbreviated Math Anxiety Scale (Carey et al., 2017)	Mid primary to mid secondary	9
Mathematics Anxiety Scale for Children (Chiu and Henry, 1990)	Mid primary to mid secondary	22
Mathematics Anxiety Rating Scale for Elementary Children (Suinn et al., 1988)	Upper primary to secondary	26
Modified version of the Betz (1978) Mathematics Anxiety Scale (Baker, 2021)	Lower secondary	10
Abbreviated Math Anxiety Scale (Hopko et al., 2003)	Mid-secondary to adult	9
Mathematics Calculation Anxiety Scale (Hunt et al., 2019)	Upper secondary and adults	26
Mathematics Anxiety Scale UK (Hunt et al., 2011)	Adults	23

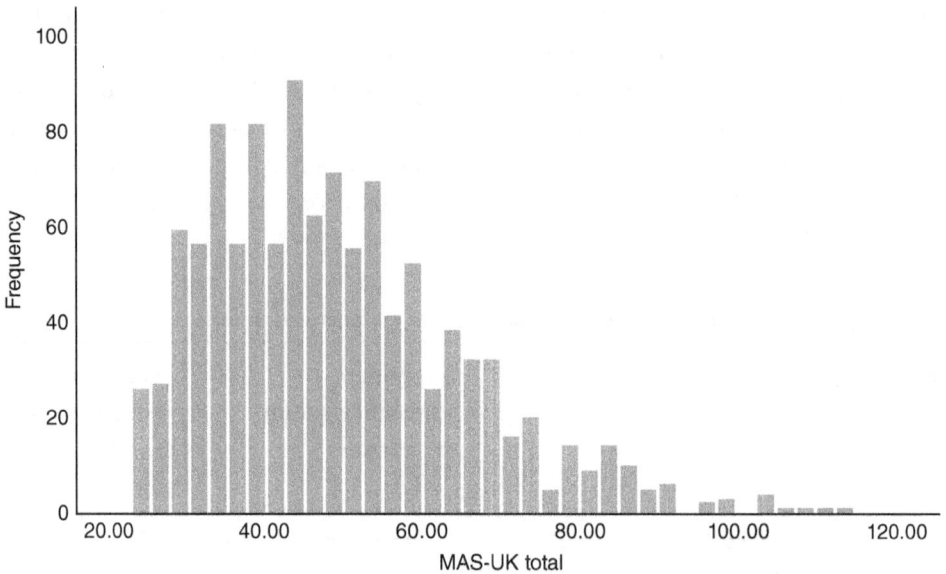

Figure 2.1 The distribution of total MAS-UK scores (Hunt et al., 2011).

The original 23-item MAS-UK was administered to over 1,000 undergraduate students in the UK and the distribution of scores (the sum of scores on the 23 items) can be seen in Figure 2.1. The maximum score is 115 and the minimum score is 23. This is a good example of how mathematics anxiety scores are typically distributed; whilst they begin to approximate a normal distribution, there is some skew towards lower levels of mathematics anxiety, with a small but notable number of high scores at the top end of the scale. The data suggests that most people experience some degree of mathematics anxiety. It also suggests that there is no obvious cut-off for what we might consider "high mathematics anxiety". Many studies arbitrarily divide participants into groups, indicating levels of mathematics anxiety, such as "low mathematics anxiety" and "high mathematics anxiety" based on a median split, i.e., those who fall lower or higher than the median value. This has some practical value, but it is important to acknowledge that it does not necessarily reflect the reality of the spread of mathematics anxiety scores across a group of learners.

Carey and colleagues (2017) carried out a study where they administered a modified version of the AMAS (Hopko et al., 2003) to over 1,700 pupils aged 8–13 in the UK. This research involved a large sample of UK learners and explored separately how pupils experience anxiety when they are learning mathematics and when they are being tested. As per the findings of Hunt et al. (2011), the overall distribution of total mathematics anxiety scores was slightly skewed, with a higher proportion of pupils at the lower end of the scale. However, upon viewing the scores for the sub-scales of Mathematics Learning Anxiety and Mathematics Evaluation Anxiety separately, the distributions were very different. The scores for Mathematics Learning Anxiety were quite highly skewed, with most learners scoring quite low in anxiety concerning learning mathematics. On the other hand, anxiety concerning mathematics evaluation was more evenly distributed, with a higher average score. The finding that mathematics anxiety was higher in the context of being evaluated or tested is very common and is likely closely linked to various features that predict mathematics anxiety, such as others' attitudes towards the mathematics learner and the mathematics learner's self-efficacy, as discussed in Chapter 1. This has important implications within mathematics education given the prevalence of mathematics tests.

In more recent years, new mathematics anxiety scales have been created with younger people in mind, particularly those within primary/elementary education. The original mathematics anxiety measures were devised with adults in mind, resulting in both content that is not relatable to a younger audience, and response options that might not be easily understood. Table 2.1 shows some of the self-report scales for mathematics anxiety that we recommend for children or adolescents. It is important to be aware that there is some variation in content and style between these scales, mostly based on the age groups that were targeted when the scales were designed. Therefore, we advise you consider each scale carefully before choosing which scale is most suitable for the age of the learners completing it.

Anxiety rating scales for young learners look very different to those for adolescents for adults. For example, the Children's Mathematics Anxiety Scale-UK (CMAS-UK) (Petronzi et al., 2018) is a 19-item scale which utilises an emoji response format (see Figure 2.2.) The items, developed through focus groups with

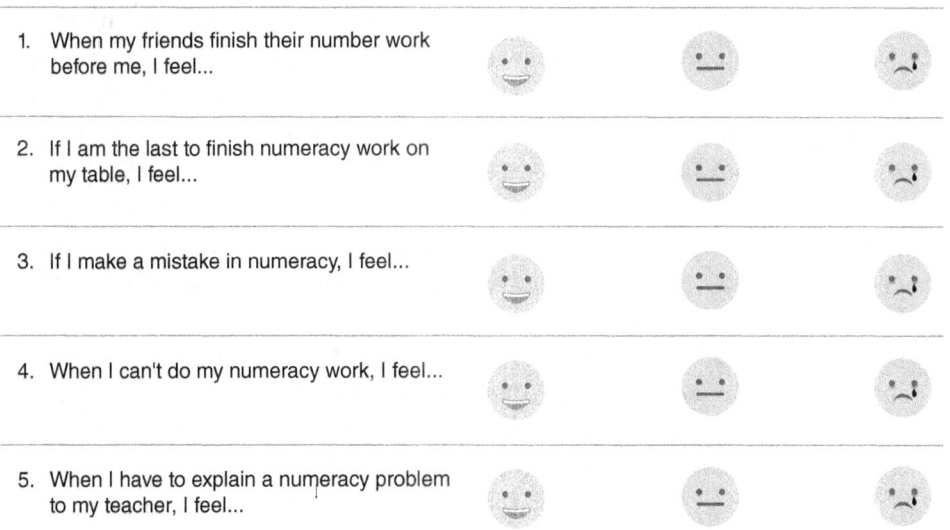

Figure 2.2 The first 5 items on the Children's Mathematics Anxiety Scale UK (Petronzi et al., 2018).

children aged between 4 and 7 in the UK, reflect a typical mathematics lesson and the types of situations a child might encounter, which the qualitative research suggested might be associated with feelings of anxiety. Items represent core concepts such as a fear of failure and negative feelings related to peer comparison. Relating these ideas to the principles of mathematical resilience, including growth mindset, value, struggle, and support may allow a teacher to consider what is and what is not causing mathematics anxiety if they find it is present in their classrooms. The CMAS-UK has proven to be a useful tool for teachers in UK primary schools, although there are some practical considerations, such as the need to support younger learners in completion of the scale owing to challenges related to reading ability and sustained attention.

Even though mathematics anxiety scales can be quite short, for example one scale proposed by Nunez-Pena et al. (2014) was just one question long, research suggests that there are many aspects or dimensions to mathematics anxiety. Some studies use factor analysis, a statistical method for exploring whether certain items group together in a way that suggests an underlying factor (measured via a subscale). We have already discussed one example that uses these ideas, that of the Mathematics Learning Anxiety and Mathematics Evaluation Anxiety subscales of the AMAS (Hopko et al., 2003). These subscales represent what are thought to be two of the main dimensions of mathematics anxiety. It may be useful for teachers to have a broader awareness of the range of factors and subscales that have been proposed as these have implications for addressing mathematics anxiety through classroom pedagogy and targeted interventions. Some scales have been shown to have a subscale relating to anxiety associated with social or everyday mathematics, beyond a formal educational setting (e.g., Hunt et al., 2011). Others have suggested the existence of anxiety associated with abstract mathematics e.g., algebra which is seen as separate from standard numerical problem solving. One self-report scale (Hunt et al., 2019) was designed to measure mathematics calculation anxiety associated with a wider range of mathematics tasks than is typically included in mathematics anxiety scales. The Mathematics Calculation Anxiety Scale (Hunt et al., 2019) includes mathematics problems from across the English GCSE mathematics curriculum (the curriculum followed by young people up to the age of 16 years in England) and is thought to be a useful measure for understanding anxiety associated with specific areas of mathematics: abstract mathematics anxiety, statistics probability anxiety, statistics calculation anxiety, and numerical calculation anxiety.

A question that we often receive from teachers is "what score [on a particular self-report scale] represents high mathematics anxiety?" It is difficult to answer this question. Since there is currently a lack of data from specific groups of learners, it is not possible to assume that the distribution of mathematics anxiety scores from self-report measures is the same for all groups of learners. This question also represents an interesting point about what the results from using the scales can be utilised for. Teachers might use the results to change the way they challenge certain learners or to help identify individual learners in need of specific support. They might also use them as part of direct intervention resources, whether that relates to mathematics anxiety specifically, or as part of a broader intervention that targets developing mathematical resilience. Whilst such scales are useful as a way of observing changes in mathematics anxiety across time, care should always be taken to avoid using scores from mathematics anxiety scales to label a learner in any particular way. Values derived from scales should always be viewed in relation to the wider group. For example, a child might score high on mathematics anxiety compared to the rest of their class, but their score could actually be fairly normal – or even low – compared to all other children in their age group.

Self-report measures are often a quick and (fairly) easy way of assessing overall mathematics anxiety among many learners in a short period of time. However, the learners who are most at risk are those whose self-report score does not quite get them into the bracket of high mathematics anxiety. After all, self-report measures are not perfect – they are subject to response bias, e.g., learners might not want to appear mathematics anxious; also, as mentioned earlier, the concept of a threshold for a high mathematics anxiety score is often ill-defined or unnecessary. Therefore, when the intention is to identify mathematics anxiety in learners for the purpose of supporting individuals with high mathematics anxiety, there is a need to employ a broader range of techniques.

Moving beyond self-report measures

Another way of identifying mathematics anxiety in learners is through observation of their behaviour. Naturally, teachers observe learners during lessons, but what we are discussing here is how one might look specifically for the types of behaviours that might suggest a learner is mathematics anxious. This approach offers another dimension to identifying and measuring mathematics anxiety beyond self-reports. It also represents a useful way of identifying learners who might be feeling mathematics anxious, especially when they are not forthcoming in speaking out or were reluctant to expose the extent of their anxiety when self-reporting. Indeed, many learners might not directly associate certain negative feelings (whether psychological or physiological) with mathematics learning or testing. Instead, they might have more of a general feeling or attitude towards the subject, such as a dislike or lack of enjoyment. Behaviours, including withdrawal, avoidance, and disruption have been observed in learners that struggle with mathematics anxiety. They can be seen as a form of communication, and they can indicate a learner's wants or needs. Such behaviours perhaps indicate that their resilience is too low to manage the stress and worry that they feel when doing mathematics.

Behaviours resulting from mathematics anxiety might also include those that indicate a certain degree of avoidance, e.g., inattentiveness or generally being disruptive. Anxiety could also manifest itself through a learner appearing frustrated or angry. They may attempt to copy from peers to avoid exerting mental effort or running the risk of individual failure. Or they might speed through mathematics work, to simply get it over and done with at the expense of making mistakes through lack of attention, which is known as the speed-accuracy-trade-off. Also, appearing withdrawn or subdued might suggest a conscious or subconscious attempt to avoid being questioned or have the extent of their mathematics work acknowledged in any way. Being withdrawn might also indicate the existence of intrusive thoughts which is thought to be a common occurrence among those with higher mathematics anxiety. Moreover, at times, intrusive thoughts can translate into negative self-talk, especially talk which hints at low self-efficacy. Therefore, teachers should note what learners say during, for example, mathematics lessons. For instance, the findings of Petronzi et al. (2017) suggest considering whether learners say things that suggest they are comparing their understanding or performance against their peers and whether they mention possible negative consequences of failure. Or maybe they comment on their perceived inability to progress. Petronzi et al. (2017) suggest it is a common theme among current or former learners who self-report as being highly mathematics anxious that being asked a mathematics question in front of others is a key trigger for their anxiety. This might be in relation to fellow learners, i.e., a fear of looking foolish if an incorrect (or no) response is given, or it might be through a fear of how a teacher might react. Considering this in relation to recognising mathematics anxiety, it becomes important for teachers to observe learners' reactions during evaluative situations involving mathematics. A no-response might be indicative of experiencing fear, rather than an absence of motivation or attention.

A major challenge when drawing links between behaviours and mathematics anxiety is whether learners behave in different ways in lessons other than mathematics. A comparison of learners' behaviours in mathematics versus non-mathematics situations is an important factor in forming conclusions about the level of a learners' anxiety that is specific to mathematics. In addition to external behaviours, in a study carried out by Hunt et al., (2017), it was observed that primary school learners' self-reported mathematics anxiety was positively related to a change in their blood pressure as mathematics problems became more difficult. As such, the physiological changes learners might experience in response to challenge sometimes go beyond what might be regarded as normal, possibly resulting in physical signs, e.g., looking flush in the face, excessive sweating, etc., that need to be acknowledged by teachers. Teachers can also aid learners to recognise and understand their own bodily changes, perhaps using tools such as the Growth Zone Model (Chapter 3); this is one of the core components of a reappraisal approach to tackling mathematics anxiety (Jamieson et al., 2016).

From mathematics anxiety to self-efficacy and resilience

Before turning to a consideration of mathematical resilience, it is worth being mindful that mathematics anxiety is not the only psychological concept which feeds into mathematical learning. Motivation, beliefs, and self-efficacy all play a part. Self-efficacy refers to the individual's belief that they can complete a certain task successfully, and research shows that self-efficacy, in terms of academic work, can be a predictive factor for academic outcomes (Owen and Froman, 1988). Moreover, both pessimism and negative attitudes in mathematics have been shown to relate to self-efficacy and can lead to motivational and cognitive deficits (Kolacinski, 2003). Therefore, it is clear to see why self-efficacy has been considered as a crucial factor within mathematics education. Importantly, as it can be increased by developing cognitive skills which also support persistence, performance, and interest, it seems important for resilience. There are scales and ways to observe and measure these constructs too, but there is insufficient space here to discuss these fully. In Chapter 1, the authors propose that students low in mathematical resilience can progress in mathematics by overcoming their mathematics anxiety with the help of teachers and resources that can empower them. Those who learn to strengthen their ability to manage their emotions and cope with situations that they perceive as threatening and challenging can enhance their mathematical resilience. Consequently, it is important to recognise and measure mathematical resilience along with mathematics anxiety among learners. This enables both learners and educators to increase awareness of a hidden dimension in the classroom, to target attention, and to identify the extent of the challenges that are faced. All these features help educators understand how best to support learners and to direct resources accordingly. Resilience can be considered on a continuum, rather than as being, for example, 'good or 'bad' or 'resilient' or 'not resilient' and there may be a number of interacting life experiences that lead a learner to have higher or lower resilience, as well as a potential genetic basis that can be activated under certain environmental conditions (Rubinsten et al., 2018).

Mathematics anxiety can be considered as multi-dimensional and similarly, as noted in the previous chapter it is proposed that mathematical resilience comprises four components: growth mindset, struggle, value, and support. The first three relate to an individual learner's self-beliefs concerning their mathematics learning. The Mathematical Resilience Scale (MRS) (Kooken et al., 2016) is a self-report scale that can be used to measure these components. This includes 24 statements (see Table 2.2 for examples) and learners are asked to respond on a scale from 1 "strongly agree" to 7 "strongly disagree". Scores can be created for the

Table 2.2 Example items from the Mathematical Resilience Scale (Kooken et al., 2016)

Example item	Sub-scale
Math can be learned by anyone	Growth
Math is essential for my future	Value
Everyone struggles with math at some point	Struggle

individual components of mathematical resilience, or a total mathematical resilience score can be created by combining the components. The MRS can be used to ascertain learners' current level of mathematical resilience, including which components might need to be targeted, e.g., one learner may have particularly unhelpful beliefs regarding the value of struggle in mathematics learning, whereas another learner might clearly experience a fixed mindset.

Currently, the MRS does not measure learners' perceptions of support. Recent work by Lau et al. (2022) highlighted the importance of the role of others in the context of learners' mathematics anxiety, demonstrating a negative relationship between learners' perception of teacher competence and the learner's mathematics anxiety. Also, people with high mathematics anxiety often talk about negative mathematics experiences with others, especially teachers (Bekdemir, 2010). Therefore, measuring perceived support could be important in the context of understanding the development of mathematical resilience.

Another scale designed to explore resilience in mathematics learning is the Academic Resilience in Mathematics Scale (ARMS, Ricketts et al., 2017). Whilst this 9-item scale is not as extensive as the MRS, it considers resilience in the context of mathematics slightly differently, although there is some obvious overlap. One of the differences is the inclusion of items related to support, e.g., "I know where to get help if I'm having trouble with math" (note that an item pertaining to seeking support was included in the original MRS as part of the development process but was later removed following various statistical analyses). Again, aside from self-reports, it is possible to look more broadly for indicators of resilience. At the macro level, selecting courses involving mathematics could reflect the development of a high degree of resilience if the learner has struggled with mathematics in one way or another. Similar observations can also be made at the micro level, for example observed perseverance when attempting mathematics problems in-class.

Measurement within action research

Some studies have utilised a mathematical resilience model to inform interventions for reducing mathematics anxiety. For instance, recently, Para and Johnston-Wilder (2023) used the Growth Zone Model (Johnston-Wilder et al., 2020) and the hand model of the brain (Siegel, 2010) to successfully reduce mathematics anxiety in Brazilian high school pupils. Others have used growth mindset-based interventions, including explaining the principles of growth mindset theories, reciting positive affirmations about mathematics learning beliefs, and reframing fixed mindset statements, to address mathematics anxiety and mathematics self-beliefs in community college students in the USA (Samuel and Warner, 2019). Furthermore, Apostolidu and Johnston-Wilder (2023) explored the effectiveness and accessibility of the Mathematical Resilience Toolkit (Johnston-Wilder et al., 2020) for building resilience, reducing mathematics anxiety, and increasing effectiveness of learning in adult learners in a UK further education college. In this case, outcomes were measured via observations of learners' verbal and non-verbal reactions to the concepts and tools introduced on the mathematics course, in addition to interviews with learners.

Observation of learners in real time provides an opportunity for teachers and researchers to note behaviour and speech in a way that is more subtle, perhaps circumventing some of the issues that interviews carry, such as general biases in responding such as inaccuracies in learners' memory or learners wanting to appear in a particular light. On the other hand, interviews provide the level of depth required to delve into the intricacies of individual experiences. In Apostolidu and Johnston-Wilder's (2023) study, interviews provided the learners with the chance to reflect on their experiences of the toolkit, which is something not always considered as part of intervention work. Including learners in this way has the benefit of (i) helping learners to consider their own thoughts and feelings, (ii) aiding learners and teachers in understanding the extent to which resilience and mathematics anxiety may have changed across time, and (iii) further developing tools based on feedback from learners.

> "The CMAS-UK is an essential part of my assessment toolkit in my role as specialist Mathematics teacher. It enables me to quickly determine if, and the extent to which, the child is experiencing mathematics anxiety. The interactive nature of the questionnaire offers the opportunity for the child to qualify their responses, both on paper and verbally, and for me to informally explore specific aspects further.
>
> Once processed, I use the scores as a basis to conduct further conversations with both the teachers and parents/carers, to validate the child's responses and determine if the child's self-reported levels of mathematics anxiety, are typically 'lived out' day to day.
>
> I also analyse the child's responses in order to use them to alert teachers of potential situations within mathematics lessons that may trigger anxiety in the classroom, or at home when completing homework.
>
> The results, discussions and recommended strategies are then recorded in my assessment report and used to inform the child's Individual Learning Plans."
>
> Gail an Schalkwyk, Specialist Mathematics Teacher.

Recently, Petronzi et al. (2023) implemented a child-friendly storybook approach in primary school classrooms in a bid to promote children's reflection on mathematics experiences and to normalise mathematics talk. In this qualitative work, children aged 6–7 years (N = 15) across two UK primary schools first completed the CMAS-UK (Petronzi et al., 2018) to indicate their level of mathematics anxiety and give further context to their discussion points. The children took part in one-to-one discussion focussed on the learners' engagement with a specially designed and written mathematics anxiety storybook. Following reflexive thematic analysis of the transcriptions of the discussions, three global themes were identified:

1. Mathematics Application: (a) counting, and (b) mathematical language;

2. Strategies: (a) social learning, (b) resilience and self-regulation; and

3. Emotive Responses: (a) perceptions of self and mathematics, and (b) success and happiness.

The findings suggest that children were able to engage with the storybook approach which integrated mathematics problems, and which normalised mathematics talk in a non-judgement-based environment. The storybook approach seemed to lead to more positive perspectives of mathematics and more resilient approaches to finding solutions. Even for those children with higher self-reported CMAS-UK scores, this approach seemed to encourage independent implementation of their individual mathematics knowledge. The children appeared to become progressively more confident as they worked through the storybook. This exemplifies the use of validated mathematics anxiety scales alongside targeted intervention approaches which are designed to support the emotion regulation and resilience of mathematics learners. As a final point, it is sometimes surprising that seemingly successful intervention work does not result in a change in mathematics anxiety scores as measured via self-report scales. However, this might be due to action researchers not having adequately sensitive measurement tools at their disposal. The question may be not so much whether a learner's anxiety has changed, but instead whether they feel better equipped to deal with their feelings of anxiety and negative self-belief.

> Andy, Head of Mathematics at a secondary school, used the MAS-UK (Hunt et al., 2011) as part of some action research he conducted and had the following to say: "What was interesting about the Mathematics Anxiety Scale UK was that it didn't just identify the students who were struggling with the basic concepts of mathematics, or suffered general anxiety issues, but it shone a light on other students across the attainment spectrum. Easy to use, it made students aware that they were not alone!"

Conclusion

In conclusion, we must first emphasise the prevalence of mathematics anxiety in many societies and the damage mathematics anxiety can do to the effective learning of mathematics. Therefore, we feel that it is useful for teachers to be aware of the range of self-report scales available to measure mathematics anxiety, especially if they are unsure if their learners experience it and are affected by it. The range available means that teachers, and others, can carefully consider the suitability of the scales' content, especially in relation to the age of their learners. Once an appropriate self-report scale is selected, the extent of anxiety present in the educator's learning environment can be measured. Such scales provide a quick and usually easy way of identifying mathematics anxiety in learners, providing a starting point for the appropriate allocation of support. However, we make clear that an over-reliance on self-reports should be avoided. Developing an awareness of typical behaviours associated with mathematics anxiety can often be helpful alongside the use of scales. Measuring mathematics anxiety is an important feature of developing and testing approaches for developing mathematical resilience. Mathematical resilience can itself be assessed in similar ways to mathematics anxiety, helping to establish the effectiveness of interventions used to develop learners' resilience in mathematics. Attempts to measure mathematical resilience are in their infancy compared to mathematics anxiety alone, and existing self-report scales are likely to be developed further as our conceptual understanding of mathematical resilience improves.

References

Alexander, L., and Martray, C. (1989). The development of an abbreviated version of the mathematics anxiety rating scale. *Measurement and Evaluation in Counseling and Development*, 22(3), 143–150. 10.1080/07481756.1989.12022923

Apostolidu, M., and Johnston-Wilder, S. (2023). Breaking through the fear: Exploring the mathematical resilience toolkit with anxious FE students. *Research in Post- Compulsory Education*, 10.1080/13596748.2023.2206704

Ashcraft, M., and Moore, A. (2009). Mathematics anxiety and the affective drop in performance. *Journal of Psychoeducationql Assessment*, 27(3), 197–205. 10.1177/0734282 908330580.

Baker, J. (2021). *You see it differently once you calm down: Developing an intervention to support learners to address their mathematics anxiety*. Ph.D. Thesis, University of Warwick.

Bekdemir, M. (2010). The pre-service teachers' mathematics anxiety related to depth of negative experiences in mathematics classroom while they were students. *Educational Studies in Mathematics*, 75, 311–328. https://doi.org/10.1007/s10649-010-9260-7

Betz, N. (1978). Prevalence, distribution, and correlates of math anxiety in college students. *Journal of Counseling Psychology*, 25(5), 441–448.

Carey, E., Hill, F., Devine, A., and Szűcs, D. (2017). The modified abbreviated math anxiety scale: A valid and reliable instrument for use with children. *Frontiers in Psychology*, 8(JAN), 1–13. 10.3389/fpsyg.2017.00011

Chinn, S. (2012). Beliefs, anxiety and avoiding fear in mathematics. *Child Development Research*. 10.1155/2012/396071

Chiu, L., and Henry, L. (1990). Development and validation of the mathematics anxiety scale for children. *Measurement and Evaluation in Counseling and Development*, 23, 121–127.

Dowker, A., Bennett, K., and Smith, L. (2012). Attitudes to mathematics in primary school children. *Child Development Research*, 2012, 1–8. 10.1155/2012/124939

Dreger, R., and Aiken, L. (1957). The identification of number anxiety in a college population. *Journal of Educational Psychology*, 48(6), 344–351. 10.1037/h0045894

Fraser, H., and Honeyford, G. (2013). *Children, parents and teachers enjoying numeracy: Numeracy hour success through collaboration*. Routledge.

Hopko, D. (2003). Confirmatory factor analysis of the math anxiety rating scale – Revised. *Educational and Psychological Measurement*, 63(2), 336–351. 10.1177/0013164402251041

Hunt, T., and Maloney, E. (2022). Appraisals of previous maths experiences play an important role in maths anxiety. *Annals of the New York Academy of Sciences*, 1–12. 10.1111/nyas.14805

Hunt, T., Bagdasar, O., Sheffield, D., and Schofield, M. (2019). Assessing domain specificity in the measurement of mathematics calculation anxiety. *Education Research International*. 10.1155/2019/7412193

Hunt, T., Bhardwa, J., and Sheffield, D. (2017). Mental arithmetic performance, physiological reactivity and mathematics anxiety amongst U.K primary school children. *Learning and Individual Differences*, 57, 129–132. https://psycnet.apa.org/doi/10.1016/j.lindif.2017.03.016

Hunt, T., Clark-Carter, D., and Sheffield, D. (2011). The development and part validation of a U.K. scale for mathematics anxiety. *Journal of Psychoeducational Assessment*, 29(5), 455–466. 10.1177/0734282910392892

Hunt, T., Marmolejo-Ramos, F., Tejada, J., Kotera, Y., Waldron, M. B. K., Correa Vione, K., Gabriel, F., Asanjarani, F., Ashraf, F., Muwonge, C. M., Popa, I-L., Sari, M. H., Bagdasar, O., Kojima, Y., and Sheffield, D. (in press). Predicting mathematics anxiety from self-beliefs and attitudes: A multi-national study. Paper in preparation.

Jameson, M. (2013). The development and validation of the children's anxiety in math scale. *Journal of Psychoeducational Assessment*, 31(4), 391–395. 10.1177/0734282912470131

Jamieson, J., Peters, B., Greenwood, E., and Altose, A. (2016). Reappraising stress arousal improves performance and reduces evaluation anxiety in classroom exam situations. *Social Psychological and Personality Science*, 7, 579–587. 10.1177/1948550616644656

Johnston-Wilder, S., Baker, J., McCracken, A., and Msimanga, A. (2020). A toolkit for teachers and learners, parents, carers and support staff: Improving mathematical safeguarding and building resilience to increase effectiveness of teaching and learning mathematics. *Creative Education*, 11, 1418–1441. 10.4236/ce.2020.118104

Kolacinski, J. (2003). *Mathematics anxiety and learned helplessness, thesis submitted for the degree of doctor of arts (DA)*. University of Miami.

Kooken, J., Welsh, M., McCoach, D., Johnston-Wilder, S. and Lee, C. (2016). Development and validation of the mathematical resilience scale. *Measurement and Evaluation in Counseling and Development*, 49, 217–242. 10.1177/0748175615596782

Lau, N., Hawes, Z., Tremblay, P., and Ansari, D. (2022). Disentangling the individual and contextual effects of math anxiety: A global perspective. *Proceedings of the National Academy of Sciences of the United States of America*, 119(7), e2115855119. 10.1073/pnas.2115855119

Luo, X., Wang, F., and Luo, Z. (2009). Investigation and analysis of mathematics anxiety in middle school students. *Journal of Mathematics Education*, 2(2), 12–19.

Nicolaidou, M., and Philippou, G. (2003). Attitudes towards mathematics, self-efficacy and achievement in problem solving. *European Research in Mathematics Education III*, 1, 11.

Núñez-Peña, M., Guilera, G., and Suárez-Pellicioni, M. (2014). The single-item math anxiety scale: An alternative way of measuring mathematical anxiety. *Journal of Psychoeducational Assessment*, 32(4), 306–317. 10.1177/0734282913508528

Owen, S., and Froman, R. (1988). Development of a college academic self-efficacy scale. Proceedings of the annual meeting of the national council on measurement in education, New Orleans.

Para, T., and Johnston-Wilder, S. (2023). Addressing mathematics anxiety: A case study in a high school in Brazil. *Creative Education*, 14(2), 377–399. 10.4236/ce.2023.142025

Petronzi, D., Staples, P., Sheffield, D., Hunt, T., and Fitton-Wilde, S. (2017). Numeracy apprehension in young children: Insights from children aged 4-7 years and primary care providers. *Psychology and Education*, 54, 1–26.

Petronzi, D., Schalkwyk, G., and Petronzi, R. (2023). A pilot math anxiety storybook approach to normalize math talk in children and to support emotion regulation. *Journal of Research in Childhood Education*, 1–19. 10.1080/02568543.2023.2214591

Petronzi, D., Staples, P., Sheffield, D., Hunt, T., and Fitton-Wilde, S. (2018). Further development of the children's mathematics anxiety scale UK (CMAS-UK) for ages 4-7 years. *Educational Studies in Mathematics*, 100, 231–249. https://link.springer.com/article/10.1007/s10649-018-9860-1

Ramirez, G., Gunderson, E., Levine, S., and Beilock, S. (2013). Mathematics anxiety, working memory and mathematics achievement in early elementary school. *Journal of Cognition and Development*, 14, 187–202. 10.1080/15248372.2012.664593

Richardson, F., and Suinn, R. (1972). The mathematics anxiety rating scale: Psychometric data. *Journal of Counseling Psychology*, 19(6), 551–554. 10.1037/h0033456

Ricketts, S., Engelhard, G., and Chang, M.-L. (2017). Development and validation of a scale to measure academic resilience in mathematics. *European Journal of Psychological Assessment*, 33(2), 79–86. 10.1027/1015-5759/a000274

Rubinsten, O., Marciano, H., Eidlin Levy, H., and Daches Cohen, L. (2018). A framework for studying the heterogeneity of risk factors in math anxiety. *Frontiers in Behavioral Neuroscience*, 12, 291. 10.3389/fnbeh.2018.00291

Samuel, T., and Warner, J. (2019). "I can math!": Reducing math anxiety and increasing math self-efficacy using a mindfulness and growth mindset-based intervention in first-year students. *Community College Journal of Research and Practice*, 1–18. 10.1080/10668926.2019.1666063

Siegel, D. (2010). *Mindsight: Transform your brain with the new science of kindness*. Oneworld Publications.

Suinn, R., Taylor, S., and Edwards, R. (1988). Suinn Mathematics Anxiety Rating Scale for Elementary School Students (MARS-E): Psychometric and Normative Data. *Educational and Psychological Measurement*, 48(4), 979–986. 10.1177/0013164488484013

Wu, S., Barth, M., Amin, H., Malcarne, V., and Menon, V. (2012). Mathematics anxiety in second and third graders and its relation to mathematics achievement. *Frontiers in Psychology*, 3, 1–11. 10.3389/fpsyg.2012.00162

Young, C., Wu, S., and Menon, V. (2012). The neurodevelopmental basis of math anxiety. *Association for Psychological Science*, 1(10), 1–10. 10.1177/0956797611429134

Tools for building mathematical resilience

Sue Johnston-Wilder and Clare Lee

Introduction

Chapter 1 discussed the idea of mathematical resilience and began to outline some of the theory and practice that makes us so sure that inclusive, supportive but challenging environments allow everyone to learn more mathematics. In this chapter, we discuss four specific tools that can help you, your colleagues and your learners to co-create environments in which everyone has the chance to thrive. Each of these tools offers a distinct way for teachers and learners to communicate about the emotional reactions that we know learners experience as they try to make progress in mathematics. We will then go on to suggest some teaching habits and pedagogical approaches which support mathematical resilience. Together these tools and techniques will help you configure a resilient learning environment where learners are aware of, and engaged with, the principles of growth, struggle, value and support, which together help learners develop mathematical resilience and protect their mathematical well-being.

The four tools

The experience of learning mathematics can seem to learners and often to their teachers as well, to present barriers that need to be overcome. Where support has not been present or appropriate, those barriers can turn into what learners perceive as insurmountable difficulties. Often it is the emotional response to mathematics which turns barriers that could be overcome with little difficulty into impossible mountains. These emotional responses must be transcended before progress can be made in learning mathematics and understanding can grow. However, learners and teachers do not always have the opportunity or enough shared language to describe and discuss the emotional side of learning mathematics. Here we discuss four tools which we know can help learners become aware of and understand their feelings and how their previous experiences may have led them to feel the way they do. The tools have been shown to be effective in allowing all learners of mathematics from primary age to adult to communicate how they are feeling and thus gain the support they need.

These four tools form a structure for work on supporting learners to manage their emotions so that they can continue to learn mathematics. They each have a role in reducing the impact of existing anxiety in mathematics lessons and empowering learners with the agency they need to progress more effectively. The four tools are as follows: the Growth Zone Model (GZM) (Johnston-Wilder and Lee, 2017), the Hand Model of the Brain (Siegel, 2010), the Relaxation Response (Benson, 2000), and the Ladder Model (Johnston-Wilder et al., 2020). The GZM enables learners and teachers to develop an accessible shared language to raise awareness of and communicate emotions whilst learning mathematics; the Hand Model of the Brain helps learners to understand why they may feel like they are 'stupid' or that they 'just can't do it' when faced with mathematics; the Relaxation Response enables learners to recover from any perceived threat and return to a calmer, more productive state; and the Ladder Model, based on Bruner's (1966) notion of scaffolding with the addition of an image, enables learners to recognise that they can progress with any mathematics when they have sufficient rungs or supportive understandings and to recognise when they may need more rungs.

These tools also have a role to play in helping teacher educators, teachers, support staff, parents and carers and learners understand better how teaching and learning approaches can be adjusted to reduce the impact of anxiety and help learners to recognise and manage their feelings. Teachers and learners who have used the tools have found them manageable and highly effective and you will find examples of the tools in action in many of the upcoming chapters. Each of the tools will now be considered in turn, followed by a discussion bringing the tools together into an effective toolkit. Each learner (and each teacher) will find some tools more helpful than others. As you read, reflect on how each tool might be used, or adapted for use, in your own context.

Growth Zone Model

The GZM is a model that can be adopted in classrooms or in support or homework sessions to help learners to name and communicate their current, otherwise hidden, feelings and emotional experiences when learning mathematics. The GZM has three "zones", as depicted in Figure 3.1, which represent ways individual learners might experience learning mathematics in the moment.

The green zone, or comfort zone, represents feeling safe and confident while learning mathematics. You could say you are in the comfort zone when you are able to use your existing knowledge without needing much support from anyone and you are not experiencing stress. This is the zone where learners feel safe and secure. They are operating within their current level of understanding but are not

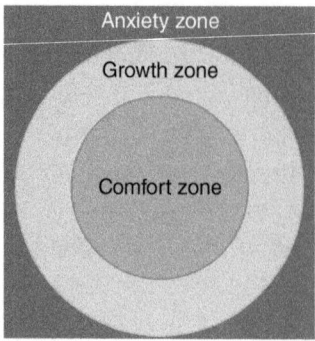

Figure 3.1 The Growth Zone Model.

challenged to grow or change. Here they consolidate their knowledge and understanding of mathematical concepts and, through intelligent practice, build fluency with mathematical ideas. While this zone can be comfortable, where a learner is reluctant to leave the comfort zone, they will limit their growth or progress in mathematical learning.

> One primary school found that several of their learners were not wanting to challenge themselves and were content with doing the same mathematics over and over again. That way they always got full marks. Their teacher explained the Growth Zone Model to her learners. She described the feelings of risk and making mistakes that are inherent in moving learning forward.
>
> Now her pupils are progressing well; rather than feeling they 'can't do it' when things feel slightly beyond them, they understand that feeling means they are growing their learning. They tell her, "I am not in amber, can I have something that will challenge me more?" Sometimes they say, "I'm turning red" and their teacher or their peers know to help them back into the more productive zones.

The red zone symbolises perceived danger or threat to well-being. If a learner is in the red zone, they are likely to feel overwhelmed and unable to think. They experience feeling as if they are in over their heads. This zone can be very stressful and can lead to feelings of anxiety and panic. In this zone, students might experience a "fight, flight or freeze" impulse which may cause them to feel angry, or to run, or to cry. Almost everyone in the red zone will experience an inability to think clearly.

The amber/orange or growth zone represents where the learner feels appropriately challenged, and that they are experiencing and learning new ideas. It is a good place to be as it is in this zone where challenge and engagement is experienced, and where understanding can grow. When a learner is in the growth zone, they might be a little nervous, or feel slightly uncomfortable, as they are learning new things. However, their feelings reflect a managed risk as they are prepared to engage with new ideas and develop new skills. When they are working in the amber zone, learners will explore ideas, make mistakes and may experience being stuck. However, if they are sufficiently supported, they will persist in their struggle to understand some new mathematical ideas. As they work in the growth zone, learners are likely to need to use coping strategies and will need to feel supported by their learning community. If that community is encouraging, listening to ideas, asking questions, and giving access to support, then the learner is likely to be able to stay longer in the growth zone and learn more (Lee and Johnston-Wilder, 2018). Managing emotions in this zone takes practice and help as otherwise learners may find themselves dipping in and out of the red zone.

The goal of the GZM is to help individuals understand the emotional experience of growth and development. It is a tool that enables learners to have an awareness of their emotions and to develop a shared language to express their feelings, thus helping them to develop Mathematical Resilience. The GZM emphasises the importance of learners taking themselves outside of the comfort zone in order to grow, but also recognises the need to balance this with self-care and support. By understanding what it means to experience the different zones, learners can better navigate challenges and opportunities for growth and development in mathematics.

Exploring the growth zone further

A learner's position in, and movement around, the growth zone in the GZM is influenced by the building blocks of mathematical resilience: value, struggle, growth mindset and support through relationships (Figure 3.2). The idea of value is related to Ryan and Deci's (2017) notion of autonomous motivation, the form of extrinsic motivation closest to intrinsic motivation. If learners see personal value in what they are being asked to do, they will balance risk with reward and are likely to be more willing to move out of the green zone into the amber zone. Struggle summarises the need for persistence and perseverance when meeting challenges; persistence involves a determination to keep working at a problem, but perseverance includes recruiting support when needed (Williams, 2014). The growth mindset (Dweck, 2006) is included because it is a vital attribute of a resilient learner and also because of the prevalence of a fixed mindset in mathematics teaching and learning. Mathematical facility can be thought of as a muscle which develops when used (Boaler, 2022). It develops most effectively when learners make mistakes and receive formative feedback focused on growth. Relatedness is included because of its importance (Ryan and Deci, 2017) in psychological safety. A learner is best placed to learn effectively where they feel trust in the teacher (Hattie, 2012) and are part of a community learning mathematics. Building group work and peer support into teaching are ways to increase relatedness and community and therefore support in learning mathematics.

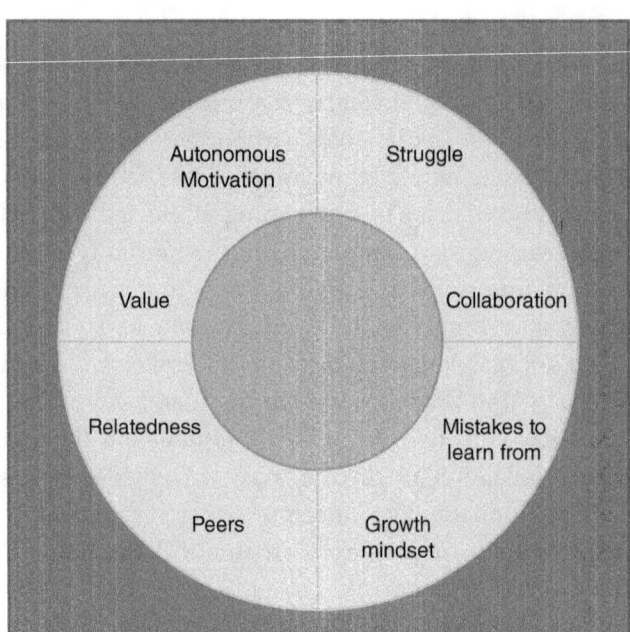

Figure 3.2 Enhanced Growth Zone Model (Para and Johnston-Wilder, 2023).

In some ways the GZM seems similar to Vygotsky's Zone of Proximal Development (ZPD) with which many teachers are already familiar. However, in Vygotsky's work the ZPD tends to be used cognitively, at least in the West, indicating the ideas and concepts that learners are positioned to be able to learn. In contrast, the GZM is focused on building the learners' understanding of the emotions they are likely to experience when working on mathematics and positioning learners to manage those emotions effectively.

Fight or flight: The Hand Model of the Brain

Siegel's (2010) Hand Model of the Brain (see Figure 3.3) can help to explain why mathematics anxiety may lead to temporary feelings of stupidity when anxiety causes panic, distress or self-defence. Where mathematics challenges are perceived as too difficult or that the support on offer is not perceived as adequate, the Hand Model of the Brain helps learners to recognise when their brain has "flipped" into threat mode, and to take appropriate action.

When the brain perceives a situation as a challenge, it activates the sympathetic nervous system, which prepares the body for action. This response is characterised by increased heart rate, blood pressure, and respiration. The brain also releases hormones such as adrenaline and cortisol, which help to increase focus and energy. These hormones cause feelings of excitement, nervousness, some anxiety and

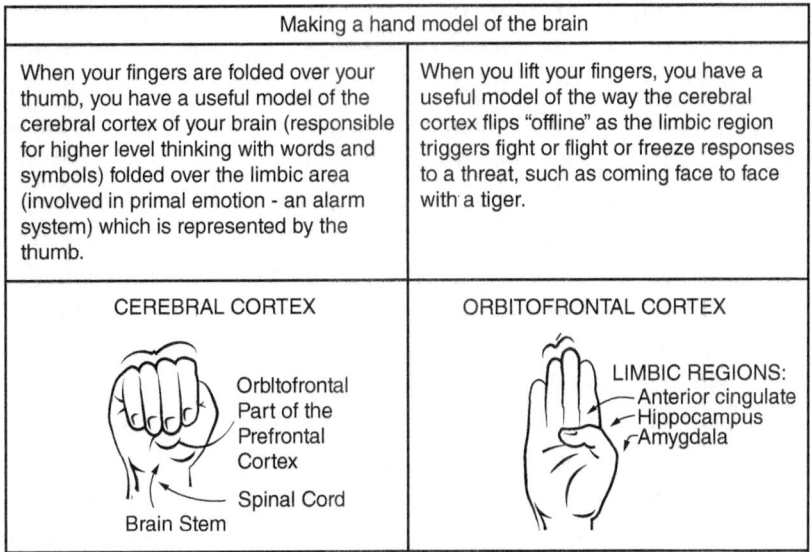

Figure 3.3 Hand Model of the Brain (Siegel, 2010).

heightened attention but do not cause feelings of severe anxiety or fear. Where learners do not understand such feelings can be experienced when faced with a challenge, they can misinterpret them, as they are similar to the feelings induced by a threat to their well-being. Where the current situation echoes a previous situation which the learner experienced as humiliating, or which led to perceived failure, then it may be appraised as a threat rather than a challenge.

When the brain perceives a situation as a threat, it activates the fight or flight response. This response is characterised by feelings of severe anxiety or fear and is accompanied by physical symptoms such as increased heart rate, blood pressure, and respiration. The brain again releases hormones to help increase focus and energy, but where a threat has been perceived the feelings of anxiety or fear block higher level thinking. The brain distinguishes between challenge and threat based on the level of control a learner feels they have over the situation, the level of predictability, and the level of support available to the learner. When a situation is perceived as controllable, predictable, and supported, it is more likely to be perceived as a challenge rather than a threat.

Siegel's Hand Model of the Brain is a model of this process which can be used to help learners recognise and manage their emotional responses. In this model the wrist represents the brain stem, and the thumb represents the primitive part of the brain, or the "alarm system", shared with all animals. The back of the hand represents the cortex, which is shared with most mammals, and the fingernails represent the prefrontal cortex, where complex human thinking, involved in reading, writing, and arithmetic, takes place.

The two hand positions in Figure 3.3 represent two contrasting ways in which the brain can process information and direct reactions. The closed fist models the brain in a situation of calm or challenge; everything is connected, and learning can happen. The open hand models the loss of regulation by the cortex known as the "fight or flight or freeze" response in which the cerebral cortex has "flipped offline" and the alarm system has taken over. Learners' capacity to think and reason logically, and express themselves verbally, is temporarily reduced. When reacting to a situation as a threat, the learner's body seeks to self-safeguard by addressing the perceived threat or escaping. The primitive brain expects fight or fleeing to result in the restoration of physical and emotional well-being. There is a video of Dan Siegel explaining his model at https://www.youtube.com/watch?v=gm9CIJ74Oxw.

The Hand Model of the Brain (Siegel, 2010) can help learners to distinguish between the productive nervousness invoked by a challenge and the kind of anxiety that interferes with mathematical thinking and leaves them feeling 'stupid' and that they 'just can't do it'. When they understand that they have perceived a threat, their immediate task is to manage their emotional distress in order to regain control. They are then better placed to find the support they may need to continue with the challenge of learning mathematics. Appropriate protective actions may include telling the teacher of their emotional state by folding and lifting their fingers so that the teacher can offer more support. Learners can manage their reactions if they are helped to learn a calming technique or given the agency to seek the kind of support they need.

> Jackie is an adult with mathematics anxiety who had agreed to take part in a session designed to build her mathematical resilience. She described how on the way to the session she had experienced nervousness but by telling herself: "Oh it's okay, it'll be fine, Sue seems kind". She'd managed to arrive at the session in a state of challenge rather than threat. When Sue suggested that Jackie had been regulating her alarm response, and explained the Hand Model of the Brain, Jackie recognised her responses and how maturity had enabled her to manage her emotions in a way she could not have done as a teenager.

Learners can experience their mind going suddenly blank, or feelings of anger or fear, due to previously acquired mathematics anxiety. When anxiety is triggered, some learners may express helplessness, saying "I can't do it", even when the task is perceived as relatively undemanding by the teacher. Other learners' behaviour may deteriorate and become more controlling, aggressive or destructive (fight) in response to their feelings of anxiety. They may become agitated and seek to run or hide (flight), or their responsiveness may decrease (freeze). Parents/carers with mathematics anxiety can experience similar emotions, and might, for example, suddenly burst into tears when asked to help a teenager with mathematics homework. A common response is to exclude themselves, saying something like "I was never a maths person either; our family is better at English".

Teaching the Hand Model of the Brain can help explain why such responses may occur. It helps learners understand that they are not 'stupid' but perceiving a threat. It also enables their response to be reframed: they are exhibiting behaviour directed at protecting the individual from a perceived threat of harm. Siegel (2010) discusses the effectiveness of naming feelings in order to reframe them. Rather than name the state of mathematics anxiety as "I can't do maths", the Hand Model of the Brain allows for a new name: "I've flipped; give me some time and I can rethink this". In a recent project (Baker, 2021), an 11-year-old boy reported that "the Hand Model of the Brain ... it sort of made more sense and you could like, understand why you get anxious, so then you could think about it and think, oh no, it's not because I'm dumb [...] everyone is like it sometimes".

The Hand Model of the Brain can also make visible a transition, from feeling challenged to feeling threatened, that might otherwise be hidden. The learner can communicate their experience promptly and effectively one-to-one and, in some classes, this has been used to check in with the whole class. For someone faced with a situation that causes psychological pain and disruption (Lyons and Beilock, 2012), who believes they are unable to do anything, whether practical or cognitive, to resolve the situation, there may seem little point in continuing to learn mathematics. Avoidance strategies keep the learner psychologically safe in the short term, but they reinforce anxiety, rather than addressing it. Some learners, when they experience anxiety behave in ways that are out of character, including angry outbursts or behaving aggressively towards others, becoming withdrawn or appearing clingy or depressed [helpless] and displaying a lack of confidence, wariness or caution. It can take a great deal of time, support and patience to help such a learner re-enter a mathematics learning environment. Whilst it can be done (Lee and Morgan, 2024) it may require the learner to have a very strong motivation – such as really needing a mathematics qualification - before they will attempt to return to learning mathematics.

The next tool can help provide both teacher and learner with appropriate responses to the threat state if and when it arises.

Relaxation Response

The Relaxation Response (Benson, 2000) popularly known as "rest and digest", is a way to manage emotions and over-ride the fight or flight reflex. This is a natural physiological response that occurs when the body is in a state of relaxation. It helps to lower blood pressure, slow down the heart rate, and reduce muscle tension. Benson (2000) showed that this state can be induced consciously. This means that it can be used as a tool by mathematics learners in the classroom, or during an examination. Having recognised that they have flipped into a threat state, if the learner can recognise that on this occasion there is no actual imminent danger, they can use the Relaxation Response to change their state in order to re-engage with the mathematics. The Relaxation Response engages the parasympathetic nervous system, reducing the body's responsiveness to hormonal messages that would increase blood pressure and heart rate; muscles and organs slow down and blood flow to the brain increases, re-enabling mathematical thought. Triggering the Relaxation Response has been shown to be effective in reducing stress, anxiety, and depression (Benson, 2000), as well as improving sleep and overall well-being and is prevalent in schools in the form of mindfulness training. It is a technique that can be practiced anywhere, anytime, and can be a valuable tool for managing stress and promoting relaxation.

Learning about the Relaxation Response has a particularly important role in the learning and teaching of mathematics when a learner already suffers from mathematics anxiety. However, experience shows that learners who already know about mindfulness need to be helped to explicitly understand that mindfulness can help in the context of learning mathematics. Mindfulness is the practice of focusing awareness, paying attention to the present moment with openness, curiosity and non-judgmental acceptance. Mindfulness research in schools has shown significant impact on attainment. Helping learners find ways to calm adverse emotional responses to tasks and to pay attention to the moment may be achieved in any learning environment by any teacher.

Within the mathematics or support classroom, and in examinations, learners can be taught what might be called *micro-mindfulness* techniques. These techniques help to trigger the Relaxation Response by practising longer out-breaths, for example, using a count of 5 when breathing in and a count of 7 when exhaling. There are learners, who are not helped by a focus on breathing. These learners can try other ways to promote relaxation, such as name three things they can hear; name three things they can see; name three different textures they can feel; or to repeat a word or phrase they have learned in a support session. Such activities can "ground" the learner in the sensory reality around them rather than in their own emotional upheaval. In cases of more extreme distress, a learner may need to leave the room, with permission, go for a walk to a safe, designated place, or seek one-to-one support. In the most serious, rare cases, a learner may need formal therapeutic intervention before further study of mathematics is sensible, but there is much that can be achieved by the anxiety-informed teacher who has learned to apply these tools in the context of learning mathematics.

Use of the Relaxation Response gives agency to the learner and the teacher to discuss emotional responses and different methods of focusing the mind on the moment. Many leaners and colleagues may already know such techniques but have not yet thought to apply them for learners of mathematics. Using the relaxation tools gives the learner the opportunity to self-manage anxiety. Micro-mindfulness can be used to support any learners who experiences their mind going blank, for perhaps the first time. Once a learner recognises they can get out of the threat state, they become more willing to engage with challenge. This has been found in primary, secondary, further and higher education and in adults coming back to mathematics after adverse prior experiences.

Ladder Model

We use the Ladder Model to show visually the communication gap that often arises between learners and formal mathematical language and ideas. In the ladder on the left, in Figure 3.4, a teacher can build a complete ladder for learners by using dialogic communication skills and their knowledge of the subject. The steps show that for learners to progress from what they understand now to what they need to understand there are intermediate ideas that must be grasped. The small steps are similar to the "small steps" referred to in much mathematics mastery material (Drury, 2018).

The ladder on the right, in Figure 3.4, illustrates that for some learners the gaps seem too big for them to take a step to learning the mathematical idea being taught. They need to fill in the gaps with much smaller steps. Some teachers may not see that gaps exist and may not have explored what can prevent learners progressing their understanding. The missing rungs could have occurred in other ways. The learner may have been absent when the understanding represented by that rung was explored or the learner may not have been able to ask the question they needed to and ended up with an incomplete understanding.

Figure 3.4 The Ladder Model, showing accessible and inaccessible maths (Johnston-Wilder et al., 2020).

> A student we will call Josie arrived late to a Saturday GCSE maths revision class close to the exams. She had been seen as a problematic student in regular lessons; the teacher was impressed she had come. Josie was struggling with the work on linear equations; it transpired she had not understood the basics of plotting co-ordinates. Sue took a small group outside, including Josie, giving each a number. We made the x-axis in chalk, stood on our number, then created human graphs by moving forwards (or backwards) according to a rule e.g., double your number add 1. Josie suddenly understood. Thereafter, she went round the class helping the other students who were stuck. She had become an asset. The activity took ten minutes.

The Ladder Model can help learners to understand that they need support to fill in any gaps, it is not a lack of capability. Developing the language of 'more rungs' helps build an inclusive classroom and reinforces the growth mindset. Explaining to learners that, if they are experiencing a jump that is too big, then all is not lost, rungs can and will be added, can make the classroom inclusive and encourage learners, and give them the agency, to take responsibility for getting the support they see that they need.

Adding rungs might involve something simple, such as another worked example, or space to ask questions as well as using examples from familiar everyday life, apparatus, body maths, or recruiting support from a peer or the internet. A resilient learner knows that it is possible to break the learning down into manageable steps, and that effective help can be recruited. They might not find the best help to fill in the rungs from the first person they ask, or the second, but effective help can be recruited eventually. Knowing that they have both the right to learn, and the ability, given manageable steps, empowers learners to take the responsibility for learning for themselves and eventually prevail.

Adding 'more rungs' where the learner recognises gaps in their understanding is not to be confused with path smoothing. Wigley (1992) identified path smoothing as a way that mathematics is often taught. In this model learners are led through a method or procedure for tackling certain mathematical problems. The learners then work through similar examples until they can do so with the minimum of error, they then revise the method later to retain the memorised method. Developing mathematical resilience demands that learners experience challenge and the good feelings that come from overcoming that challenge and really understanding the mathematical idea being explored. Path smoothing removes those good feelings and often replaces them with anxiety because the method must be memorised and retained. The Ladder Model emphasises a connectionist view (Askew et al., 1997) where mathematical understanding is seen as involving understanding many connected ideas, and if a vital connection is missing then that can compromise effective learning.

> Molly, wanting to train as a teacher, had learned about mathematical resilience; she was working as a teaching assistant. She told the maths teacher that she was re-taking GCSE maths and was invited to seek support whenever she wished. When she later asked him why 1/3 is 0.3 recurring; he said, 'it just is'. From her work on mathematical resilience, Molly knew that was not a resilience-building response and asked someone else who helped her understand.

Nurturing a safely challenging learning environment

Every learner who wants to learn mathematics should have the opportunity do so in an environment where they feel welcomed, challenged and able to grow their understanding. They should also feel safe and in control. Using the four tools that have been discussed previously will allow this environment to be one where the learners can communicate how they are feeling in ways that allow them to feel well supported in progressing their mathematical understanding. There will be a brief discussion here of ideas that will help to set up an environment enables helps learners to build mathematical resilience. The future chapters in this book will deal with this in much more detail.

The environment within which mathematics is learned requires thought from teachers because mathematics can be a difficult subject to learn. Learners will require perseverance as there will be barriers to overcome. They will need to struggle towards understanding and making mistakes will be part of learning. Sometimes these aspects of learning mathematics are hidden from the learners because some teachers believe that mathematics is all about remembering and applying procedures and then it becomes 'easy', or because a teacher has a fixed mindset and therefore is reluctant to admit vulnerabilities which may reflect badly on their own intelligence or for many other reasons. If the struggle that is part of learning mathematics is hidden, when learners experience difficulties, as they will, the difficulties can surprise them and cause anxiety. Their emotions may overwhelm then and cause them to decide they just 'can't do it'. When teachers hide or minimise struggle it can undermine the trusting relationship that is needed between a teacher and a learner. Mathematics often requires a great deal of thinking and puzzling out, which can be hard work – but this is not a bad thing!

Using a tool such as the GZM is only the start. A learner might recognise that they are in the green zone or the red zone, but it is then up to the teacher to respond, valuing the learner's feedback and ensuring they have ways to move forward in that moment. The teacher's responses might openly be connected to some aspect of the tools, for instance by giving learners a choice of a range of tasks with different levels of scaffolding or difficulty, or they might be used more subtly, for instance when the teacher changes their questioning approach, or sets up a group work task to promote mathematical learning in a challenging but supportive way.

Where teachers model behaviours and have access to a range of techniques which promote an inclusive, collaborative environment their learners can feel challenged, and able to challenge themselves to increase their learning, but always feel supported. These ideas are discussed next.

An inclusive environment

There are many ways to develop a learning environment which supports the growth of mathematical resilience, and what is effective will vary between teachers, classes and schools. However, an inclusive environment will always be one where the learner can start from where they are, and they are supported in making progress. To be sure that appropriate challenge is offered, and the right support given formative assessment (Black et al., 2003) will be used by both the teacher and the learner themselves. When both teacher and learner know what the learner does or does not understand then the next steps will be clearer.

Inclusive working environments

- Emphasise this is a classroom where every learner should feel supported. They should feel able to work hard and understand the mathematics but also to ask for whatever help they need.

- Use tasks where everyone has a role, such as problem-solving scenarios, playing games to increase fluency and low threshold, high ceiling activities.

- Always ask for how an answer was reached not just for the answer, so that when working together it is everyone in the group's job to ensure everyone understands the mathematics, not just to get an answer.

It will also be an environment where people listen to one another. Modelling active listening as the teacher is one of the best ways to help learners know how to listen to one another and it will help with formative assessment (Lee, 2006). Active listening is more than just paying attention to what someone says, although that is important. You show you are listening, and valuing what the learner says, by making eye contact and clearly acknowledging what has been said. Once the learner has had time to make their point, you respond appropriately to what has been said. If you respond with a judgemental response, even if it is positive, you will close down the conversation. Instead, you might ask for more clarification, where they were thinking of going next or if they had thought of doing something you know may help. Extending the dialogue in this way will help all the learners know that you, as their teacher, are interested in them and are willing to take their ideas and questions seriously. You will know how to help and support their learning and learners will see the value in what they are doing as it is overtly designed to help them make progress. Discuss active listening with the class and perhaps develop a set of rules for listening in their groups if they seem to need help doing this. Sometimes people can find learning hard because they have not been shown how to listen and respond appropriately. Helping students listen to one another is a good way to make the learning environment more inclusive and supportive.

A challenging environment

An environment that nurtures learners' mathematical resilience is both challenging and supportive. It focuses on understanding ideas and connecting those ideas (Askew et al., 1997) across areas of mathematics and to the real world so that the value of those ideas becomes more obvious to the learners. Every learner is offered carefully considered challenges, but also the support that is appropriate is discussed as that may be different between learners.

> **Increasing the challenge**
>
> Not everyone can change the content of the scheme of work they are expected to teach but everyone can increase the challenge in their classroom. Here are some ideas:
>
> - Ask the learners what they have found easy in the lesson and why? Then ask what they have found hard and why? Ask them for ideas about how the hard problems or tasks could be made easier.
> - Ask the students to work together to make a chart linking the ideas in the lesson to other ideas they have learned and to ideas outside the learning environment.

Increasing the challenge in a learning environment is not just about doing the same standard exercises or examination questions about more advanced or complex mathematical ideas. It is about exploring ideas fully and checking a full understanding of the ideas by applying them in different situations. It is about asking the learners to make links and connections between other areas of mathematics and asking them to explore how ideas could be applied in life outside the classroom.

A supportive environment

Support is a vital ingredient in a learning environment that nurtures the learners' mathematical resilience. Without support the challenge can become too much triggering anxiety and panic, or what can be called 'being in the red zone'. Where a learner finds that they need 'more rungs', using the Ladder Model, then they will need to be offered support to fill in the gaps.

Most learners will obtain support by working with another learner; just talking through the ideas is often all the help someone needs. But sometimes that will not work, and the pair will need more help, and some people will find talking to another difficult. Make it clear that seeking help is normal and natural and whether that help comes from looking up the ideas in a textbook, looking online, working on a whiteboard so that 'failed' experiments can be rubbed out, whatever works for you is fine.

Increasing the support available

- Activate all learners to support each other – there are more of them than you. Instigate the rule that you only ask the teacher when you have asked someone else and you both don't understand.

- Discuss what help means. It doesn't mean supply the answer as that deprives the other of the good feelings that come when you succeed with something difficult.

- Ask each learner to note what they think helps them understand and progress with mathematics. Make sure they can access what they need.

- Make sure each learner has a designated "buddy" who they can chat to about any barriers they encounter. Change the buddies regularly to emphasise that everyone is required to help each other.

- Have digital technology available for searching for help. Discuss how that may be used for support.

Learning ways to independently get support will be important when they are outside the learning environment or when others also need help. Becoming skilled in obtaining appropriate support is an important part of being resilient.

A classroom that develops mathematical resilience also nurtures the learners' skills so that they can offer appropriate support to one another. Some teachers find that their learners do not understand how to help one another in mathematics (Lee and Ward-Penny, 2022). Instead of helping one another understand the mathematical concepts, learners can think that supplying the 'right answer' is helpful. When the feeling of successfully understanding and gaining control over mathematical ideas were discussed the learners understood they should help one another get these positive feelings of success after struggle, not just give the answers. Knowing that working at understanding mathematics requires struggle and that making mistakes is inevitable when you are working at the right level of challenge is part of become a resilient learner of mathematics. It is true to say, if mistakes are not being made then the learners are in their green or comfort zone and should seek more challenge to grow their understanding.

Learners will need to be in the green zone often during their mathematical journey as there they become fluent in using and controlling mathematical ideas and comfortable with ways of working. However, whenever they are learning and growing the depth and breadth of their mathematical understanding, they will need to move into their amber or growth zone. Here they will feel uneasy as they are trying new things and they will make mistakes and get things wrong. Talk to your learners about this. Every mistake made is a chance for learning to happen, so that the same mistake is not made again. Sharing mistakes is a way to help others learn to not make the same mistake. Mistakes are not an occasion for shame but rather a learning opportunity and resilient learners want to learn.

> **Making 'making mistakes' part of learning**
>
> - Work on white boards – stating that they will be working together, and they will be making mistakes as you really want to challenge them. Making mistakes will show they are learning.
>
> - State clearly that exercise books are learning books and will therefore have lots of mistakes and wrong turns in them.
>
> - Instead of asking for who got the right answer, ask who has a mistake they made to show us, so that we can all learn from it.
>
> - Always take an interest in the mistakes that have been made and how they can help everyone learn. If there are no mistakes, increase the challenge of the work and tell the learners why.

Bringing it all together– The power of 'not yet'

In this chapter, we have discussed four specific tools that will help you and your learners understand that learning mathematics is an enterprise that requires some resilience but that it can be worthwhile. Learning mathematics has value to learners, as discussed earlier in this book those with mathematics qualifications, on average, earn more than those without. It also has value in allowing learners to stay in control of budgets and expenditure and understand and evaluate what others want them to believe. Perhaps most importantly mathematics has value in practicing thinking, reasoning and puzzling things out. It gives positive feelings when barriers are overcome, and a learner understands that they can use and control previously perplexing mathematical ideas.

One of the big ideas in building mathematical resilience is helping learners acquire a growth theory of learning. One child who understood the growth theory of learning redefined the red zone as the "not yet zone" (Anon, 2019), an area available to be explored in due course when the necessary precursors were in place. She was empowered to ask her mother for a slightly more accessible task, one that would put her in her current amber or growth zone. The power of "yet" should not be underestimated. Whenever a learner says, "I can't do it!" ask them to rephrase it as "I can't do it YET!" and then ask what help and support they might need to access in order to progress.

We have discussed how the tools facilitate direct communication between teachers and learners about the emotional reactions that we know students experience as they try to make to progress in mathematics. We have also discussed some day-to-day ideas for making a classroom a resilient learning environment in which the kind of communication facilitated by the four tools can be used to build resilience in mathematical learning. Anxiety and fear are acknowledged, and learners are helped to understand why they might feel that way and how they might diminish those feelings and return to being able to learn. The mathematically resilient learning environment is one in which the learners can approach a mystifying set of symbols with curiosity and perseverance, and a conviction that they have the support available to gain an understanding and overcome any barriers.

The rest of the book will provide suggestions and practices from different voices all of which will continue to show how the tools and approaches can be and are being used in different contexts. The next chapter in the book will discuss more about building a challenging and supportive learning environment.

References

Anon (2019). How talking about maths suddenly became easier – The Toast Model from a parent's perspective. 6 March 2019. https://www.mathsontoast.org.uk/how-talking-about-maths-suddenly-became-easier/ accessed 6 October 2023.

Askew, M., Brown, M., Rhodes, V., Wiliam, D., and Johnson, D. (1997). *Effective teachers of numeracy: Report of a study carried out for the teacher training agency.* London: King's College, University of London.

Baker, J. (2021). *'You see it differently once you calm down': Developing an intervention to support learners to address their mathematics anxiety.* Ph.D. Thesis, University of Warwick.

Benson, H. (2000). *The relaxation response.* New York, NY: William Morrow and Company.

Black, P., Harrison, C., Lee, C., Marshall, B., and Wiliam, D. (2003). *Assessment for learning: Putting it into practice.* Buckingham: Open University Press.

Boaler, J. (2022). *Mathematical mindsets: Unleashing students' potential through creative mathematics, inspiring messages and innovative teaching* (2nd edition). New Jersey, USA: Jossey-Bass.

Bruner, J. (1966). *Toward a theory of instruction.* Cambridge: Harvard University Press.

Drury, H. (2018). *How to teach mathematics for mastery.* Oxford, UK: Oxford University Press.

Dweck, C. S. (2006). *Mindset: The new psychology of success.* New York: Ballantine Books.

Hattie, J. (2012). *Visible learning for teachers.* London: Routledge.

Johnston-Wilder, S., and Lee, C. (2017). Addressing the affective domain to increase effective-ness of mathematical thinking and problem solving. In *IMA and CETL-MSOR 2017: Mathematics education beyond 16: Pathways and transitions, 10–12 Jul 2017,* University of Birmingham.

Johnston-Wilder, S., Kilpatrick Baker, J., McCracken, A., and Msimanga, A. (2020). A toolkit for teachers and learners, parents, carers and support staff: Improving mathematical safeguarding and building resilience to increase effectiveness of teaching and learning mathematics. *Creative Education*, 11, 1418–1441. 10.4236/ce.2020.118104

Lee, C. (2006). *Language for learning mathematics – Assessment for learning in practice*. Buckingham: Open University Press.

Lee, C., and Johnston-Wilder, S. (2018). *Getting into and staying in the growth zone*. NRICH.

Lee, C., and Morgan, J. (2024). Remembering learning mathematics – We can run but we can't hide. *Teacher development* [in press]

Lee, C., and Ward-Penny, R. (2022). Agency and fidelity in primary teachers' efforts to develop mathematical resilience. *Teacher Development*, 26(1), 75–93.

Lyons, I., and Beilock, S. (2012). Mathematics anxiety: Separating the math from the anxiety. *Cerebral Cortex*, 22(9), 2102–2110. 10.1093/cercor/bhr289

Para, T., and Johnston-Wilder, S. (2023). Addressing mathematics anxiety: A case study in a high school in Brazil. *Creative Education*, 14.

Ryan, R. M., and Deci, E. L. (2017). *Self-determination theory: Basic psychological needs in motivation, development, and wellness*. New York, NY: The Guilford Press. 10.1521/978.14625/28806

Siegel, D. (2010). *Mindsight: Transform your brain with the new science of kindness*. Oneworld Publications.

Wigley, A. (1992). Models for teaching mathematics. *MT*, 141.

Williams, G. (2014). Optimistic problem-solving activity: Enacting confidence, persistence, and perseverance. *ZDM*, 46, 407–422.

PART 2
Teaching for mathematical resilience

4 Building a resilient mathematical learning environment

Robert Ward-Penny and John Thomas

Introduction

Does mathematical resilience matter in my teaching?

There is a widespread problem across today's mathematics classrooms: quite simply, many learners do not enjoy the experience of mathematics lessons. In the results of the 2019 Trends in International Mathematics and Science Study (TIMSS), 50% of English Year 9 learners reported that they did not like learning mathematics (Richardson et al., 2020, p. 142). Not only was this result higher than the international average, but it had increased from previous reports in 2015 and 2011 (48% and 42%, respectively). This finding is even more concerning as the connection between emotion and cognition becomes clearer (for example, Immordino-Yang and Damasio, 2007).

It can be challenging for busy mathematics teachers to pick up on learners' emotional disengagement and psychological strains, especially if they have not had similar experiences when learning mathematics themselves. Learners' internal struggles can fade away into the background of a large class or be camouflaged by good grades and compliant behaviour. They can be expressed as quiet disaffection (Nardi and Steward, 2003) or through diverse means. In their discussion of mathematics anxiety, dos Santos Carmo, Gris and dos Santos Palombarini (2019) offer a list of escape-avoidance strategies including frequent delays, refusal to participate, prioritising other subjects and activities, damaging materials and even completing work quickly (p.407). Even the most inclusive and sensitive teacher will likely recognise some of these behaviours, perhaps arising from their learners' previous experiences as well as attitudes towards mathematics found in most cultures. We would likewise suggest that it would be very difficult for a teacher today to read through the items of a mathematics anxiety rating scale

DOI: 10.4324/9781003334354-7

(such as those described in Chapter 2) and not see the behaviour of some of their learners within the statements.

Mathematical resilience, then, is relevant and important in all our classrooms, and teachers have a pressing and shared responsibility to develop and sustain resilient learning environments.

Questioning routines

No teacher goes out of their way to construct an unhappy learning environment - at least, we would hope not! However, it can take time, thought and effort to build classroom practices that uniformly promote participation and actively work against exclusion. Many traditional, inherited routines can fall short of these goals.

Consider, for example, how you set up questions in the classroom. When you ask questions, do you expect them to be answered as quickly as possible in order to move the lesson along? Do you ask for hands up, insist on hands down, or look for volunteers? How do you respond to a correct answer, and what do you say when a learner gets it wrong? It is likely that you use a mixture of approaches, but a balance can emerge that values pace over participation and frames classroom questions as a competition, rather than as part of a conversation.

For example, imagine a class where the same four capable learners strive to be the first to answer every question. Although enthusiasm is a great thing to see in mathematics classrooms, over time this kind of situation can isolate other learners who are not as speedy, and the absence of visible struggle can give both the teacher and other learners a false impression of how a lesson is progressing. To address this, the teacher might introduce an element of random selection, whilst also building in wait time (Black et al., 2003, pp. 32–33) and making sure that incorrect attempts at answering are respected and used as learning opportunities where possible.

Some habits are harder to spot. Over time, questioning practices can turn into codified routines for teachers and learners. If you stand behind one of your learners and watch them working, do they immediately assume that they are doing something (mathematically) wrong? If you are talking to a learner and ask, "are you sure?" are they more likely to think that you are evaluating their thinking, or that you are trying to signal that they have made a mistake? Would it make a difference to say instead "do you think you could convince your friend that you are right?"

As teachers aspire for their learners to develop mathematical resilience, they develop a greater pedagogical awareness of how classroom actions can affect learners' senses of autonomy, competence and belonging. You may choose to adopt, adapt, or even abandon different routines after reflecting on your practice in your own context.

Inviting participation and displaying ownership

The tone of a classroom environment can be informed by (and reflected in) small practices as well as big ones. For example, I (Robert) have picked up the habit of writing the digits of the date as a mathematical statement. The date on which I am writing this, 11th April 2023, could be written as $(1 + 1) \times 4 = 2^3$ whilst tomorrow might end up as $\left(\frac{12}{4}\right)! = 2^0 \times 2 \times 3$. The 'rules' here are very flexible: although I don't let myself use any other digits, I'm quite happy to bring in square roots and trigonometric ratios, skip the '20' part of the year, or even stretch the definition of 'statement'; I expect that the 24th of July, 25 will be going on the board as a right-angled triangle with integer sides!

Originally, I had hoped that this routine might help learners revise a few topics, such as the order of operations, and reinforce the meaning of the equals sign. However, over time I noticed that if I was slow to write up the date, or absent, some of my learners would volunteer their own formulations. What had started as an intellectual exercise for me became an inclusive but voluntary practice through which learners could start the lesson seeing their contributions acknowledged and shared.

Another way in which learners can see themselves reflected in their learning environment is through display work. Take a moment to look around your classroom (or call to mind a classroom in which you have taught or learned mathematics). Do the walls feature posters, learners' work, or a combination of both? If there are examples of work, would your current learners recognise the names on this work, and have they been chosen to showcase creativity and effort, or neat writing with no mistakes?

These are leading questions, but they lead to a serious point. Professionally made posters can be fantastic for demonstrating key concepts and brightening a room, but without examples of learner work to balance them out, a classroom can appear impersonal. Similarly, if the only samples of learner work are flawless, faded contributions from ten years ago, they can push learners to valorise presentation over progress. Having honest and recognisable examples of mathematical work on some of your walls and display boards can help learners to feel that they belong in their classroom space.

Path smoothing versus productive struggle

A resilient learning environment is intended to be warm and inclusive, but that does not mean that learners should find everything easy and effortless. The goal of developing mathematical resilience does not ask teachers to apologise for the inherent difficulties of learning mathematics, or require them to limit its cognitive challenge. Indeed, it encourages teachers to normalise struggle, and warns against the dangers of making things too simple.

One approach widely adopted when teaching mathematics is 'path smoothing'. Wigley (1992) explains this term by describing a teacher who leads learners through a method and then gets them to practice similar exercises:

> *The key principle is to establish secure pathways for the learners. Thus it is important to present ways of solving problems in a series of steps which is as short as possible, and often only one approach is considered seriously.* (pp. 4–5)

Path smoothing is an enticing strategy: it paves the way for learners to move through questions and tasks with little apparent difficulty, it can be easier to implement in demanding classroom contexts, and it aligns with the design of many textbook series and contemporary approaches to teaching mathematics (Blair and Hindle, 2019). However, path smoothing can disadvantage learners. Atomising the mathematical experience into small steps can limit opportunities for conversation, conjecture and creativity, and when teachers flatten out difficulties, they can leave learners ill-prepared for challenge and independent work (Foster, 2013).

A teacher aiming to develop mathematical resilience recognises that there is an especial value in challenge, and much to be learned through experiencing difficulties and working through them:

> *being stuck is an honourable and useful state because that is when it is possible to learn about mathematics, about mathematical thinking, and about oneself.* (Mason, 2015, p. 101)

Such a teacher would try to avoid immediately telling their learners what to do, or how to resolve a difficulty, but would start by asking probing and supportive questions, such as "why do you think that?" and "could you try a few values and see what happens?", or even afford the learners permission and time with suggestions like "why don't you play with that for a while?" (Warshauer, 2015). They recognise that the practice of perseverance can support creative mathematical thought (Williams, 2014).

A learner experiencing challenge should also have access to strategies and resources beyond the teacher which might address their difficulties. These can range from simple prompts to additional resources and pedagogical strategies, but together they address another tenet of teaching for mathematical resilience: that every learner should have access to appropriate support. For example, a teacher might:

- Display a poster with several prompt questions, such as "can I draw a picture of this?", "have I seen something similar before?" or "can I break the problem down into smaller parts?" which can encourage the learners to work through these questions before asking the teacher for help.

- Consistently have resources out which learners are allowed to use during regular lessons without asking, such as mathematical dictionaries, algebra tiles or tablet computers with access to mathematics software.
- Establish that it is acceptable (and even encouraged) for learners to work with their desk-partner if they are stuck, or move tables into groups to promote the expectation of collaborative working (see Chapter 6 for more on group work and support).
- Put two or three sticky notes on each learner's desk at the start of each lesson; the learner loses one every time they ask the teacher a question and cannot ask any additional questions. This encourages learners to think about their questions before they speak, and to access other means of support.
- Set up out-of-class avenues for support which involve peer coaching, for instance a lunchtime homework club run by sixth formers for younger secondary learners.

It can be difficult for both teachers and learners to break the habits which bypass struggle in the mathematics classroom, and an atmosphere of 'positivity and perseverance', as described in Chapter 1, will need time to develop. However, an overuse of path smoothing can limit learners' involvement and learning, and in the long run it can be productive to, in the words of the US mathematics educator Dan Meyer, "be less helpful" (Meyer, 2010).

Choosing and adapting tasks

One way of integrating productive struggle and a growth mindset into practice is to involve your learners explicitly in your differentiation. For instance, after explaining a new idea or method through some whole-class work, you could offer learners a selection of four worksheets of progressive difficulty, and ask them to aim to complete two. Learners would then have the choice of where to start, whilst still meeting a minimum expectation; they could also adjust their choice of sheet if the first one turned out to be too easy or uncomfortably challenging. This approach can be deployed elsewhere: if your only resource is a worksheet of 50 questions, you might have three different starting points, or give learners the choice of whether to do every question or every other question.

A second way is to choose a task which is both accessible and challenging, such as a 'low threshold, high ceiling' task. These are tasks where "everyone can get started, and everyone can get stuck" (NRICH, no date). For example, an upper-primary or lower-secondary class might tackle the following task, based around Pick's theorem.

Start with a piece of square dotted paper. Try drawing some 'one-dot' shapes: shapes with straight edges, where all of the vertices are on the dots, and there is one dot caught in the middle of the shape (see Figure 4.1). Look at the areas of these shapes. What do you notice?

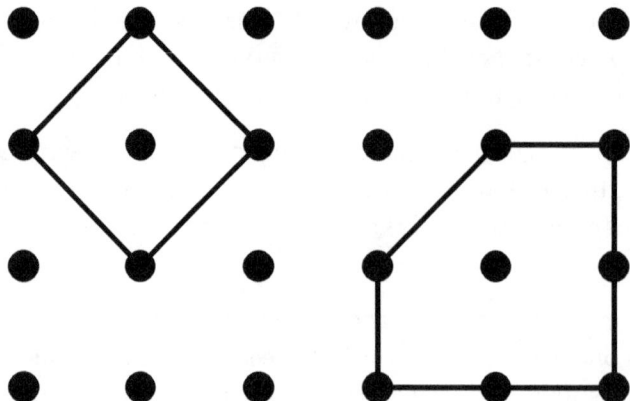

Figure 4.1 Two one-dot shapes.

This task is relatively easy to begin; learners can 'collect' shapes and find their areas by breaking them into squares and right-angled triangles. However, it develops quickly. What stays the same and what changes? Can they find a pattern or a sequence? Could they write this using a word equation, or using algebra? There is also potential for some sticky moments: how would they find the area of the one-dot triangle in Figure 4.2?

The task rewards further exploration. Could we repeat this investigation for a two-dot or a three-dot shape? Would a no-dot shape be possible? Is there a 'pattern of patterns' which could tell us the area of any shape, given the number of dots inside the shape and the number of dots on its perimeter?

The scope of this problem ranges from learners calculating simple areas by counting squares and half-squares to expressing a three-variable relationship with formal algebra. In this way it affords opportunities for learners to see that everyone can grow their mathematics from where they are, and that everyone will experience barriers, whatever level they are working at. Related task types, such as 'rich tasks',

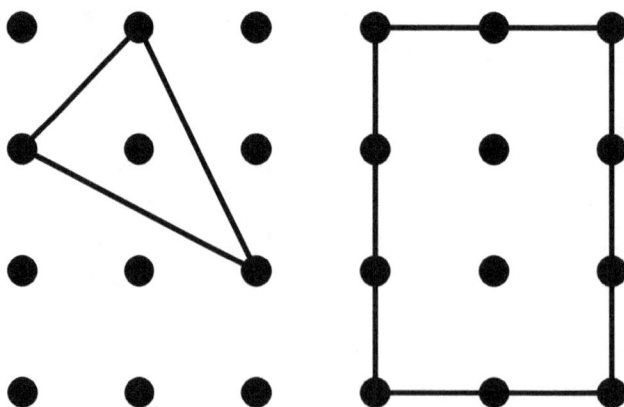

Figure 4.2 A one-dot triangle and a two-dot rectangle.

can also furnish learners with this awareness; for a second example, please see 'Diffy' in Chapter 6.

Explicitly involving learners in developing their own resilience: A case study

As an experienced but busy head of department, John is very aware of how mathematical resilience must support and fit in with many other aspects of school practice. In this section he provides a brief account of how his department worked together to improve results and learners' affect by building inclusive learning environments.

At the time of writing, I (John) am leading a large department in a selective boys' school with a mixed sixth form where all students continue to study mathematics to the age of 18 as they work towards the International Baccalaureate Diploma. Having worked in a variety of different schools, I am aware that there is a popular assumption that most selective school learners are quick to settle, happy to learn and ready to succeed. Our reality is very different. A large-scale survey of learners' attitudes and feelings towards mathematics in our school showed us that, although our learners arrived in Year 7 with slightly more positive than typical attitudes, their levels of anxiety quickly increased to expected levels as they moved through the school, and they were worryingly high in the sixth form. Additionally, following the introduction of the revised grades of 9-1 being awarded for GCSE in mathematics, our results dropped significantly indicating that there were gaps and inconsistencies in the way that we, as a department, approached teaching the subject. It was clear that something had to change.

Strategy for moving forwards

My department decided to focus on building resilience as a way of improving learners' affect and attainment. Our starting point was that the results achieved by our learners were fundamentally linked to their experience throughout the school. By focusing on resilient behaviours in class, we began the process of adapting our practice so that we could ensure that all learners were given the toolkit they needed to approach their aspirational targets.

Our approach to teaching and learning is now based on three principles or strands. All the teachers in the department aim to work at the intersection of these three strands (Figure 4.3).

Quality first teaching

Quality First Teaching (QFT) was an approach introduced by our school's special educational needs department in 2022, drawing on government guidance. Although resilience is not a central theme of QFT, the goal of ensuring that all

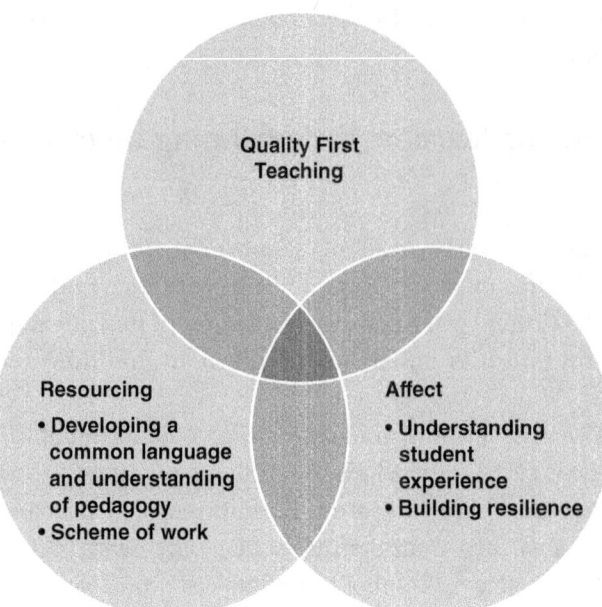

Figure 4.3 A three-strand approach for developing departmental practice.

learners, irrespective of need, are able to access the curriculum in all lessons sits well alongside the ideas of growth and value in mathematical resilience. The principles behind QFT are that lessons and learner experiences are developed via a plan-do-assess-review cycle, forming the main teaching approaches aimed at improving lessons for all learners. In our mathematics department, we have collaborated to address mathematical resilience in our cycles, believing that all learners should be equipped to be resilient in their learning rather than assuming that they will develop the skills independently.

As we have worked to provide learners with the tools they need to become resilient learners, we have found that we naturally fall into a system where we regularly assess and review our progress. This review comes in several forms which are undoubtedly common in many schools including:

- Regular check-ins with the relevant subject coordinator.
- Regular low-stakes, online mini-tests where learners get instant feedback on their performance and teachers get timely information about how well their classes are progressing.
- Lessons built into the schedule where teachers can revisit concepts that were poorly understood initially.
- Short department development meetings before school one day per week.
- Department meetings where staff work collaboratively on ideas to improve further the way that we work together.

- Student voice activities so that we can gather feedback on how the work is perceived by learners.

Resourcing

The second strand of our strategy was 'resourcing'. This started with creating a well-structured scheme of work but grew to include other ways of ensuring that learners' experiences through the key stages had a high degree of consistency. We aimed to ensure that learners do not experience changes in their teacher as a hurdle in their learning journey. One aspect is a continuing discussion about how we use words such as "minus" and "negative" across the department with the aim that correct terminology is used consistently (for example, "multiplying by the reciprocal" when referring to dividing fractions). This allows learners to develop the vocabulary needed for higher-level discussions as they move through the course. We have found that time spent on developing the use of the correct language pays off as learners start the process of exploring and explaining their work. As well as a focus on developing consistent use of language through CPD sessions, we chose to use aspects of Rosenshine's Principles of Instruction (Sherrington, 2019) as the basis of what a standard mathematics lesson will look like in our school. As a department, we now aim for a shared understanding that all lessons need to have high-quality and inclusive questioning, clear instruction, high expectations of learners' contributions, and clear scaffolding and modelling of solutions; much of this is strongly consonant with the parallel aim of developing mathematical resilience. We felt it essential that all staff were comfortable demonstrating the creative and risk-taking aspects of the subject so that learners would become equally comfortable with trying ideas out and fixing errors as they go. It is also important to note that this expectation of a typical lesson is not so prescriptive that teachers are tied into a particular resource or script and that all are able to express their own preferences within the broad framework outlined.

Our curriculum in years 7 to 9 (ages 11–14) is rooted in the International Baccalaureate Middle Years Programme. While learners still have regular, traditional tests in mathematics, we also use more open-ended tasks based on the low threshold, high ceiling principles discussed earlier in this chapter as part of the department's assessment regime. These assessment tasks normally come at the end of a sequence of lessons and include ideas such as "Who is the average boy in your form group?" and exploring the use of indices in writing very large and very small numbers. In line with Rosenshine's principles of instruction (Sherrington, 2019), we make sure that scaffolded instructions are provided but, crucially, as learners move through the school that scaffolding is reduced and eventually removed.

It is a fundamental belief within the department that by ensuring a degree of continuity in expectations, and having a common approach to questioning, modelling, and scaffolding, along with regular reviews and staff development, we are doing everything that we can to allow learners to become resilient learners. In line with the principles of mathematical resilience as laid out in Chapter 1, the structure of our curriculum reinforces the message that providing challenging but appropriate work, adopting a growth mindset and linking this with an understanding that help is available, is the best way to ensure high levels of progression and attainment.

Affect

Finally, we looked at learners' emotional responses to mathematics. If we were to be successful in improving the overall experience of learners learning mathematics, we needed to ensure that we understood how learners felt about mathematics so that we could then work on building their resilience and therefore their ability to cope with setbacks and disappointments during the course.

We began by defining what we meant by resilience in mathematics and made heavy use of the idea that mathematical resilience is based on an understanding that struggling is a common experience (explaining that "if they're finding it hard it's not them, it's because maths can be challenging"); that there is value in working at getting better in mathematics; that there is no shame in seeking assistance when the work gets more difficult; and that with practice, they can get better at doing mathematics. The model used was based on my MA research which looked at how teacher expertise has a direct influence on learners' experiences, as illustrated by the cycle in Figure 4.4.

Students joining our school at age 11 have generally had more positive prior experiences in mathematics and have become accustomed to achieving success with (sometimes) very little effort. However, defining the experience of struggle as a fundamental aspect of being a resilient learner in mathematics was new to teachers and learners. This was further developed when the Growth Zone Model (see Chapter 3) was introduced to teachers and learners, alongside work on the idea of managing cognitive load and working memory, to represent struggle and growth in a meaningful way. It is now common within the department to hear teachers talking about not working in the comfort zone and identifying topics and questions

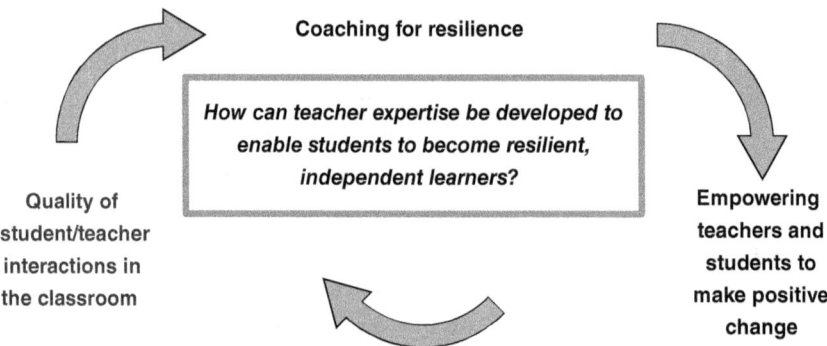

Figure 4.4 Teacher empowerment and development cycle.

that result in learners attempting to work in the panic zone. More importantly, we are continuing to develop a list of strategies that everyone can use to help the learners work in the growth zone. The list includes simplifying the way in which we present new ideas, the sequencing of our schemes of work, the way that we use independent learning tasks, and providing learners with "quick fixes" that they can use when they start to find the struggle overwhelming.

This is still very much a work in progress, but the initial indications are that the changes are positive. Observations show that learners are much more willing to participate in class; the quality of learners' work is slowly improving; all staff are more aware that those learners who seem to be disengaged may be using avoidance tactics to mask their anxiety; sequencing and modelling in lessons has improved significantly.

When learners were invited to comment on the range of strategies that we had introduced to our learners, the Growth Zone Model stood out as one that learners seemed to associate with more readily:

This helped me to understand that being uncomfortable is okay and that something good actually comes out of it since you grow and learn things in uncomfortable situations.

This gave me context as to how to develop my maths abilities and also gave me the confidence to take a step back when I start to panic. I often think of this image when I start to struggle with questions and it helps me to classify if I just need to push through and grow or revisit prior knowledge to avoid panic.

It made clear that taking time to slowly come out of the comfort zone and into the growth zone would help me improve at maths at a good and comfortable pace. It showed that pushing too hard all of a sudden does more harm, by putting you into the panic zone, than good.

The final comment is a good example of a learner who has clearly paid close attention to the ideas of managing cognitive load in their own learning and making good use of the ideas behind the Growth Zone Model as a way to understand how their own negative response can limit learning.

Impact of changes

We have a high degree of confidence that the changes we have made are making a difference but, equally, we know that there is still more to do. Schools are unrelentingly busy places, and it is not surprising that a department's focus can get a little stuck on how to progress through the huge amount of content that is present in a typical mathematics curriculum and the demands of high stakes external assessments in the subject. However, by taking a step back and spending time reviewing – and renewing – our focus on equipping our learners to manage their emotional responses to learning mathematics, our approaches to teaching and learning have developed and we still find that we are able to meet the demands of the challenging curriculum.

For us, the next step will be to revisit and renew the focus on resilience building as, without reinforcement, it is becoming clear that, even within an anxiety-informed department, the drive to focus on just the development of pedagogy with little attention paid to affect can quickly become overbearing and likely to result in more negative experiences for those learners who find the subject difficult. It is reassuring that, in a system with high levels of accountability, our external examination results have shown a significant improvement while still allowing learners the space and time to develop as resilient, independent learners.

Developing resilience in and through assessment

Assessment is a critical element of many stages of education and the high-stakes examinations associated with mathematics cast a long shadow over many aspects of teachers' practice and learners' experience. Pragmatically, teachers may be limited in how much they can control system-wide assessment practices.

Nonetheless, where teachers do have control, they can marshal activities and approaches to help their learners develop mathematical resilience, ready for formal summative examinations. Instead of simply mirroring the form of high-stakes assessment, teachers can help build their learners' confidence and memory over time with frequent, low-stakes tests (Ofsted, 2021) or, as in the previous case study, through more substantial tasks.

A resilience-focused teacher should make sure that assessments are accessible. This includes being aware of the pitfalls of overusing uniform tests where lots of items relate to mathematics that the learners have not yet met. One approach which some teachers use involves taking a few minutes to colour-code a test paper with traffic light colours before beginning to write. This can help direct learners' attention to where they might succeed at the start of an assessment, and what they might try next.

Likewise, if timed practice of skills is considered necessary, it should be handled with care. For example, learners could be asked to complete as many questions as they can in three minutes (and perhaps beat their previous best); this diversifies the nature of success, and would arguably cause less stress than insisting that the whole class must keep pace with a teacher reading out a new question every six seconds.

More widely, it is difficult to overstate how much formative assessment (Black et al., 2003) contributes a critical part of the narrative surrounding learning. Assessment *for* learning resonates strongly with the growth principle of mathematical resilience, as every learner is given support through feedback to develop their mathematical understanding from where they are, and constructive comments promote autonomy by inviting learners actively to take part in moving their learning forwards. Feedback can further support learners' self-efficacy, growth and interest in mathematics (Rakoczy et al., 2019), in support of the value strand of mathematical resilience.

Conclusion

Today's mathematics teacher must balance many demands in their planning and practice. They are asked to enter the classroom with an eye on attainment goals, behaviour management, school-specific policies and much more. It would be understandable to see building a resilient learning environment as an undue extra, or one stress too far on teachers' time and attention. We offer three points which address this concern before concluding.

First, a lot of good practice with differentiation, assessment and so on, is already consonant with the principles of developing mathematical resilience. Explicitly integrating elements of growth, struggle, value and support into teaching and learning can strengthen and enhance existing pedagogy, with the aim that resilience is no longer just a sometime side-effect but a resource which is deliberately cultivated in learners over time.

Second, the goal of developing mathematical resilience supports many of the other goals mentioned previously. There is no way of guaranteeing perfect behaviour from every learner, but it also stands to reason that if mathematics is perceived as something that is engaging but not threatening, and as something that is of great value but not intimidating, then that will positively influence learner behaviour overall. A similar argument could be made connecting resilience with attainment.

Third, there are many ways to improve your own teaching for resilience, and every teacher can select, develop and evolve approaches which work in their own institutional context, and which recognise their own situation. You may be in a position, like John was, where you can set up a comprehensive programme. However, it is more likely that you will start with smaller changes. Perhaps you could integrate one low threshold, high ceiling task once a term across a key stage, start awarding 'mistake of the day' to a productive error in your class (Lee and Ward-Penny, 2022, p. 85), or just fine-tune how you word questions. It does not take massive changes to become more attuned to learners' mathematical well-being; simply adjusting routines can promote mathematical resilience by involving everyone, making struggle a visible part of learning, collectively respecting that it takes time to learn mathematics, and openly valuing effort and progress, not just attainment.

We hope that you have found some of the ideas and discussion offered earlier helpful. In time, perhaps we can reverse some of the findings offered in the introduction to this chapter, and bolster cohorts of learners who are cognitively and affectively better equipped to engage with the important and enjoyable work of learning mathematics.

References

Black, P., Harrison, C., Lee, C., Marshall, B., and Wiliam, D. (2003). *Assessment for learning – Putting it into practice*. Maidenhead, UK: Open University Press.

Blair, A., and Hindle, H. (2019). Models for teaching mathematics revisited. *Mathematics Teaching*, 268, 37–40.

dos Santos Carmo, J., Gris, G., and dos Santos Palombarini, L. (2019). Mathematics anxiety: Definition, prevention, reversal strategies and school setting inclusion. In: D. Kollosche, R. Marcone, M. Knigge, M. G. Penteado, & O. Skovmose (Eds.), *Inclusive mathematics education*. Cham: Springer. 10.1007/978-3-030-11518-0_24

Foster, C. (2013). Resisting reductionism in mathematics pedagogy. *The Curriculum Journal*, 24(4), 563–585. 10.1080/09585176.2013.828630

Immordino-Yang, M. H., and Damasio, A. (2007). We feel, therefore we learn: The relevance of affective and social neuroscience to education. *Mind, Brain and Education*, 1(1), 3–10. 10.1111/j.1751-228x.2007.00004.x

Lee, C., and Ward-Penny, R. (2022). Agency and fidelity in primary teachers' efforts to develop mathematical resilience. *Teacher Development*, 26(1), 75–93. 10.1080/13664530.2021.2006768

Mason, J. (2015). On being stuck on a mathematical problem: What does it mean to have something come-to-mind? *LUMAT*, 3(1), 101–121. 10.31129/lumat.v3i1.1054

Meyer, D. (2010). Math class needs a makeover. [TEDxNYED Talk]. Posted May 2010. Available at https://www.ted.com/talks/dan_meyer_math_class_needs_a_makeover (Accessed May 2023).

Nardi, E., and Steward, S. (2003). Is mathematics T.I.R.E.D.? A profile of quiet disaffection in the secondary mathematics classroom. *British Educational Research Journal*, 29(3), 345–367. 10.1080/01411920301852

NRICH (no date). Low threshold high ceiling. [Webpage]. Available at: https://nrich.maths.org/8769 (Accessed May 2023).

Ofsted (2021). Review research series: Mathematics (assessment section). Available online at https://www.gov.uk/government/publications/research-review-series-mathematics/research-review-series-mathematics#assessment (Accessed May 2023).

Rakoczy, K., Pinger, P., Hochweber, J., Klieme, E., Schütze, B., and Besser, M. (2019). Formative assessment in mathematics: Mediated by feedback's perceived usefulness and student's self-efficacy. *Learning and Instruction*, 60, 154–165. 10.1016/j.learninstruc.2018.01.004

Richardson, M., Isaacs, T., Barnes, I., Swensson, C., Wilkinson, D., and Golding, J. (2020). *Trends in International Mathematics and Science Study (TIMSS) 2019: National report for England*. UK: Department for Education. Available online at https://www.gov.uk/government/publications/trends-in-international-mathematics-and-science-study-2019-england (Accessed May 2023).

Sherrington, T. (2019). *Rosenshine's principles in action*. United Kingdom: John Catt Educational Limited.

Warshauer, H. K. (2015). Productive struggle in middle school mathematics classrooms. *Journal of Mathematics Teacher Education*, 18, 375–400. 10.1007/s10857-014-9286-3

Wigley, A. (1992). Models for teaching mathematics. *Mathematics Teaching*, 141, 4–7.

Williams, G. (2014). Optimistic problem-solving activity: Enacting confidence, persistence and perserverance. *ZDM Mathematics Education*, 46, 407–422. 10.1007/s11858-014-0586-y

Understanding the power of a coaching approach

Debbie Inglis and Sue Johnston-Wilder

Introduction

Learning mathematics is a journey across both a spectrum of mathematical areas, and within each mathematical task. The coaching approach also represents a journey, starting from where the learner is and supporting them to develop a range of tools, strategies and personal resources to achieve a solution.

The International Coach Federation (2023) defines coaching as "partnering with clients in a thought-provoking and creative process that inspires them to maximise their personal and professional potential." In the context of this chapter, 'clients' are thought of as learners and the 'process' as the process of learning mathematics, including any part of the learner's journey. The person they are partnering with in this coaching relationship could be a teacher or tutor who has trained in coaching skills, or a specialist coach who knows about the mathematical resilience framework of tools discussed earlier in Chapters 1 and 3. We will refer to both teacher and specialist coach as 'the coach' in the first part of this chapter. It is important to note that the role of the coach is not to know the specifics of someone's job or role (and in this case the mathematics content), but rather to create a psychologically safe space in which the individual learner can articulate their perceptions of the situation, consider and discuss options, then identify a plan for moving forwards towards a solution (Gardiner and Kearns, 2012).

The coach will use a specific coaching mindset and group of skills to support and guide the learner. Importantly the coaches' role is not to tell the learner what to do. This can feel frustrating, particularly for those coaches who have a good knowledge of the subject matter (in this case mathematics) and who want to share this knowledge as a first level of help. Instead, the coach needs to put their knowledge to one side, resist the urge to give the learner advice, and ask questions to find out what is going on in the learner's mind.

In this chapter, we start by sharing the mindset and skills of a coaching approach, and how to set up a coaching approach in a learning environment. Then we will focus on how to coach for mathematical resilience in older learners and adults with mathematics anxiety, before exploring the power of coaching to overcome personal barriers to learning.

Coaches can apply the mindset, skills and tools of coaching in a wide range of situations (Grant and Cavanagh, 2011). In the education sector alone, this includes:

- Peer coaching to support learning and professional development;
- Coaching colleagues on wellbeing topics, such as workload;
- Enhancing conversations with parents, such as getting to the nub of issues quickly, and finding positive and productive ways forward;
- Self-coaching to increase personal resourcefulness;
- Plus, of course, supporting learners.

Introducing a coaching approach

Coaching can be applied in one-to-one and small group situations, both in person and at a distance via phone or online. A coaching approach can be used as part of the role of a teacher working with a whole class. The mindset and skills of a coach, can enable a facilitator of learning to become a more effective communicator, develop better relationships, and expand their personal development toolkit.

The coaching mindset

There are several core elements that create a coaching mindset. For the purposes of this chapter, we focus on the following four:

1. Having unconditional positive regard (UPR).
2. Focusing on the learner, their experiences, thoughts and feelings.
3. Creating a collaborative conversation built on openness and trust.
4. Believing that the learner's self-knowledge improves their performance.

We look at each of these in turn.

Having unconditional positive regard (UPR)

UPR is a concept in humanistic psychology, based on the work of Carl Rogers (1957), that refers to the acceptance and validation of another person without judgement or evaluation. In the coaching mindset, UPR means that the coach accepts and values each learner as a unique individual, regardless of their abilities or behaviour. Coaches are non-judgemental and do not hold preconceptions about the learner's potential, but rather believe in the learner's capacity for self-determination, growth and development. If they do find themselves making judgements, it is important that the coach lets these thoughts go and does not allow them to influence how they work with the learner. A coach who practises UPR creates a positive, supportive and inclusive learning environment for learners. This can lead to greater engagement, motivation, and perseverance in the learners, as well as improved self-esteem and self-worth. UPR does not mean that the coach is always positive and avoids dealing with learners' negative behaviours. They set boundaries and consequences for negative behaviours, while maintaining an overall attitude of non-judgement and acceptance towards the learner. This approach helps learners to take responsibility for their actions, learn from their mistakes and improve their behaviour and learning.

In summary, for a coach, unconditional positive regard is a mindset and approach towards learners that is accepting, non-judgemental and focused on the learner's growth, development and well-being.

Focusing on the learner, their experiences, thoughts and feelings

Under pressure of time, there is a temptation to help a learner who is 'stuck' to get to the answer by telling them what to do. This can lead to frustration and overlooking the process of learning. A coaching approach encourages the coach to resist telling the learner what to do or giving them the answer. Instead, it promotes a position of curiosity and adopts a range of largely open questions to explore the learner's current thinking. As a result, there is less of a tendency to make assumptions about what is going on for the learner, which can hamper rapport and result in the learner feeling that their opinions do not count, or that they are not being listened to.

Whilst mathematical knowledge is important in helping the learner, learning is the result of a range of mathematical experiences, mistakes and understanding. Everyone possesses their own unique view of the world, and understanding the learner's own view and comprehension better equips the coach to provide the best guidance and scaffolding. So, there is a case to be made here for redefining what is meant by 'helping'. A coach may have the best intentions when they share their learned strategies and knowledge, but, with a coaching approach, the focus must be on helping the learner to help themselves.

In terms of the learner's feelings, often the first thing that is observed with learners who are struggling is strong emotion. This could range from a total shutting down and apparent lack of emotion to more extreme anger borne out of frustration, fear or anxiety. This has been discussed in Chapter 3 where the tools to help identify the underlying causes of strong emotions were introduced. The next section offers another mindset element that can help here, and later in this chapter you will be offered coaching strategies to support learners who find themselves in this situation.

Creating a collaborative conversation built on openness and trust

Within this coaching mindset element, it is useful to think of a collaborative conversation as a discussion where knowledge and ideas are shared and combined to achieve the desired outcome; in this case, the solution to a mathematical problem, or figuring out useful ways of working. However, the roles of teacher and learner do not naturally suggest equality in the status of ideas that tends to indicate a collaborative approach. There is a natural hierarchy based on the status or role of the teacher and their increased knowledge compared to that of the learner. So, creating a collaborative conversation in practice requires some thought and some particular actions.

It is important for the coach to foster an open environment, either 1-to-1 or in a group, where the learner feels safe to share their thinking, try things out, explore ideas and take risks which might not always result in the correct answer. This environment will lead to learners feeling more comfortable and confident, and safer to make mistakes and ask questions without fear of ridicule or embarrassment, leading to greater learning and growth. In the following Coaching Skills section, we also share some ideas on ways to build rapport that will positively contribute to this type of collaborative conversation.

Further ways to create a safe space are:

1. **Being aware of and sensitive to cultural and individual differences**
   Being aware and respectful of different backgrounds, experiences, and perspectives.

2. **Being inclusive**
   Making sure that everyone feels included and welcome, regardless of their background or current abilities.

3. **Being available and accessible**
   Including being present, responsive and accessible to participants' needs and concerns.

4. **Being adaptable**
   This includes being flexible and open to changing approaches to meet the needs of participants and to making sure that the space is safe and comfortable for each individual.

Believing that the learner's self-knowledge improves their performance

This core mindset element may not come as a surprise, considering the earlier point where the attention is on the learner's experiences. This element focuses specifically on the learner developing knowledge of themselves, how they learn, what skills and resources they have, what they can do when they get stuck or anxious, and so on. Enhancing the learner's knowledge of themselves in practice is a process of reducing the time spent telling the learner what to do, and increasing the time spent asking questions which are designed to help the learner discover (or re-discover) what they already know. Helping learners figure things out for themselves creates an environment where the learner can experience the joy and increased confidence which comes with having a key part to play in their own learning.

Self-knowledge also builds resilience. A resilient learner will be more able to identify sources of support when they need help and know how to ask for it. They are also in a better position to employ self-safeguarding and maintain a position in their growth zone, addressing any periods of anxiety or avoidance of challenge.

A key part of self-knowledge is the learner's values, what is important to them. Learners are motivated by their values, and demotivated by anything that appears to conflict with those values. For example, when it comes to career choice, one learner may want to be a nurse and can see the benefits of a mathematics qualification. Alternatively, another learner may want to work with people, or want to get a manual job, and is unable (yet) to see the relevance of school mathematics. Conversely, learners who say they want a job that, in their mind, does not need a mathematics qualification could be motivated to say this because of their fear of mathematics. We return to values and motivation later in the chapter, when exploring how to set up a coaching approach in practice.

The coaching skills

As you read through this list of skills involved in a coaching approach, you will probably recognise them as ones that you already have developed to some extent as a teacher:

- Active listening
- Open questioning
- Building rapport
- Offering feedback for growth
- Setting goals

But each skill takes on a specific identity within a coaching relationship.

Active listening

Starr (2021) offers 4 levels of listening: Cosmetic; Conversational; Active and Deep.

When you are using **cosmetic listening**, you are not really listening at all. Your attention is elsewhere; in the classroom, it might be on another group of learners rather than the one in front of you.

When you are using **conversational listening** your time is divided into 3 parts:

1. listening to the other person,
2. thinking about what you will say in response,
3. and then speaking.

As a result, in this level of listening two thirds of your time is potentially spent thinking about yourself, not the other person.

When you are using **active listening**, you focus your attention on the speaker. You might ask questions, but they will be from a point of interest and curiosity. Your attention is much less on yourself, and your aim is to engage the speaker, drawing out their thoughts and feelings in a way that feels supportive and positive for the speaker.

Using **deep listening** takes this further. It invokes intuition and a level of listening that means you get a greater sense of the other person. You can 'hear' what they are *not* saying as much as what they are. This is less likely to be reached in a busy learning environment, although it is worth being aware of deep listening so that you can use it for the occasions that may happen during longer one-to-one periods with a learner.

Both active and deep listening are skills needed by a coach. When choosing to adopt a coaching approach with a learner, the aim would be to actively listen from the start. This may take some practice, for both the coach and the learner, especially when the coach, as teacher, is used to being the one doing most of the speaking in a learning environment, but will also depend on how much the learner is used to being listended to.

Open questioning

Coaching-style questions are generally open and start with one of the five W's or one H, namely: What, When, Where, Who, Which and How? The question 'Why' is avoided as it provokes a feeling of having to justify thoughts and actions and can feel judgemental for the learner, particularly when they are in a position of fear or anxiety about a mathematics question, or mathematics in general. Note that you can still ask a 'Why' question with a mathematics anxious learner without saying the word itself. For example, you can change: "Why did you choose that method?" to: "What was your thinking here?"

Coaching style questions should generally be open and not leading; "Could you do it like this?" is leading as it is suggesting a method. Whereas "How might you approach this problem?" invites discussion and different possibilities. The reason for asking open questions is to gather as much information as possible in order to help the learner. There are several different coaching question models that support this particular skill, and the third section, "Setting up a coaching approach", of this chapter offers details of one particular model, the Egan (2002) model, that has been used to good effect with mathematics anxious tutors and learners.

Building rapport

Rapport is a feeling of connectedness with another person and can come from already having things in common with the other person or from building a relationship by creating connections, ideally in an authentic way. Rapport leads to trust, which leads to the learner being more willing to share their thoughts and feelings.

Three key factors contribute to rapport:

1. A non-judgemental approach: the learner will be more likely to talk openly if they feel they are not being judged (see UPR discussed previously)

2. Active or ideally deep listening: the coach can build trust more easily and the learner will communicate more if they feel they are being listened to.

3. Using the learner's language: the learner will be more likely to engage in the process of learning if the coach asks questions which use some of the learner's words and phrases. This can be called "mirroring", as your words mirror theirs. This can include asking, "Which part in particular is difficult?", when the learner describes the task as 'difficult', rather than the coach changing 'difficult' to a similar word such as 'tricky', which might have a very different meaning for the learner. The use of 'mirroring' back the learner's words and phrases is a key part of rapport building, and shows the coach is both listening and valuing the learner's thoughts or feelings.

Offering feedback for growth

Feedback can be perceived negatively by the recipient. Even the thought of receiving feedback can sometimes result in concern or anxiety. Much feedback is of a summative nature: marks and grades are given, which may be an attempt to help the learner know 'where they are' and whether they are progressing towards certain goals. Marks and grades are known to be ego-involving (Black et al., 2003); learners can feel labelled and as though they have no control over what mark or grade they get. Such summative feedback is rarely received as positive. Formative assessment focuses on the process of doing mathematics rather than the learner or the outcome; it assesses what has been done well and what can be improved or learned from any mistakes. Because of the focus on what can be done, or next steps, formative feedback is not ego-involving and is usually much better received by the learner. Unless it is accompanied by a mark or grade, in which case if it is a good mark the learner will often ignore the formative feedback as they are OK, and if it is a poor mark the learner will tend not to look at any other feedback due to the negative emotion it engenders.

The coaching approach has ongoing feedback inherent within its supportive and developmental process. Rather than offering opinions, coaches feed back on what they are noticing with the learner, in a non-judgemental and factual way. A gentle way to approach offering feedback, particularly for the mathematics anxious learner, is to ask permission to offer some formative comments. In the coaching relationship they can, of course, say 'no', but when you have fostered an open and honest conversation, and built trust, feedback is more likely to be openly received. For example, you might say, "I've noticed something about how you're working that I'd like to feed back to you; is that OK?" And your piece of feedback might be, "I noticed when you did that calculation, you hesitated for a while before using the calculator, and I wondered what made you pause."

The feedback statement can be followed up with a question to further explore the problem, the learner's current understanding, and their awareness of personal and external resources. A follow-up question might be, "What were your thoughts as you were holding the calculator?". This shows the coach's curiosity and refers to a fact not an assumption. It picks up on a key moment in the learning process where the learner hesitated. Exploring this hesitation could uncover a learner's limiting beliefs or unhelpful thoughts. They may have a question but thought it was inconsequential, so held back from asking it.

Setting goals

When a coach and their client work together, they usually try to overtly identify what they are working towards from the outset. When a teacher or tutor and a learner work together, the goals may be tacit, for example "to do better at mathematics" but trying to be clear about what the learner wants to work on more explicitly may help build a good coaching relationship. Examples of goals might include:

- To gain greater confidence in group settings,
- To develop productive time management skills,
- To develop resilience strategies to avoid panic or burnout and stay longer in the growth zone.
- To identify how I can help myself get the support I need to understand the mathematics for my GCSE examination.

Whatever the goal, the coach-client relationship starts with where the client is now, and works towards an agreed point in the future. This may also be applied to the teacher-coach relationship, with a slight difference. Learners may have a goal for the term or year which is set for them by their teacher or by institutional or government targets. Nevertheless, it is still possible for the learner to feel they have some control over what they are working towards. This control comes in the form of the teacher helping the learner create their own smaller targets that still lie within wider educational goals. Questions to help the learner create mini goals include:

- How would you like to feel about this mathematics topic by the end of the lesson today or by the end of the week?
- What would be a really good outcome for you from today's lesson?
- Which skill do you want to get better at this week or term?

From this point onwards, the coach can be mindful of the learner's own goals in light of the wider goals and can be sensitive to these when supporting the learner.

Setting up a coaching approach

Having established what a coaching approach is, the chapter now focuses on the practicalities of developing a coaching approach in different mathematics learning contexts.

At the core of this process is the Egan Model (Egan, 2002), a model that can be used to support learners, ranging from mathematics anxious to mathematics confident. It can be applied 1-to-1 and in groups with specialist coaches trained in mathematical resilience, and 1-to-1, in groups and classes with mathematics teachers using in coaching skills (differentiated here as 'teacher-coaches').

The Egan Model

The Egan Model is drawn from the book 'The Skilled Helper' by Gerard Egan (2002). The model leads both coach and learner to resist focusing on 'the answer', but rather to allow time for the learning process to happen and to foster the development of resilience. It proposes three stages to use when helping someone with a problem. Using these three stages can help prevent diving in too quickly with a solution that may not be helpful and will not have come from the learner themselves. The 3-stage Egan Model supports the learner from exploration of their starting point, when presented with a mathematics problem or question, through to taking action and finding the solution. It can also help enthusiastic mathematics teachers refrain from focusing on the solution, instead their focus is on the process and what the learner brings to the problem or question.

The three stages are: Explore – Options – Actions

Exploring the current situation

At the start of this process, it is important that the teacher-coach or specialist coach creates a safe space using active listening, open questions and unconditional positive regard to enable the learner to explain and explore their situation. In this space, the coach will have empathy with the learner, will summarise the conversation using the learner's words and phrases, and will check understanding to eliminate assumptions and misunderstandings. If the learner experiences the coach as a genuine person, they are more likely to trust them. Rogers calls this congruence (Rogers, 1957).

Options

The coach invites the learner to brainstorm possible strategies and ways of finding support, asking open questions and occasionally putting additional options 'on the table', without telling the learner what to do.

Actions

The coach invites the learner to select from the options and to make a plan. The agreed plan ideally would be SMART: specific, measurable, achievable, relevant and time bound. The coach supports the learner in setting such goals in line with the learner's own values and constraints and will support reviewing these goals at a subsequent session.

Adapted to blend with the Growth Zone Model in Chapter 3, the Egan model encourages specialist coaches or teacher-coaches to start by helping the learner to identify which zone they are in: red, amber or green. Mathematics anxious learners are likely to be in the red zone and finding thinking very difficult. Contrast this with a teacher-coach or mathematics specialist, who is likely to be in the green zone, and who thus needs to draw on empathy and a coaching approach to bridge the gap and to support the learner's agency. In groups or classes, learners may indicate their zone by showing red, amber or green cards. The teacher-coach or specialist coach can then tailor their next step accordingly.

The following section details the three stages and provides examples of questions and other coaching elements related to each stage.

EGAN STAGE I: EXPLORE

This stage is about exploring what is going on for the learner from the outset; how they feel about a mathematics problem or task and their ability to engage with it. The coach and learner start by using the Growth Zone Model from Chapter 3 to find out how the learner(s) is(are) feeling; are they in the red, amber or green zone?

Questions the coach might use if the learner is in the **red zone** include:

- Where is the feeling in your body?
- What shape/colour is it?
- What's happening in your head?
- What do you notice about your breathing?
- If you were an animal/film/song, what would that be?

The overall purpose with these questions is to help the learner simply notice what is going on for them, labelling emotions where possible, and without judgement. The role of the teacher-coach or specialist coach is crucial here. They need to resist 'rescuing' the learner and remember that another way to define 'helping' them is to help them to help themselves. This builds the learner's resilience and adds to their collection of personal resources.

What the teacher-coach or specialist coach *can* do at this stage, in addition to asking questions in a non-judgemental way, is to offer the learner some strategies. It is important to note that the learner should still feel in control here. The coach might encourage this by saying, "There are a range of strategies that I can share here that could help you. You can choose the ones you would like to try, or they may help you to think of your own."

Some examples of strategies that the coach could offer the learner include:

- Give them permission to step away from the situation, this is important if they start to panic or get anxious. They can indicate this with a 'Stop' sign such as the flipped position of the hand model of the brain.

- They could go for a walk.

- A way to remember that everyone makes mistakes and that is OK because we can learn a lot from mistakes.

- Taking time to think and not rush.

- Starting by making notes on paper; getting thoughts out of their head can help to 'unjumble' them.

The coach could then follow up the offer of these ideas by asking, "What would be most useful here?" or asking if the learner has an alternative strategy to offer. After trying out different strategies, and making a note of what works, remind the learner they can use their chosen approach to move from red to amber more quickly and easily the next time they find themselves in a similar situation.

Examples of questions you might ask when the learner indicates they are in the **amber zone**:

- What do you notice here about this question/task?

- What do you already know that can help you here?

- Which words or phrases are familiar/unfamiliar?

- How could you find out what they mean?

- What is your current understanding of this task?

- Which bits do you need to clarify?

- What do you need to find out?

- How do you prefer to work?

- What resources or support are available to you?

- Who could you ask?

Some learners may indicate that they are in the **green zone**, which also needs to be explored. Is every part of this question or task comfortable to them, or are they simply confident about having a go at it? Once they begin the task, they may move into the amber zone, but they also may decide that they already know the answer and understand the mathematics used.

Here are some questions to ask:

- How will being in the green zone help your learning?
- How will being in the green zone hinder your learning?
- What would make this more challenging for you?
- What would make the task more exciting or interesting?
- What if … .?

EGAN STAGE 2: OPTIONS

Once there is some clarity about the learner's starting point, the learner and coach can move on to identifying the learner's options, as they work through the question or task.

Examples of questions at this stage are:

- What are all the things you need or want to find out, or clarify, before you start? List them if it helps.
- How might you find out what you do not know yet? How else? (Repeating 'How else?' a few times can help uncover more creative ways of working)
- How might you approach this task?
- What are your options?
- What resources or support can you access? (resources, people, websites, etc)
- What are the pros and cons of each option?

If the learner displays signs of moving into the **red zone**, as they work on the task, the coach can offer the previously mentioned supportive strategies and ask the red zone questions. Supporting the learner back into the **amber zone**, and completing the Options stage, is key before continuing with Stage 3.

EGAN STAGE 3: ACTIONS

There are two elements to this stage: developing strategies and action planning.

Examples of questions to support developing strategies are:

- Which strategies are best for this situation?
- Which strategies best fit your way of learning?
- What are the pros and cons of using this strategy?
- What will help you achieve this task?

- What will hinder your achieving the task?
- How will you minimise the things that may hinder you?
- What do you need to do?
- How can you get started?
- What can you do right away?
- What could you do next?

Examples of questions to support action planning are:

- What actions will you take to achieve the task?
- By when?
- Who could help you?
- What are all the things you need to do to complete the task?
- Do they have to be done in a specific order? If so, what is the order?
- What should be done straight away? What can wait?
- Where can you get the resources or support you need?
- What will you do if you get stuck?

As you can see, there are lots of questions to choose from, and you do not need to ask all of them. Success of this model will depend on actively listening to the answers and tailoring subsequent questions to what you hear. Remember to keep the questions open, curious and not leading, as much as possible.

Silent coaching

This is a powerful adaptation of the coaching model, which supports the coach's ability to work with several learners, or indeed a whole class, at once. It involves asking a set of open questions (5–10 is usually ideal) which are progressive in nature, guiding the learners to think through a particular topic, such as a recent experience in a lesson. The coach asks one question at a time, giving the group time to silently record their thoughts, rather than saying them out loud.

A useful structure for 'silent coaching' follows the Egan model, and includes questions on:

- Exploring something that has just happened in the lesson
- Considering possible options to manage this and move forward
- Identifying next steps

This model means that several learners can be involved at once, and it allows for a range of individual answers which, depending on the question, could all be valid. The Egan Model detailed previously provides examples of questions that could be used here. Here are three examples of when a teacher could use this approach in a lesson:

1. At the start, following the introduction to a mathematical problem, and the teacher-coach wants to give the learners thinking time.

2. During a lesson, when a group of learners are stuck, and the teacher-coach wants to help them pause, reflect on what they already know, and consider options for moving forward.

3. At the end of a lesson, as part of the reflection or 'what have we learned?' process.

You could also follow-up a 'Silent Coaching' exercise with partner work, where learners share one or more of their answers verbally with a partner.

Coaching training

Many of the ideas and approaches shared in this chapter have been part of two successful training programmes. These programmes have been run successfully with groups of FE mathematics teachers, mathematics anxious teachers and teaching assistants, and older learners around England over recent years, by Sue Johnston-Wilder, Debbie Inglis and colleagues to help develop mathematical resilience (Johnston-Wilder et al., 2013).

One programme is known as *Developing Coaching Skills for Mathematics Teachers* and includes STEM teachers. The other programme, known as *Coaching for Mathematical Resilience*, is aimed at mathematics anxious adults, including teachers and support staff, and older 16–18 learners, including learners who are re-taking their GCSE examinations having failed that examination at least once already and learners who thought they had left mathematics behind by choosing advanced study subjects such as psychology, geography and chemistry.

For the specialist STEM teachers, a level of patience is required, as the emergent teacher-coach adapts their default position from 'rescuer' to 'helper'. There is also an element of preparation required, before going into a mathematics learning environment, in order to adopt a coaching mindset. Developing both the coaching mindset and skills take practice, but the rewards speak for themselves. The mathematics specialist teachers reported that employing the coaching approach resulted in learners reporting having:

- more autonomy
- increased confidence
- feelings of being trusted

- a sense of gratification
- feelings of satisfaction
- and rich rewards

(Johnston-Wilder et al., 2013)

The mathematics anxious participants learning to coach themselves and their peers reported:

- feeling encouraged
- there was a 'we can do that together' attitude
- questioning to find out where they were
- efforts to establish a relationship
- feeling supported

(Johnston-Wilder et al., 2013)

> Joe (aged 17) took part in a *coaching for mathematical resilience* programme. He had failed GCSE maths and was due to resit that examination. One lunchtime, he was sitting next to a friend studying maths A-level who was stuck. Joe used his coaching strategies to help his friend make progress. Joe was delighted.

Self-coaching and continued learning

Once the coaching mindset and coaching skills are understood, as well as the Growth Zone Model, they can be applied in a self-coaching model. Learners can recognise when they are in a 'red zone' state and think of the options they have which will help them to move back to the green zone. They can also recognise when they are in the green zone, when things feel too easy for them, they are in danger of feeling bored and could benefit from more challenge.

Learners can be helped to help themselves become 'unstuck' in any area of learning. You might suggest initially that they think in terms of being kind to themselves, recognising what is happening and removing judgement, before considering their options and subsequent actions. They can set themselves personal development goals, asking themselves some probing questions to ensure clarity, identify their personal motivation for their goal, set timescales and break a larger goal into smaller chunks, so that it feels more doable.

The key things to remember are the coaching mindset and the skills. As you explore using these, you will learn from the feedback you receive from the learner and how it feels to coach. The learner's feedback may not always be verbal; you can usually tell how good a question is by how it is received by the learner. Does it make them think? If they do not initially respond, do not be tempted to dive in with another question, or rephrase the one you have just asked. Observe the learner. Are there clues that they are thinking? Is it a comfortable, thinking silence, or are they looking at you wanting another question or some other sort of support? If in doubt, wait or ask. As with all learning, it is often in the doing where progress is made.

Conclusion

In line with the wider philosophy of this book, a coach for mathematical resilience is someone who adopts a coaching approach to help individuals or groups develop the skills and mindset necessary to persevere and succeed in mathematical learning. This can include supporting learners in developing a growth mindset, exploring strategies for dealing with frustration and setbacks, and providing opportunities for learners to develop their confidence and self-efficacy in mathematics. The coach will also aim to promote the belief that, with effort, perseverance and support, anyone can achieve success in mathematics.

In this chapter you have been introduced to the mindset and skills of coaching. Using Egan's (2006) concept of the skilled helper, you have been offered ways to help learners of mathematics overcome the barriers they may have previously developed and succeed in their learning. You have been shown how to support and guide a learner using the coaching mindset skills and understand that the coaches' role is not to tell the learner what to do. The coaches' role is to help the learner explore their feelings, values and beliefs about their current situation, to decide what options might be available and which may work for them and then to encourage the learner to make a plan for action.

References

Black, P., Harrison, C., Lee, C., Marshall, B., and William, D. (2003). *Assessment for Learning- putting it into practice.* Maidenhead, UK: Open University Press.
Egan, G. (2002). *The skilled helper: A problem management and opportunity development approach to helping, 7th edition.* Pacific Grove, CA: Brooks Cole.
Egan, G. (2006). *Essentials of skilled helping: Managing problems, developing opportunities.* Pacific Grove, CA: Brooks Cole.
Gardiner, M., and Kearns, H. (2012). The ABCDE of writing: Coaching high-quality high-quantity writing. *International Coaching Psychology Review,* 7(2), 237–249.
Grant, A., and Cavanagh, M. (2011). *Designing positive psychology: Taking stock and moving forward.* New York: Oxford University Press.
International Coach Federation (2023). https://coachingfederation.org/about

Johnston-Wilder, S., Lee, C., Garton, L., Goodlad, S., and Brindley, J. (2013). Developing coaches for mathematical resilience. In *2013 ICERI 2013: 6th International Conference on Education, Research and Innovation*, 18–20 Nov 2013, Seville, Spain.

Rogers, C. (1957). The necessary and sufficient conditions of therapeutic personality change. *Journal of Consulting Psychology*, 21, 95–103.

Starr, J. (2021). *The coaching manual.* London UK: Pearson Education Ltd.

Working with groups of learners

Robert Ward-Penny

Introduction

Group work is an important part of education. It models an important feature of many community and work environments, and can lead to better outcomes; research has shown that cooperative learning approaches tend to have meaningful, positive effects on both achievement and affect in mathematics (Capar and Tarim, 2015).

Group work also aligns well with the principles of teaching that enable leaners to develop mathematical resilience. If teachers intend for learners to see themselves as part of a community within which they can seek and offer support, it follows that opportunities to work with others will feature in the learning environment. However, much widespread practice in school positions learners differently. Whenever learners are listening passively to an explanation, copying down an example, or working through problems from a textbook, they are chiefly working as individuals, cut off from their peers. If this isolation is persistent, it not only excludes the many potential cognitive and social benefits of collaborative working (see for instance Swan, 2006) but can also lead to "quiet disaffection" (Nardi and Steward, 2003). Teaching for mathematical resilience has the potential to rebalance classroom practice, by reminding teachers of the importance of communication and collaboration.

Context, however, is key. The same group work activity which generates mathematical discussion and promotes resilient behaviours in one set of learners might lead to anxiety, conflict and behaviour management problems in another. Teachers can also be limited by outside pressures and practical concerns, even down to the physical space and furniture available. There is no single, infallible way for any teacher to ensure constructive group work in the mathematics classroom, and it can take time to set up. Sometimes teachers must take the time to teach and refine collaborative practices that will encourage learners to develop their mathematical resilience.

This chapter examines and suggests how you as a teacher might use group-based activities to develop classroom cultures, promote mathematical well-being, and support the growth of mathematical resilience in your learners. However, it is not intended as a to-do list, nor as a gold standard, and your own professional judgement will be essential as you reflect on the following discussions and work to provide positive and inclusive learning experiences.

Building constructive groups

The principles of teaching for mathematical resilience suggest that group work should ideally be set up in a way that supports personal growth, generates and endorses struggle, safeguards support within learning and accentuates the value of mathematics. These intentions can steer your practical and pedagogical decisions. For example, how should you group your learners: with their friends, with learners of similar attainment, or in some form of mixed grouping?

Friendship groups might naturally lead to the respect and support associated with developing resilience and are sometimes favoured by the learners themselves (for instance Nardi and Steward, 2003, pp. 353–354). Working in this way can support the development of resilient practices, but friendship groups can limit challenge and growth, isolate individuals who do not fall neatly into a group, and even contribute to behaviour management issues. Attainment-based grouping might seem necessary in some curriculum settings, but it can quickly lead to unhealthy comparisons which undermine the ideals that mathematics is for everyone, and that all learners are valued parts of a learning community. Research (for example Wiliam and Bartholomew, 2004; Marks, 2014; Boaler, 2022) has repeatedly demonstrated how ability grouping can have negative effects on learners' performance and affect. Mixed grouping offers some advantages: it can better reflect how everyone can progress in mathematics from where they are, illustrate how it is usual for everyone to experience struggle at some point, and give all learners a chance to support each other in some manner (Barclay, 2021); however, without mutual respect, guidelines and appropriate tasks, mixed groups can devolve into hierarchies.

The success of a group often depends on personal qualities which are not always visible, as well as changeable social factors; this is discussed further in Chapter 16. It may also be that learners have additional educational needs which must be considered. Some learners struggle with the interpersonal demands of working in groups and may need alternative or finely structured provision. Others might prefer to work with a group of peers using a different common language, so that they feel supported and more able to participate.

Another option is to build in some element of randomisation. Liljedahl (2014) makes the case for using fully random groups in the mathematics classroom and gives a thought-provoking account of one class of 15- to 16-year-old learners where the frequent and open use of random grouping led to the learners becoming more amenable, enthused and collaborative; as "the class coalesced into a community, their reliance on the teacher as the knower diminished and their reliance on themselves and each other increased" (p. 142).

Changing the way or ways in which you group learners can quickly have a significant effect. Whilst writing about teachers working to develop mathematical resilience, Lee and Ward-Penny (2022) shared an account from a primary school teacher who had moved away from attainment-based seating to an adapted form of mixed grouping; learners could choose to sit alone, with a partner, in a group, or at the teacher's table if they felt they would need extra help.

> *We very quickly noticed a change in the atmosphere of our maths lessons. Because the children are now sitting in mixed attainment groups, there seems to be more animated discussion all around the room, rather than it being focused on one or two tables ... we have found that children who wouldn't historically be involved are now putting their hands up and joining in. One hypothesis for this is that they feel more confident now because other children on their table are answering questions, which didn't happen before* (pp. 85–6).

What is possible will vary significantly across and between schools, and the benefits and risks of each approach need to be balanced. Group structures can also flow and develop as the culture of the classroom grows. Nonetheless, it is perhaps easiest to connect the principles of mathematical resilience with some form of mixed or random grouping practices.

Setting up groups in the classroom

There are several other practical questions surrounding group work in the classroom. How might groups be seated in the classroom? How big should each group be? Again, the principles of mathematical resilience do not offer definitive answers, but they do highlight many potential benefits and concerns.

For instance, when a classroom has tables that are all arranged in rows with everyone facing the front, there is an implication that the learner's attention should be directed forwards towards the teacher. Working with anyone other than a neighbour or desk partner can also be physically awkward. When desks form small islands around the room, it is usually easier for learners to talk with their peers, and access support from different parts of the classroom learning community.

Desks will not be the only type of furniture which can have an impact on group work; some classrooms are fortunate enough to have large whiteboards or whiteboard tables where all members of the group can gather and share their thoughts in writing. This process can be emulated with large pieces of paper on regular desks, but whiteboards enable learners to suggest, refine and erase their ideas without worrying about leaving a permanent record.

The most effective group size will depend heavily on the specific task, but it will also be influenced by learners' previous experiences. If learners have not become used to taking an active part in lessons and have not yet developed the habits of listening to and valuing each other's contributions, a sudden shift to working in a large group is perhaps more likely to generate anxiety and confusion than resilience. In some learning contexts it can be useful, even necessary, to build towards larger group sizes and scaffold positive learning behaviours.

One relevant classroom tactic which can be used to introduce working with others gradually is 'think-pair-share'. The teacher sets a question, such as "if I multiply two numbers together, is the answer always bigger?" They then ask everyone to think about it individually for a short period of time in silence. Next, they invite learners to discuss their thoughts in pairs. After a short time, pairs are brought together into small groups to settle on an agreed answer to be made ready to share with the whole class. This approach is intended to encourage reflection, revision and teamwork, but it also develops resilience; the stepped approach is inclusive without being threatening, since learners get to share, amend or even abandon their ideas before being 'put on the spot'. They can develop their skills of talking about, listening to and considering one another's ideas, and collaborating on their mathematics in a small-scale, low-stakes situation.

Another approach is to tackle expectations directly, by setting out or negotiating a set of 'ground rules' before starting on group work involving larger groups of learners. Not everyone arrives in the classroom having already learned the unwritten rules of speaking and listening at home (Littleton and Mercer, 2013); you may in turn have to be aware of diverse cultural practices and expectations. A related strategy is to assign roles within the group; for instance, one person is responsible for taking notes, another for presenting to the class at the end. A third idea is to reflect explicitly on both the mathematics and the group work after the fact, perhaps getting each team to nominate their MVP (most valuable player) with an explanation of why they were valued. Rules and roles for effective group work might look very different across age groups, but the principle is to aim to offer the optimal amount of structure to support and activate learners.

One further practical consideration is resourcing. The support principle of teaching for mathematical resilience holds that learners should know how to get help from a variety of sources, and you might choose to set up your classroom to include extra age-appropriate resources such as mathematical dictionaries, access to mathematics software or the internet. This kind of 'choice-affluent' set-up can help maintain collaborative work without teacher intervention when groups become stuck or confused (Pruner and Liljedahl, 2021).

Selecting and appraising group tasks: Three examples

Today's mathematics teachers can choose from a plethora of tasks, games and activities, drawing from the internet and published resources. Many of these are designed for learners to work together, but they can lead to very different learner outcomes, ranging from collaboration and cooperation to competition or even just coexistence. Once again, mathematical resilience offers a critical lens which teachers can use to evaluate and balance their choices.

For instance, one common activity in secondary mathematics classrooms at the time of writing is a live online quiz. Using software such as Kahoot, Blooket or Quizizz, learners can answer questions in real-time to test their recall and gather points. These quizzes can be fun and exciting and focus the attention of a class, but they do very little to help some learners construct mathematical resilience. Even if learners are working in pairs or small groups to arrive at their answers, having to work at speed and then being graded so publicly is likely to engender anxiety in some or possibly in many (Ashcraft and Krause, 2007). A teacher who wants to develop mathematical resilience in all their learners might choose to handle such quizzes carefully and use them economically, after reflecting on the culture of their particular classroom. Competition can be fun for some, but comparison is too often the thief of joy – even if you have managed to come in second place! When online quizzes are used, small deliberate changes such as hiding the bottom of the leaderboard from public view, turning off streak bonuses so mistakes are not penalised as harshly, or encouraging learners to compete with their previous personal bests can make a difference.

Another familiar type of activity is a card sort or jigsaw, such as a Tarsia puzzle. These hands-on resources typically have learners starting with lots of labelled cards which they can rearrange to match questions with answers (for instance functions with their derivatives) or to make connections between different sets of objects (such as decimals and percentages, or between tables, graphs and equations of straight lines). Some excellent examples of these for secondary-aged learners are available for free within the Improving Learning in Mathematics (Standards Units) resources.[1]

If there are plenty of cards to go around, this kind of task can certainly inspire cooperation, with many hands making light work; the need to make group decisions can encourage discussion and the use of mathematical vocabulary. However, none of this is guaranteed. Overly repetitive or simplistic content can reduce a card sort to being a worksheet in disguise; some learners might also be tempted to sit back and let others do the work, particularly if the activity has been set up as a race, so that their contributions might be seen as 'getting in the way'.

Activities like these can be strengthened by thinking through both the content and intent of the task. The activity is more likely to provide appropriate challenge, and involve meaningful struggle for everyone, if the questions cover a range of difficulty. Another approach is to include space for learners to produce some questions and answers themselves; simply adding two or three blank cards to a pack can enable learners to personalise a task in a satisfying way. One more visible extension can be used with jigsaws in the shape of equilateral triangles that have blank spaces around the edge. If the learners glue the jigsaw pieces onto some thin card, they can move on to add new elements around the edge which match up when the triangle is folded up into a tetrahedron – adding a new dimension to their learning! It is impossible to guarantee cooperation and resilient working, but adaptations like this are arguably more likely to promote the development of mathematical resilience.

A third type of activity is an extended group task. As a simple example, you could challenge each group of learners to redecorate your classroom, given a set budget. Depending on how you scaffold it, this task could require skills involving real-world measuring, scale drawing, research, finance and presentation. Extensions could involve greater levels of realism or additional constraints. The multi-faceted nature of this type of task allows learners to explore and value their own different strengths and contributions within the project, and the realistic aspects emphasise the value of mathematics. The different ways in which learners can succeed might also help the teacher to move the project away from being a competition: instead of declaring a clear winner, they might share and praise the most accurate plan drawing, celebrate the most creative response, and congratulate the group that ended up closest to being on budget. Some of the tasks offered by Bowland Maths[2] for lower secondary-school learners fit this style.

Lotan (2003) uses the term "group-worthy tasks" to describe tasks which are "as close as possible to genuine dilemmas and authentic problems ... teachers delegate intellectual authority to their students and make their students' life experiences, opinions, and points of view legitimate components ..." (p. 72). It may be that your own teaching context limits the use of extended group projects that meet all of these criteria, but even shorter, occasional tasks with some of these characteristics can help to promote a broader and more inclusive approach to working mathematically together.

Team games and talking mathematics

Discussion is an important part of collaborative work, but it can be challenging to encourage, manage and evaluate talk in the mathematics classroom. Teachers often urge their learners to keep talk 'on-task', but the boundaries between conversational modes are not always clear and, just like in the workplace, some 'off-task' classroom chatter might serve an affective purpose, and ultimately support collegiate mathematical collaboration. Another important question about talk in the mathematics classroom is how mathematical it is (Ingram and Watson, 2018). Even when the talk is 'on-task', it is possible for communication to become focused as much on turn-taking and the practical side of completing a task or challenge as it is on the mathematics. A teacher who is working to establish a positive environment and develop mathematically resilient learners must make a lot of judgements about talk, both in planning and in the moment.

One approach to managing talk, mentioned earlier, is to prepare your learners with a discussion on roles and expectations (Littleton and Mercer, 2013). However, some tasks are designed with built-in structures which can support or even by-pass this process. For instance, the tasks contained in *Mathematical Team Games* (Lucas, 2014) have each been summarised in 12 cards. These cards contain everything that the learners need, from the opening instructions to the information required to reach the solution. A set of cards is cut out, shuffled, and distributed amongst each team; Lucas suggests a team size of three or four. Theoretically, there is no need for the teacher to do anything else from this point, and if you insist that learners cannot show each other their cards, this requires everyone in the group to contribute verbally to the discussion as the group shares their information and works through the mathematics (Figure 6.1).

If learners are already working in a sufficiently encouraging and inclusive environment, this kind of approach to structuring communication can promote co-operation and positive working behaviours; the teacher is entrusting the learners with some extra responsibility, and although there is room for individuals to negotiate their roles within each group, there is also a guarantee of some level of minimum participation. However, if any learners are too uncomfortable or in some way unable to talk easily with the others in their group, this task could be a source of anxiety for them. This would likely be made even worse if the task was set up as a timed race to the answer. Once again, balance and professional judgement are paramount in establishing and maintaining an effective learning environment.

A different approach is to structure classroom discussion by positioning your learners as young mathematicians. Mathematically rich starting points can allow learners to approach mathematical practice by speculating, guessing at patterns, testing conjectures and justifying their thinking. For instance, Steward (2012) shares

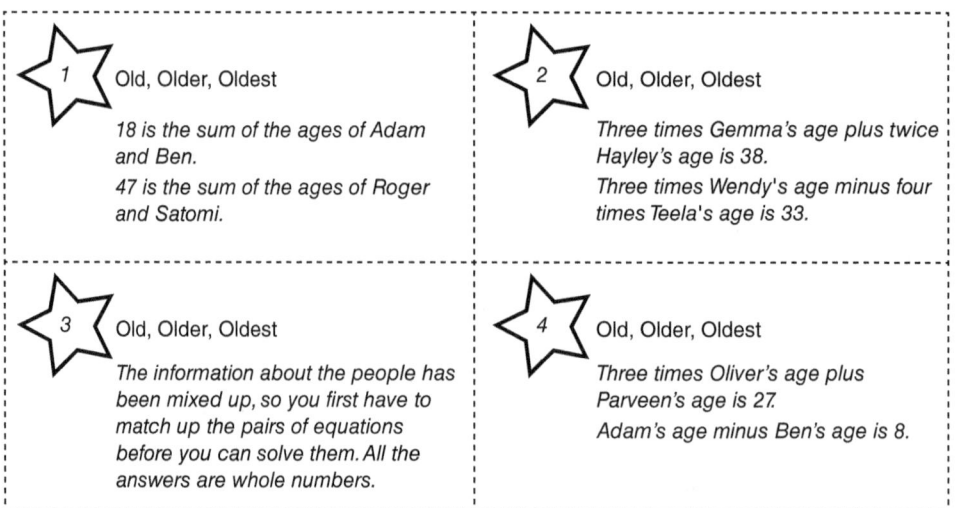

Figure 6.1 Cards from a page of *Mathematical Team Games* (Lucas, 2014).

4	15	11	3
11	4	8	1
7	4	7	10
3	3	3	3
0	0	0	0

Figure 6.2 A 'Diffy' with four steps.

a mathematical task called 'Diffy' which has been presented in various forms and places over the years. Start with four positive numbers and write them in a row. Underneath, write the (absolute) differences between each pair of numbers: the first and second, the second and third, the third and fourth, and finally the fourth and first. Repeat this process until you get to a row of zeroes (see Figure 6.2).

This problem is very easy to access – it only requires a knowledge of subtraction – and it is also quick to lead to conjecture and exploration. If you invite your learners to work in small groups and come up with one 'interesting question' each, they might offer up some starting points such as:

- Will this process always stop at 0 0 0 0? If so, what is the longest possible number of steps and when does it happen?
- What is the shortest possible number of steps and when does it happen?
- Does it make a difference if you can use negative numbers at the start?
- What happens if the starting numbers follow a pattern?
- What happens if you start with the first four square numbers? The cube numbers? The prime numbers?
- What happens if you start with more (or fewer) than four numbers?

Learners can choose a challenge, then go off to work either independently or in small groups, ready to report back every fifteen minutes with either conjectures, justifications, or new questions to add to the board. These can naturally give rise to further mathematical thinking, as learners explore the domain of relevance of a statement, notice underlying structures or come up with counterexamples to their peers' conjectures. (Although there is scope here for formal algebraic proof, the purpose of the activity is not to prove or disprove every statement before the end of the lesson; rather, it is to provide a rich starting point for mathematical talk, co-operation, and collaboration.)

This type of work can line up well with the factors of mathematical resilience, as learners become more aware of growth and struggle and ideally gain experience of working within a supportive learning community. This is perhaps especially evident when learners are taking their first steps in justifying conjectures. Mason, Burton and Stacey (2010) talk about justifying in terms of three steps of conviction: being able to convince yourself, being able to convince a friend and being able to convince a sceptical (internal) enemy. This process offers one possible structure to sharing thoughts in the classroom, and a way to value learners' first steps towards mathematical proof. After a learner justifies their reason for believing in the truth of a conjecture, they might be asked to nominate a 'friend' and then a 'sceptical enemy' to comment on their thinking. Another way in which the teacher can illustrate the principle of support is by 'cross-pollinating', that is selectively sending a member of one group into another group to observe, ask questions, and then return to their own group with a new insight or suggestion.

However, successful group work which allows space for this level of mathematical discourse can require you as a teacher to change some aspects of your own behaviour in the classroom (McCrone, 2005). Thinking carefully about whether, when, and how to intervene (Webb, 2009) will be important, as will ensuring that you model good practice yourself. Francisco (2013, p. 436) writes that "just asking students to work in groups may not be enough", going on to say that:

Teachers also need to pay more attention to the culture of "doing" mathematics that they may be promoting in their mathematics classrooms. This can be the difference between students developing positive or negative views and dispositions toward mathematics, which can have a positive or negative effect on their achievement.

This brings us back to developing mathematical resilience.

Conclusion

This discussion has touched on some of the many cognitive and social advances which can be linked to group work in the mathematics classroom, but there are still others. Swan's (2006) research with learners in further education, for instance, suggests that collaborative work might also lead to improved recall over time. However, it would be disingenuous to conclude by implying that any of these gains are automatic or guaranteed. All the potential benefits of group work will be mediated by the context and culture of your own classroom.

A similar caution can be made with respect to mathematical resilience. Group work is undoubtedly a powerful and proactive way to afford learners healthy and mathematically meaningful experiences. Collaborative tasks have a clear potential for learners to feel connected to, and valued by, the wider learning community. However, if learners are unexpectedly faced with unclear and inappropriate extended tasks, forced into badly constructed groups, and then suddenly put on the spot to report back, they are more likely to experience anxiety than develop resilience. You will need to use your professional judgement to select approaches and cultivate practices which match your situation and activate your learners.

Group work offers learners ways not only to develop mathematical resilience, but also to display it. It is no small thing to observe your learners experiencing autonomy, contributing and drawing from their learning community, and engaging in a healthy way with mathematical challenges. Look out for these moments, reflect on how they have come about, and celebrate them.

Notes

1 At the time of writing these are available online through STEM learning at https://www.stem.org.uk/resources/collection/2938/teaching-activities-and-materials (Accessed May 2023).
2 Available at https://www.bowlandmaths.org.uk/ (Accessed May 2023).

References

Ashcraft, M., and Krause, J. (2007). Working memory, math performance, and math anxiety. *Psychonomic Bulletin and Review*, 14(2), 243–248. 10.3758/bf03194059

Barclay, N. (2021). Valid and valuable: Lower attaining pupils' contributions to mixed attainment mathematics in primary schools. *Research in Mathematics Education*, 23(2), 208–225. 10.1080/14794802.2021.1897035

Boaler, J. (2022). *Mathematical mindsets* (2nd Edition). Hoboken, NJ: Jossey Bass.

Capar, G., and Tarim, K. (2015). Efficacy of the cooperative learning method on mathematics achievement and attitude: A meta-analysis research. *Educational Sciences: Theory and Practice*, 15(2), 553–559. 10.12738/estp.2015.2.2098

Francisco, J. (2013). Learning in collaborative settings: Students building on each others' ideas to promote their mathematical understanding. *Educational Studies in Mathematics*, 82, 417–438. 10.1007/s10649-012-9437-3

Ingram, J., and Watson, A. (2018). But are students communicating mathematically? *For the Learning of Mathematics*, 38(2), 19–21.

Lee, C., and Ward-Penny, R. (2022). Agency and fidelity in primary teachers' efforts to develop mathematical resilience. *Teacher Development*, 26(1), 75–93. 10.1080/13664530.2021.2006768

Liljedahl, P. (2014). The affordances of using visibly random groups in a mathematics classroom. In: Y. Li, E. Silver, & S. Li (Eds.), *Transforming mathematics instruction: Multiple approaches and practices*. New York, NY: Springer. 10.1007/978-3-319-04993-9_8

Littleton, K., and Mercer, N. (2013). *Interthinking: Putting talk to work*. Abingdon: Routledge. 10.4324/9780203809433

Lotan, R. (2003). Group-worthy tasks. *Educational Leadership*, 60(6), 72–75.

Lucas, V. (2014). *Mathematical team games*. St. Albans: Tarquin Publications.

Marks, R. (2014). Educational triage and ability-grouping in primary mathematics: A case-study of the impacts on low-attaining pupils. *Research in Mathematics Education*, 16(1), 38–53. 10.1080/14794802.2013.874095

Mason, J., Burton, L., and Stacey, K. (2010). *Thinking mathematically* (2nd Edition). Harlow: Pearson Education.

McCrone, S. (2005). The development of mathematical discussions: An investigation in a fifth-grade classroom. *Mathematical Thinking and Learning*, 7(2), 111–133. 10.1207/s15327833mtl0702_2

Nardi, E., and Steward, S. (2003). Is mathematics T.I.R.E.D.? A profile of quiet disaffection in the secondary mathematics classroom. *British Educational Research Journal*, 29(3), 345–367. 10.1080/01411920301852

Pruner, M., and Liljedahl, P. (2021). Collaborative problem solving in a choice-affluent environment. *ZDM – Mathematics Education*, 53, 753–770. 10.1007/s11858-021-01232-7

Steward, D. (2012). MEDIAN Don Steward mathematics teaching: Diffy.

Swan, M. (2006). *Collaborative learning in mathematics: A challenge to our beliefs and practices*. London: National Research and Development Centre for Adult Literacy and Numeracy and National Institute of Adult Continuing Education.

Webb, N. (2009). The teacher's role in promoting collaborative dialogue in the classroom. *British Journal of Educational Psychology*, 79, 1–28. 10.1348/000709908x380772

Wiliam, D., and Bartholomew, H. (2004). It's not which school but which set you're in that matters: the influence of ability grouping practices on student progress in mathematics. *British Educational Research Journal*, 30(2), 279–293. 10.1080/0141192042000195245

7 Helping individual learners to make mathematics manageable

Janet Kilpatrick Baker

Introduction

In this chapter I describe the lessons learned from my research on making mathematics learning manageable. The participants were eleven-year-old learners in a state secondary school located in a central Midlands market town in the United Kingdom. A similar approach has subsequently been used with adult learners, with the same high success rate. The Growth Zone Model has been described in detail in Chapter 3. As this model is the foundation of my research, the reader is advised to read Chapter 3 before continuing with this chapter.

My research revealed that learner autonomy and emotional awareness are key requirements which enable learners to use the Growth Zone Model, and therefore to learn effectively. In this chapter I will explain these requirements in detail, and then discuss the actions that learners can take in order to manage their emotions. I will also explain why I think that this needs to be an individual approach, led by learners and based on their different needs.

To begin, however, I explain that the Growth Zone Model (Figure 7.1) is a dynamic model, where learners steer themselves back into growth, and stay there.

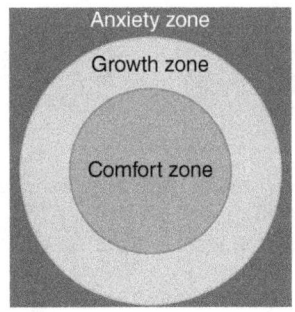

Figure 7.1 The Growth Zone Model.

Autonomy and awareness

The Growth Zone Model is a useful tool to help individual learners, as it makes learning mathematics a more manageable experience. This is because the Growth Zone Model responds to the specific needs of the learner in a dynamic way. Just like turning the handlebars on a bike to stay on the path, mathematics learners should be regularly, or ideally constantly, steering themselves back into the growth zone. Once they become aware that they have slipped into the comfort zone, learners can use the ideas in the Growth Zone Model to creep out of comfort by looking for a greater challenge. If they recognise that they have gone the other way and ended up in the anxiety zone, then the Growth Zone Model prompts them to step back from the mathematics and recover their equanimity.

Making mathematics manageable is all about helping learners to blossom in the growth zone. Blossoming implies that the learner widens their growth zone to accept the range and variety of emotions that can be experienced when learning mathematics. However, to be able to do this, they firstly need to be aware of their emotions, and secondly need the autonomy to take the steps that are necessary to manage or act on their emotions. The skills of awareness and autonomy are key to resilient learning and can be learned.

The model shown is a representation of the factors which promote the ability to blossom in the growth zone, or in other words to be a resilient learner of mathematics. I call it the 3As model –Autonomy, Awareness, Action (Figure 7.2).

Some learners naturally look for challenges, and relish the experience, as Dweck describes in her books (Dweck, 2000; 2017). These are people who have developed a growth theory of learning, however most people, at least in mathematics, hold a fixed theory of learning and, to these, challenges can seem like a chance to fail and feel humiliated, rather than a chance to learn. Other learners will have been cosseted in the comfort zone by teachers who, understandably, do not want their pupils to suffer the mathematics anxiety that they experienced. However, as has already been explained in Chapter 3, this is actually an unhelpful approach. Not offering challenge and support to meet that challenge means that learners do not build their resilience and miss out on the satisfying experience of solving a mathematical problem. Once they develop autonomy,

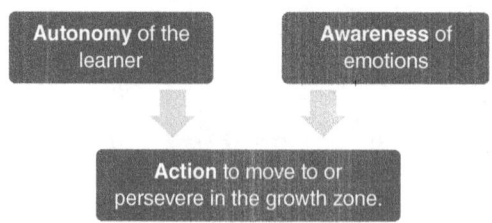

Figure 7.2 The three As Model: Autonomy, Awareness and Action.

these learners can be encouraged to step out of the comfort zone, face challenges and build resilience.

In order to become mathematically resilient, learners need to know that they have the autonomy to take action to meet their personal emotional needs. By autonomy I mean the knowledge that they **can** do what they **need** to do, to continue to learn mathematics effectively. This is slightly different from agency and self-efficacy. Autonomy is an essential psychological need that is an important prerequisite for motivation (Ryan & Deci, 2017). Once learners have a degree of control of their decisions, they feel more able to take risks and make mistakes, as they can commit or refuse, to please themselves rather than someone else. Agency is very similar but subtly different in character – for me, agency is situation dependent, so whilst a consistent assumption of autonomy is a psychological need, agency is a description of the possibilities and restrictions within a particular environment or situation. Self-efficacy is quite different from the other two as it is a measure of the individual's perception of their ability to complete a particular task. For example, a learner may feel that they are able to solve a GCSE word problem, but unable to solve a quadrilateral equation, and therefore have different levels of self-efficacy depending on the task.

The agency that a learner may experience in a school context, where the teacher may want to encourage independent engagement but must balance this with behaviour management of not one but many learners, is different to their agency when learning outside school. When working independently outside school, learners may have the freedom to step away from the textbook or computer and make themselves a snack or have a chat with a friend. Within school their agency seems inevitably more limited, but their autonomy need not be. I use the term autonomy to reflect the general understanding of the learner in all situations that they have choices and can choose the best way to progress their own learning. It is important to make it explicit that learners have some choices in all situations – maybe fewer choices inside the classroom, but certainly choices to help them find the best way to learn.

Within the classroom, school and teacher expectations will vary. Take a moment to think about a classroom situation with which you are familiar – what options are available when learners need more time to process a new concept or stop and think about the best way to steer themselves back into the growth zone? How can these choices be made explicit to the learners?

Once the learner understands that they are in charge of their learning, the decision to pause to review their emotional state becomes a sensible one. Previous chapters have described the impact of anxiety on learning, and the importance of realising the need to resolve the anxiety before continuing with learning. One adult learner very effectively described this as a shift in assumption, from feeling that they were procrastinating to realising that they were in fact being productive. However, many learners may continue to feel that they are wasting time. They may have unrealistic expectations of themselves will need support in trusting that they are making their own learning process as productive and effective as it can be.

Awareness

The 3As model prompts learners to realise that it helps to be aware of the emotions that hinder learning and address them, rather than ignoring them and plodding on. This is important because mathematics anxiety acts as an emotional handbrake (Riley, 2023). Once learners are aware of these emotions, they can take a little time to release the handbrake and learn effectively, rather than struggling on.

However, this can be very hard as learners are so used to ignoring their emotions. To help learners overcome this in a classroom situation, I included the growth zone model in a toolkit I developed to support learners. This toolkit is a small, zipped pouch containing laminated cards that could fit easily in a blazer pocket (a blazer is often worn as part of a school uniform in the UK) and was co-designed with learners.

The first card was an image of the growth zone model and came with a small glass pebble. Some learners chose to keep the toolkit in their pocket for their personal and private support, and some shared the growth zone model as a way of communicating to themselves and their teacher, as Lynden explained: 'the way I use it is I put this on the table. I've told Miss about it. And I put the pebble wherever I am in the lesson, and so she can analyse it and give me the correct ... treatment, can I say?'

To help learners become more aware of the emotions that could be hindering their learning, I firstly asked them to tell me about recent experiences of learning mathematics, and how they felt about it. For some it took a little time to draw out specific identification of emotions, rather than saying something like they felt stuck, but I was able to capture useful words that could then be related to the different zones of the model.

For the comfort zone, learners described feelings such as

'I felt more comfortable.'
'I wasn't feeling too anxious.'
'I wouldn't have anything to worry about.'

For the growth zone, many learners could not identify feelings but rather described situations when they engaged with the learning. However perhaps the best description was:

'it's challenging but I can manage it anyway.'

For the anxiety zone, learners described:

'It made me feel anxious.'
'Felt angry, angry at homework, angry at myself for not getting it.'
'It was too frustrating.'

Once learners can associate specific feelings with the zones on the growth zone model, they can then identify their current position on the model, and once they know their position, they will be more able to do something about it.

Action

Once a learner is aware of an emotion that is holding them back and understands that they are able to change how they are thinking and what they are doing, then they can take action to steer themselves back into the growth zone. These actions will be described in more detail shortly, but at this point two aspects of the growth zone model should be noted.

The first is a reminder that the growth zone model is very much a dynamic model, where the intention is that the learner is regularly, or continually, steering themselves towards growth. To continue with the analogy of learning to ride a bicycle, initially these turns will be, or feel, quite significant, and require a lot of thought and effort. However, once a learner has found that it is possible to recover from the anxiety zone, then the next experience requires a little less effort.

The second aspect is that, whilst the teacher can support the process, it must be led by the learner. Emotions are intensely personal, and only the learner will know how they are feeling and what they need to do about it. The teacher, parent or coach can encourage autonomy and awareness, and can suggest appropriate actions to recover, but will only ever be the middleman in the process. The learners need to control their own learning process.

Staying in or returning to the growth zone

Here I will go into more detail about the actions that the learner can take to stay in or return to the growth zone. First, I will look at the process of recovering from anxiety, then I will cover the process of creeping out of comfort, then lastly, I will look at how to stay in the growth zone.

To recover from **anxiety**, the first step is for the learner to understand that their reaction to an uncomfortable mathematical situation is not reflective of their ability, but rather a perfectly normal response to a perceived threat, as has been explained in Chapter 3. When the fear response is triggered by a perceived mathematical threat, it really helps learners to understand that this is a biological process developed by evolution to keep them safe. It is not a disaster, or the end of the world, as it is perfectly possible to recover equanimity and continue learning. There are various steps to the recovery process. The first is to stop actively engaging with mathematical learning, as when in the red or anxiety zone engaging is both ineffective and unhelpful. The second is that the learner should remind themselves that they are not dumb, but that they have slipped into the anxiety zone, no-one can think when they are anxious and they just need to get themselves back on track. The third step is to deploy the action or way that works best for them to trigger their own recovery response. This may be different depending on the situation – for example, whilst a stretch and a hot drink can work in many situations, it will not often be accepted in a classroom situation. Examples of ways to trigger the recovery response that can be used in a classroom include deep, diaphragmatic, box, or 5/7 breathing, doodling, and looking away from the textbook or screen for a short time.

This shift in view from perceiving an experience of the anxiety zone as a huge problem, to seeing it as a temporary problem that can be easily rectified, can be illustrated from my personal experience. I came to realise that the shift from anxiety zone back to the growth zone was more like wandering into the sea when walking on a beach rather than falling off a cliff. I was practicing being aware of my emotions when I realised that I could pull myself back by pausing, reassuring myself, and calming myself down. It took time for me to build self-efficacy, and to remind myself that I can make that change. As learners experience more and more instances where they recover from the anxiety zone, they fear the red zone less and become more confident to push the boundaries of their growth zones.

When the learner needs to creep out of **comfort** zone, it can be helpful to consider why they have ended up there in the first place. It may be because they have previously experienced too great a challenge and not enough support, causing them to feel that they are falling behind. This is a horrible experience, vividly described by the following learners. Chris explained that, because she failed to understand the mathematics and then did not manage to produce the expected work, she expected a reprimand from the teacher: *'I just didn't want to get told off for not doing enough work because I can't help not doing enough work when I don't understand it.'* Erin and Indigo expressed concern over appearing to be different to their peers: *'well, I still don't, really, ask for help, because everyone else in the class is like 'oh yeah I get it' so I don't really want … to get up and be different from everyone else and ask for help'* (Erin) and *'Bit annoyed cause I couldn't do as well as the other people'* (Indigo).

No wonder these learners have decided that it is not worth bothering with mathematics and that any effort they make will all be in vain.

Being in the comfort zone can also result from a teacher making the learning too easy, and lacking challenge. This is an understandable response if the teacher themselves has mathematics anxiety, or wants to prevent it in their learners, but in fact is an unhelpful approach. The learning, and the interest, is in the struggle to work out the answer, and when learners understand that a little struggle, or manageable challenge, is part and parcel of being in the growth zone, they can stay longer (Lee and Johnston-Wilder, 2018) in this optimal place for learning. As long as the learner is protecting themselves from harmful levels of anxiety, they can be supported to carefully creep out of their comfort zone.

At this point many learners may be asking why they need to creep out of the comfort zone – it is after all, a comfortable place to be. The answer to give is that it is not the best place to learn, and learning mathematics is important because it has value for both careers and other situations. So, if a learner is reluctant to creep out of the comfort zone, they will need to take small steps at first. Small successful challenges can build self-confidence and ensure that it is always possible to come back to comfort, if they need to do so.

So far, we have considered the moves of recovering from anxiety and creeping out of comfort. These moves require the learner to take action. However, once in the growth zone, it is important to stay there, and this will also require action from the learner. One of my learners described being in the growth zone as feeling *'like the challenge is manageable'* and this description captures the essence of the zone. But how can learners get to the place where they feel the challenge is manageable, and what can teachers, parents, support staff and others do to help them?

One way is to meet the needs of learners who prefer an ALIVE teaching approach rather than a TIRED or TRIED one. TIRED is the acronym developed by Nardi and Steward (2003) to encapsulate the 'quiet disaffection' they observed in mathematics classrooms. The letters stand for tedium, isolation, rote and cue following, elitism and depersonalisation. Johnston-Wilder et al. (2016) further developed this concept by highlighting that not all learners fall into the category of becoming disaffected by this approach and reordered the letters to form the acronym TRIED, to acknowledge that while the approach does not work for all, it does works for some learners. They further proposed a new acronym, ALIVE, to describe a more inclusive way of teaching mathematics, with the letters standing for accessible, linked, inclusive, valued and engaging.

Accessible learning involves explaining concepts at an appropriate level. This is akin to the learner trying to climb a ladder with spaces between the rungs that are an achievable stretch. When learning to climb, it is important to have a reach that is challenging yet manageable. This photograph shows a 3-year-old learning to climb. Because the distance between the rungs, or in this case, the hand- and footholds on the slope, were just about achievable for her at this stage in her development, she was able to climb the slope and gained a great deal of satisfaction in the process (Figure 7.3).

Had the hand- and footholds been farther apart, she would not have been able to climb, and could have become frustrated and disaffected. Similarly in the mathematics classroom, if the steps are too challenging for the learner to achieve, then no learning can take place. However, this situation can be addressed by more rungs being inserted, so that the stretch is achievable, as illustrated in the following image (Figure 7.4).

Several of my learners have described situations suggesting that their teacher did not scaffold the learning appropriately. Here is Blake: *'The teacher explains it to me. But I don't exactly get it cause she's a bit complex the way she explains it.'* And Lynden said: *'it's been awful. At the start of the year, I was ok cause we were doing stuff that I knew, and we kept expanding that a little bit. But all of a sudden, we're jumping out of nowhere like, y equals something on a grid, or*

Figure 7.3. Ada learns to climb. Copyright: R. Baker-Frampton.

stuff that I never even actually did in primary school and it's just that massive jump that made me feel uncomfortable. And it makes me anxious.'

It is important to note at this point that it cannot be the sole responsibility of the teacher to decide on the size of the gap for a learner. Assuming that the learner has the autonomy and resilience to do so, they are in the best position to decide on the optimal distance of the next step, and therefore should be supported to make their own decisions on a variety of learning options. Mathematical resilience has been explored in previous chapters of this book. The key aspects here are that the learner understands that a degree of struggle is to be expected, that it is expected that you have to work an answer out rather than know it straight away and, and lastly, that mathematics is of great value to their future, both in terms of their careers and their everyday lives.

One learning option may be to ask for help. Learners will need to work out their best resource in terms of support. Parents can be wonderfully supportive and enabling, but equally can have understandings that are out of date. Teachers can often

Figure 7.4 The Ladder Model (Johnston-Wilder et al., 2020).

be the best help but may also be remote and hard to contact, friends can support but may also judge, YouTube can be very enlightening but can also be unreliable and misleading. Additionally, the best advisor for one situation or challenge may not be the best for all. Learners should try lots of ways to get support and think about the way they prefer and the ones that work best for them. This way they can self-safeguard and avoid ending up in the anxiety zone, ensure that the rungs in their learning ladder are spaced so that the challenges are manageable and maintain motivation through a sense of autonomy.

I will finish the chapter by briefly considering the needs of learners who may not feel that they are falling behind, and may not be easily spotted in the classroom, but none the less can have mathematics anxiety. Here I will discuss those who achieve high results in mathematics, but still experience mathematics anxiety. For learners who easily understand the concepts covered in mathematics classes and quickly complete exercises and homework, the comfort zone is relatively large. The following image shows Jack's experience of the growth zone model, and his words explain that he feels he 'is learning sometimes' (Figure 7.5).

Finley explains this in a little more depth, stating that she feels in the comfort zone because of repetitive lessons that do not offer a variety of challenge.

Jack and Finley are examples of higher achieving pupils who are used to a lack of challenge in their lessons and tasks that are similarly lacking in challenge. They are often rewarded but the awards are for achievements that take little effort on their part. This may cause them to develop a fixed mindset (Dweck, 2000) and they may proceed through school never learning how to tackle anything that really presents a

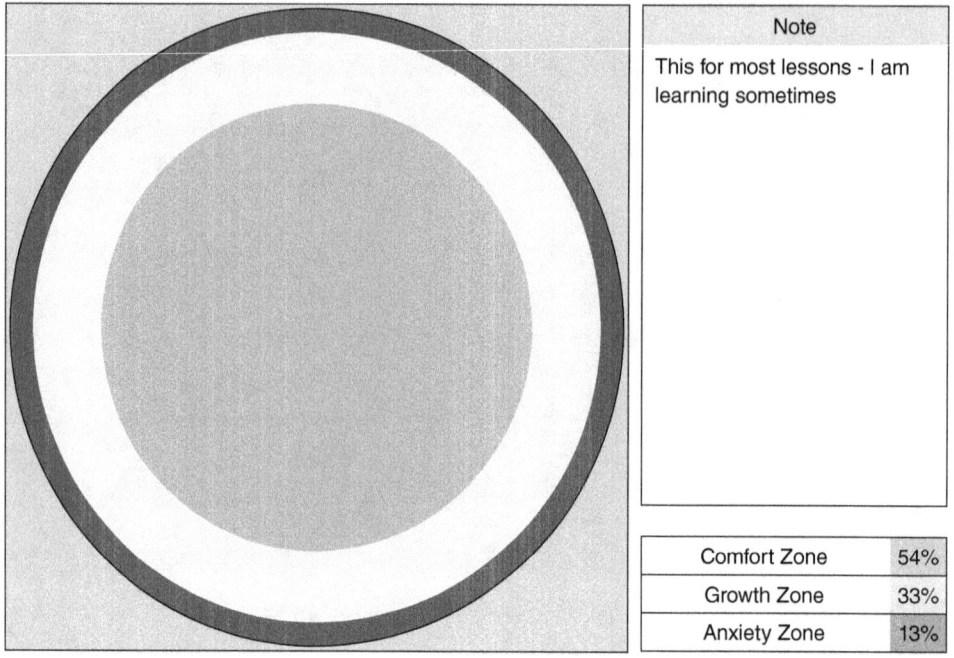

Figure 7.5 Jack's Growth Zone Model.

new challenge. When they do meet mathematics that is difficult for them to understand, which may perhaps be as late as when studying mathematics at university level, they can feel the challenge presented means they cannot learn this type of mathematics (Ward-Penny et al., 2011), and may develop mathematics anxiety. To address these small growth zones, the response is clearly to present a greater level of challenge. As a teacher myself, I know that this is not always easy, but there are many ideas available online on for example, the Nrich website, (https://nrich.maths.org/), the Bowland Maths website (https://www.bowlandmaths.org.uk/) and Jo Boaler's Youcubed website (https://www.youcubed.org/).

Conclusion

I hope that I have prompted you to think about learner autonomy and emotional awareness and that you now have enough information to try using these ideas in your own practice. I strongly believe that learners should be given both the tools and the opportunities to actively manage their emotions as they are learning anything but especially mathematics. You will want to consider the different needs of your different learners, they will not all need the same support and direction, but they will all need support if they are to meet the challenge of learning mathematics. You may well choose to create toolkits for your learners like the one I describe. If so, encourage them to personalise them with the prompts and actions that they have worked out work for them.

References

Dweck, C. S. (2000). *Self-theories: Their role in motivation, personality, and development.* Psychology Press.

Dweck, C. S. (2017). From needs to goals and representations: Foundations for a unified theory of motivation, personality, and development. *Psychological Review*, 124(6), 689–719. 10.1037/rev0000082

Johnston-Wilder, S., Pardoe, S., Almehrz, H., Evans, B., Marsh, J., and Richards, S. (2016). Developing teaching for mathematical resilience in further education. ICER2016, 9th International Conference of Education, Research and Innovation, Seville (SPAIN).

Johnston-Wilder, S., Baker, J., McCracken, A., and Msimanga, A. (2020). A toolkit for teachers and learners, parents, carers and support staff: Improving mathematical safeguarding and building resilience to increase effectiveness of teaching and learning mathematics. *Creative Education*, 11, 1418–1441. 10.4236/ce.2020.118104

Lee, C., and Johnston-Wilder, S. (2018). *Getting into and staying in the growth zone.* NRICH.

Nardi, E., and Steward, S. (2003). Is mathematics T.I.R.E.D? A profile of quiet disaffection in the secondary mathematics classroom. *British Educational Research Journal*, 29(3), 345–367. 10.1080/01411920301852

Riley, A. (2023). *Maths anxiety – Help secondary students overcome their fear of maths.* https://www.teachwire.net/news/maths-anxiety-help-secondary-students-overcome-their-fear-of-maths/

Ryan, R. M., and Deci, E. L. (2017). *Self-determination theory: Basic psychological needs in motivation, development, and wellness.* The Guilford Press: New York, USA. https://doi.org/10.1521/978.14625/28806

Ward-Penny, R., Johnston-Wilder, S., and Lee, C. (2011). Exit interviews: Undergraduates who leave mathematics behind. *For the Learning of Mathematics.* 31(2), 21–26.

Working with groups of support staff

Georgie Ford and Sue Johnston-Wilder

Introduction

Support staff within the educational field are integrally involved in teaching, learning and progression. In recent times this role has evolved to include numerous, arguably more complex, dimensions of the learning environment in schools and colleges. There is a growing body of evidence to suggest that learners' well-being and learning are synergistic and that the quality of learner relationships with support staff and teachers is critical to positive outcomes (Littlecott et al., 2018). There is some recognition that a learners' relationship with a member of support staff can differ both cognitively and emotionally to a learner-teacher relationship, but minimal research has investigated this concept. This chapter explores the way that these differences can contribute to an increased self-efficacy and resilience in mathematics.

Arguably, there is an urgent requirement across education to provide further training for support staff working in a mathematics environment. This is not a requirement to further develop mathematical skills but to help support staff manage and alleviate any mathematics anxiety which they suffer themselves, and to equip them with the knowledge of how to support the development of mathematical resilience in their learners. If they use the behaviours described within this book such as listening, growth mindset, curiosity behaviours and modelling resilient behaviours, they will help their learners be better prepared to learn mathematics. All learning requires a certain amount of resilience, this chapter aims to empower support staff to engage with, and understand further, the construct of mathematical resilience.

To acquire mathematical resilience, support staff can be trained in a way that helps them understand that both parties require a willingness to discuss, reflect and persist in times of challenge and difficulty. The roles and responsibilities given to support staff varies across educational sectors and geographical areas. Institutions take a varied approach to how they define the role of the support staff they employ. This chapter recognises that support staff can be employed to support on an individual basis whilst others move around the classroom helping anyone that appears to need support. The principles discussed within this chapter are appropriate to both operational styles.

There is a significant space for further research into how support staff themselves see their role. Support staff who see themselves as having one purpose, that of increasing examination grades, are unlikely to build the connection and strengthen the resilience required in STEM subjects such as mathematics. Organisations have a cultural responsibility to create a transparent vision for support staff based upon fundamental human principles such as resilience and connection.

Additionally, for staff supporting mathematics, it is critical to self-explore and understand their own attitudes and cognitive schemas which have resulted from their previous mathematical experience. Schemas (Woolfolk, 2007) are a well-researched phenomenon but the application of these from a staff perspective is less examined. Schemas are unconscious internal rules that dictate our behavioural responses to stimuli and situations (Cockcroft et al., 2022). In mathematics, schemas are frameworks and outlines for solving problems. They are a cognitive heuristic that facilitates our understanding of the environment and the people within it. Learners will consistently build new schemas of their relationship with and expectations of support staff. For example, the attitudes and emotions of support staff towards mathematics are contagious. If humans are exposed to certain emotion in someone else, they are likely to mimic it, consciously and unconsciously (Hatfield et al., 1993). If support staff can create and demonstrate a resilient approach, giving encouragement to overcome barriers and authentically communicate emotions, these behaviours will form a schema that develops resilience in learners.

If support staff have acquired negative or maladaptive schemas of their own mathematics experience, there is a significant probability that these will translate into the mechanisms of their support. This can result in a maladaptive support style. Similarly, if support staff perceive their role to be enabling learners to achieve better examination grades, an unconscious schema may be created preventing connective relationship behaviours. Positive connections are arguably the most vital concept in forming supportive academic relationships. The creation of a safe relationship is the first critical step in a learner support relationship and should be prioritised over any attainment assessment or functional mathematics working. This can be seen as mathematical safeguarding (Johnston-Wilder et al., 2020), which is a crucial requirement in developing a mutual psychologically safe relationship. Support without the existence of a connective relationship is unlikely to encourage the development of skills to overcome barriers, which are a common experience in mathematics.

Humans hold expectations of how a person in a certain role will behave and mathematics learners are no exception. Their schemas will inform them how they expect support staff to behave. If support staff do not understand and explore learner expectations and bias, there could be conflict between the goals of the two people seeking connection, which will impact their relationship and create disconnection, rather than the empathy and self-awareness which is known to build mathematical resilience. Positive relationships are dependent on both staff and learners being aware of their emotions and authentic interpersonal conversations (David, 2016).

Goleman's (2004) concept of emotional intelligence suggests that emotional capabilities are not innate talents but learned capabilities. This implies emotional capabilities can be developed in supportive relationships. Goleman (2004) acknowledges self-awareness as a crucial principle of emotional intelligence. Mathematical resilience includes learners having and feeling a sense of value and worth (Johnston-Wilder et al., 2020) and this is an element that support staff can encourage by promoting authentic emotional literacy in conversations and dialogue.

Support staff who maintain an awareness of the emotions of learners and are able to ascertain when the learners are experiencing high levels of fear or anxiety will be able to help learners manage these negative emotions. It is known that emotions such as fear trigger survival behaviours, often leading to the experience and behavioural outbursts of anger. This is not conducive to learning and can create more mathematics anxiety for many people. Support staff can help create exit strategies for those learners who experience such emotions which can allow the brain to de-escalate, the arousal levels to drop and the learner to re-engage.

Support staff – A paradoxical role

Topics such as mental health, well-being and psychological safety have consistently remained in the educational spotlight for at least the past ten years (Mackenzie and Williams, 2018; Waller et al., 2018). Policy makers and educational researchers agree that these concepts should have parity with academic achievement; teachers and support staff are encouraged to prioritise these elements in the same way they do academic achievements. However, operationalising these roles can pose a challenge to education. The various descriptors of the roles of support staff range across teaching assistants, coaches, specialist support workers, emotional literacy support assistants and STEM support workers. This variety creates varying perceptions of the role and schema to work from within the classroom. For example, those appointed as specialists in their field often reinforce the perception that their role is to implement specialist knowledge and enhance the learner's skills and consequently academic attainment in one specific subject area.

With the importance mandated for STEM subjects by governments across the world, the focus of the mathematics specialist often reduces to an attainment focus, not one of emotional connectivity and empathy, which are often perceived as softer skills or the role of another body such as Welfare Services. The disparity and variation in what is required of support workers in mathematics suggests that the acceptance of an embedded coaching model is imperative. Without this, there are different groups of people sharing the support title but not necessarily the same work. Felux and Snowdy (2006, p. ix) note, "Different titles exist for this position—math coach, math specialist, math support teacher, math resource teacher, and more. And just as there isn't consistency with my colleagues' titles, there isn't much consistency with their responsibilities". To demonstrate a holistic approach to mathematics support, there are examples from education in which all these titles work together.

Supporting with connection – A framework

The next section of this chapter focuses on the support worker–learner connections that are crucial to stimulate the development of mathematical resilience and allow the schemas and experiences of mathematics learners to regenerate. In 2001, a funded study by the Learning and Teaching Support Network (Lawson, Croft, and Halpin, 2003) set out some common features of successful mathematics support, and, whilst this is dated historically, these key findings remain paramount at a human emotive level. These include:

1. The provision of a non-threatening and supportive environment promoting autonomous and choice-led solution-focussed learning.

2. A non-judgmental acceptance of the learner's needs (however basic).

3. Support staff who are not perceived as part of the assessment process (Lawson et al., 2003).

These person-centered strategies can be incorporated into a support coaching model which follows a purpose, strategy, observation and goals model.

Purpose phase

The initial meeting between learner and support worker face-to-face will usually occur in their first lesson at the start of a new term. For a support worker, this often takes the form of a public encounter with minimal opportunity to personalise the situation and begin to build familiarity with the individual. A public encounter at the beginning of term is, from a humanistic angle, flawed if staff are to achieve productive, emotionally authentic relationships and communication. A new academic term for many learners is laden with worries and anxieties about a new environment, classroom and teaching staff. This is exacerbated for learners in a transitional phase of education who find themselves in a novel environment, facing the prospect of significant unfamiliarity in the routines and mores in this environment. The transitioning learner often suffers further feelings of insecurity in a mathematics environment, due to adverse prior experiences and a lack of familiarity with the likely content and how it may be presented.

In situations which trigger anxiety, working memory is consumed with cognitive affective components, such as worry, and thus it is unlikely that a learner is able apply themselves effectively or solve, for example, an algebraic equation (Pelegrina et al., 2020). Research advises that anxiety concerning mathematics especially impacts upon tasks that require working memory. This is like asking somebody with no notice to perform on stage to a stranger. The brain instinctively engages in the fight or flight response and the learner is flooded with adrenaline and concerned primarily with judgements on their performance (Buckley, 2016). Learners often experience this arousal as threat. Support workers can aid the learner in reframing this threat as challenge. Support workers who know and understand the prior experiences of the learners they are supporting will be able to anticipate these negative reactions and help the learners manage and control them. Support workers can be encouraged to build an affective profile of their learners using, for example, observing learners and creating notes for each one.

Support workers can use tools such as the Growth Zone Model to enable learners to differentiate between perceived challenge and threat, and to reframe some perceived threat as challenge. This tool will help support workers encourage learners to label their emotions and recognise and name their current feelings.

The concept of psychological safety is only achieving maturation within the educational world in recent times, and it is enormously exciting that this book introduces it to mathematics. It is highly rated by large companies, such as Google, for the creation of human cohesion and group safety. A scoping exercise across mathematics support workers demonstrated that a significant proportion have not encountered this term as part of a support worker role, nor do they know how to create a psychologically safe environment for their learners. Emotional and psychological safety is the lynchpin of effective mathematics support. If learners working with support staff are not feeling safe, this will increase arousal and the activation of the central nervous system and survival response, and learners will not be well placed to learn.

A central tenet in building learners' confidence in being able to use and control mathematical ideas is that they know with absolute certainty that they will not be judged on a response but rather they will receive growth-promoting feedback. Many young people have been subject to public humiliation within a mathematics classroom, and this contributes to a negative schema preventing wider participation. Feeling psychologically safe is the difference between a confidence- and resilience-building working relationship and one that lacks connectivity and consequently productivity. More crucially, feeling psychologically safe allows the learner to be able to explore mathematical ideas, even if that means making mistakes, with their support worker. This is a significant part of the learning process and can empower learners to reframe a mathematical process. Support workers should be able to access continuing professional development within their workplace that guides and develops their abilities to create psychological safety on a group and individual level. Support workers should be able to create brave spaces in which risk-taking and mistakes are encouraged and accepted.

Experiencing psychological safety permits learners to act as their authentic selves without fear of negative or judgmental consequences. The chances are, if a learner has prior negative experiences with mathematics, they will enter the support relationship feeling psychologically unsafe. It is documented that the higher the levels of unfamiliarity that humans experience, the further the requirement for psychological safety advances (Educational Support, 2021). There is minimal evidence-based research about the role of psychological safety in the mathematics support worker – learner relationship, nonetheless it is sensical that the creation of this is critical to the quality of that relationship. Unpredictability is a factor that makes our need for safety rise. The effect of uncertainty can be mediated through models such as the Growth Zone Model (Johnston-Wilder et al., 2020). Time spent in the comfort zone will increase feelings of psychological safety, which will encourage learners to spend time in the growth zone. Working in this way creates autonomy and independence in learning as learners begin to understand when they have reached the red zone, how to communicate this without fear of judgement and how to gain the support they need to return to the comfort or the growth zone. Psychological safety is the tool that will allow learners to voice that they need to step back or take a breath, key principles of autonomous learning.

The foundation of psychological safety can be achieved through a simple familiarisation phase. As a rule of thumb, humans and their brains benefit from predictability. This means the effort required by us to process a situation is minimal. Humans like to know what they will encounter next including sensory information such as smells, sounds and textures. This enables us to experience calm and for the brain not to activate the fight or stress response which can result from a perception of danger. Therefore, an ideal first stage of a support worker-learner relationship would be to meet and encourage engagement with a learner or group of learners in advance of any mathematics classes and share understanding of the Growth Zone Model (Johnston-Wilder et al., 2013). This will help enable connectivity, trust and familiarity to be built and allow the learners' brains to build schema and realistic expectations of the support they will be offered. Contact and familiarisation begin to reduce unpredictability, creating safety, and belonging. The first meeting is also an opportunity for learners to disclose worries and fears and a chance for the support worker to be empathetic to any prior experiences that may predict behaviour in future mathematics lessons.

This is not to suggest that the support worker takes the role of a therapist or counsellor but that they create necessary trust mechanisms to create psychological safety in mathematics. Empathy is the skill that drives and creates authentic connections and signals to the learner that they are in a safe place. This is a misunderstood skill arising from emotional literacy techniques and can be confused with sympathy which often inadvertently creates disconnection within a relationship (David, 2016). Support workers should not feel as though they have to say they understand a learner if they do not. Instead, they can validate that mathematics can be challenging and that they are there to listen and support. Phrases can include *"I can see that this is challenging for you, and that seems frustrating. Can we try another way? Or what help might we need?"*

This phase is highly effective if it occurs within a mathematics environment and can help learners associate a positive experience with a physical environment. Physical environments stimulate many unconscious responses and behaviours and can contribute to the activation of the fight/flight response. Operating at this level is the red zone of the Growth Zone Model (Johnston-Wilder et al., 2020). The body perceives danger and behaves in an appropriately reactive manner. Behaviours in this zone often include fleeing, anger, hiding under furniture or outbursts that originate from the emotion of fear; responses that often result in removal from a classroom. The emotion cannot necessarily be controlled, but the behavioural response can.

When support workers understand that there are mechanisms by which the learner can let their physical body know that it is safe, learning can continue despite the perception of a threat. These mechanisms include regulatory breathing exercises, connecting in a meaningful way and movement. If learners do not inform the physical body that it is safe, the automatic responses to threat will continue and will often cause burnout (Nagoski and Nagoski, 2019). Support workers can assist learners to differentiate between the stress (physical feeling) and the stressors in the mathematics classroom by exploring the specifics of the triggers that instigate the physical stress response. Specific CPD opportunities can be created for further training in this area.

Expectation and construction explorations

Familiarisation offers an opportunity for the learner to explore their own expectations and purpose in engaging with the support worker–learner relationship. When learners are paired with a specialist support worker, expectations may form cognitively well in advance of meeting; this is what our brains are designed to do. The learner may predict that the support worker will be able to make mathematics easier for them or teach them new ways to do mathematics that will not feel as hard. For a learner who has experienced prior mathematics trauma, expectations can often be built on hope and may not be realistic. Support workers should recognise that the learner will form subconscious expectations, resulting from prior support relationships and interactions, and understanding and exploring these expectations collaboratively will help establish a new, safe and supportive relationship, and helps ensure that the learner does not feel complex emotions, such as shame or embarrassment, later.

Support workers who have access to training programmes that equip them with the knowledge of how emotions affect learning will be better prepared to support learners of mathematics. No two support relationships are similar and there will be differences in engagement and collaborative learning; it is useful to acknowledge this with learners. Mathematics as a subject evokes many individual differences in humans, both learners and support workers, and its learning can be accompanied by the most significant of emotions and anxious responses. Behavioural difficulties often arise when learners do not feel safe enough to communicate a problem or feel embarrassed to speak up. The following vignette demonstrates this communication malfunction.

Support B

"I am employed as a learning facilitator in mathematics in a secondary school. This role is fairly new but appeared to be support work with the ultimate goal of facilitating learning. I was informed that it was an exciting role, as I could create it as I went along due to nobody being in post prior to me. On the first session, I was introduced as the learning facilitator and the lesson began. It did not occur to me at this point that learners would not cognitively understand or be able to decipher the term facilitator. As the term went on, it became clear that learners perceived me much more as an assistant teacher and required instruction consistently but just on a 1:1 level. Any attempts to encourage or create autonomy often ended in conflict-style situations or learners walking out of the classroom. I realised quickly that expectations should have been explored and a shared understanding created. Due to learners not understanding the term facilitator, resilience was weakened as they had misunderstood my role in their learning."

This scenario is not unusual in mathematics support work roles and provides a rationale and evidence for the presence of a familiarisation phase with a defined structure and exploration, such as presented in Table 8.1. This suggested model is not generalisable to every individual; it is designed to be transferable, for the support worker to decide on the appropriateness of each stage and plan according to the contextual needs of the learner.

The purpose of this phase is to create an affective profile of the learner and to understand how best to approach the supportive relationship and use strategies that promote autonomy and choice. Support workers must also be clear on the purpose of their role. Research has explored the conceptualisations of educational staff involved in mathematics learning and there is a correlation between the views and beliefs of support staff and instructional behaviour; the same applies to teachers (Thompson, 1984). Therefore, encouraging groups of support staff in mathematics to become acutely self-aware of their own schemas and experiences related to mathematics will enable them to offer more and more effective support. Schemas and experiences build and become stored in the brain from primary school age and so it makes sense that adults possess numerous pockets of information about our mathematics experiences. Culturally, mathematics teaching has also changed significantly since most adults were in school and it is important to recognise this.

Table 8.1 Prompts for support familiarisation phase.

Question	Purpose
PURPOSE What is your expectation of me as a support worker? What do you think my job is? How do you perceive my role?	Schema exploration and an opportunity to share mutual expectations of the relationship. Opportunity to correct misperceptions.
OBSERVATIONS What has your experience of mathematics been previously? Do you have any anxieties about mathematics?	Identify the mathematics persona already built. How is the student approaching mathematics? Negatively or with optimism? If experiences are negative, what does safety look like for a student?
STRATEGY How do you feel about having a support worker? Do you enjoy mathematics? What do you think are your mathematics needs?	Some students experience difficult emotions about having a support worker or worry that their peers will judge them.
SMART GOALS What are your goals? These are often not concentrated on mathematics ability but softer skills.	Collaborative, shared goal setting. Ability to share with class teacher. Explore "softer" goals. To be revisited as part of a regular evaluation cycle between student, support worker and class teacher. How are the goals measured and adapted?

> Georgie: Reflecting on my own experience as a Year 5 pupil in a middle school over 20 years ago, the teacher would encourage every learner to stand up at the end of the session and individually point to all learners with the label of good or stupid based on his perceptions of performance within that lesson. Understandably, this created a schema of negativity towards mathematics. Later, I worked with a specialist tutor to assist me with getting through a GCSE.

There are many adults in education with negative experiences in mathematics and as a mathematics support worker, it is likely that there will be things that will trigger personal emotional responses; for some, it is algebra, and for others multiplication tables. This does not mean that people with adverse prior experiences cannot be effective mathematics support workers, quite the opposite. What it does mean is that support workers should ensure they are aware of how these prior experiences may, consciously or unconsciously, affect their support style. Being able to self-safeguard and modelling bravery and safeguarding behaviours, e.g., asking for time to work on a problem, are critical tools in the support worker repertoire and toolkit of strategies.

Support workers who work to refine their own belief that any learner can progress in mathematics, including themselves will help their learners develop a growth theory of learning. An effective methodology to explore and refine this belief is via learning communities and working groups with other support workers. There are often subconscious cultural and historical beliefs that exist in the minds of support workers that suggest only those with a logical brain can progress and achieve in mathematics. This appears to arise from a cultural stereotype suggesting that mathematics is a logical ability – one that you either have or do not have. These stereotypes can sit in the unconscious field and be triggered with little awareness, leading some support staff to the belief that certain learners do not possess mathematics ability and instead must be creative learners who will thrive in subjects such as Art. Exploring these constructions enables self-awareness and can enable any resulting problems to be solved. Becoming self-aware of any stereotypes often means the thought can be recognised when it occurs and be corrected. Support workers are well placed to encourage growth mindsets and can encourage their learners to think that mathematics is like a muscle; the more it is used the more it can grow. Learners who believe mathematics capability is fixed and rigid will require support to understand that they can learn and improve through perseverance (Dweck, 2000). The support worker can holistically use the Growth Zone Model (Johnston-Wilder et al., 2013), emotional naming, literacy skills and connection to encourage the belief that mathematics ability is not fixed. This chapter and the wider book encourage support workers to encourage curiosity, autonomy, and courage in their learners. The importance of trust, relatedness and a significant adult cannot be underestimated within this subject domain.

Mathematics is stereotyped as a male domain (Forgasz et al., 2001) and research provides evidence in support of this ideology. This is a further example of a construction requiring open exploration and refining, ideally within a community of support workers to encourage the belief that any learner of any gender has the potential to succeed in this field. This type of learning community amongst mathematics support workers is a vehicle to allow transformational learning and the refining of schemas (Mezirow, 2009). This community approach enables the development of a shared identity, vision and goals whilst ensuring that support staff are using evidence-based strategies. A familiarisation process such as the one described in this chapter is simply the start of the mathematics support process; there is a need for continual growth of the support worker community through collaboration and professional learning opportunities. Professional learning opportunities of value in this domain can aid in the creation of new meaning schemas and should not just be concerned with mathematics but rather building an emotionally literate community of support workers.

Solution focus and integration

With these stages implemented, the relationship between support worker and learner should move towards a place of psychological safety, preventing cognitive worries and red zone experiences from impeding mathematics progression. These person-centered elements require maintenance throughout the remainder of the relationship and should be revisited on at least a termly basis. It is useful to check in with the learner at the onset of every session to build emotional self-awareness and labelling skills. For example, a support worker could ask:

"Which zone are you in about maths today?"

"Now you have given the zone, can you label and describe your feelings?"

The justification for this is that a whole spectrum of emotions can be experienced prior to a mathematics lesson including complex emotions such as fear. Learners are often not adept at recognising and accurately labelling their emotions and this can lead to ineffective problem solving. As a support worker you can encourage a deeper understanding of the emotions that they are trying to manage in a mathematics environment. Red zone emotions such as anger can display as rage, or annoyance or frustration, and correctly labelling the learners' feeling can change how they view a problem. Identifying emotions as often as possible in relation to mathematics will help the learner move towards a solution-focused mindset and towards autonomy. It is also helpful to enable learners to differentiate between the mathematics stress and the stressor by encouraging the use of specific language and communication instead of "I don't know" or "I'm just stressed". Physical bodies need to feel safe, and support workers can be vital in encouraging a relaxation response and embedding these behaviours in a classroom.

Conclusion

This chapter is intended to provide guidance into the human factors present in the complexity of the relationship between a mathematics learner and a support worker. The strategies discussed throughout this chapter are designed to be simple mechanisms to create a relationship in which mathematics resilience can be strengthened and the cognitive components of mathematics worry can be addressed to form a mutual understanding. By shifting the focus of the support away from mathematics progression and abilities to understanding the relationship and factors that interplay in learning mathematics, support workers can understand that often the right answer is not the most important focus. Rather strengthening resilience and redefining existing negative schemas of mathematics should be the priority in the supportive relationship.

References

Buckley, S., Reid, K., Goos, M., Lipp, O. V., and Thomson, S. (2016). Understanding and addressing mathematics anxiety using perspectives from education, psychology and neuroscience. *Australian Journal of Education*, 60(2), 157–170. 10.1177/0004944116653000

Cockcroft, J. P., Sam, C., Berens, M., Gaskell, G., and Horner, A. J. (2022). Schematic information influences memory and generalisation behaviour for schema-relevant and -irrelevant information, *Cognition*, 227, 105203.

David, S. (2016). *Emotional agility: Get unstuck, embrace change and thrive in work and life*. 1st ed. London: Penguin.

Dweck, C. (2000). *Self-theories: Their role in motivation, personality, and development*. Philadelphia, PA: Psychology Press.

Education Support (2021). *Teacher wellbeing index*. (Accessed 21 February 2021). twix-2021.pdf (educationsupport.org.uk).

Felux, C., and Snowdy, P. (Eds.) (2006). *The math coach field guide: Charting your course*. Sausalito, CA: Math Solutions.

Forgasz, H., Leder, G., & Kaur, B. (2001). Who can('t) do maths—Boys/girls? Beliefs of Australian and Singaporean secondary school students. *Asia Pacific Journal of Education*, 21(2), 106–116. 10.1080/02188791.2001.10600197

Goleman, D. (2004). *Emotional intelligence and working with emotional intelligence*. 1st ed. London: Bloomsbury Publishing Plc.

Gonzalez Thompson, A. (1984). The relationship of teachers' conceptions of mathematics and mathematics teaching to instructional practice. *Educ Stud Math*, 15, 105–127. 10.1007/BF00305892

Hatfield, E., Cacioppo, J. T., and Rapson, R. L. (1993). Emotional contagion. *Current Directions in Psychological Science*, 2(3), 96–100. 10.1111/1467-8721.ep10770953

Johnston-Wilder, S., Lee, C., Garton, E., Goodlad, S., and Brindley, J. (2013). Developing coaches for mathematical resilience. In *6th International Conference of Education, Research and Innovation* (pp. 2326–2333). Seville: IATED.

Johnston-Wilder, S., Kilpatrick Baker, J., McCracken, A., and Msimanga, A. (2020). A toolkit for teachers and learners, parents, carers and support staff: Improving mathematical safeguarding and building resilience to increase effectiveness of teaching and learning mathematics. *Creative Education*, 11, 1418–1441. 10.4236/ce.2020.118104

Lawson, D., Croft, T., and Halpin, M. (2003). *Good practice in the provision of mathematics support centres*, 2nd edition. Birmingham: LTSN, Maths, Stats and OR Network.

Littlecott, H. J., Moore, G. F., and Murphy, S. M. (2018). Student health and well-being in secondary schools: The role of school support staff alongside teaching staff. *Pastoral Care in Education*, 36(4), 297–312. 10.1080/02643944.2018.1528624

Mackenzie, K., and Williams, C. (2018). Universal, school-based interventions to promote mental and emotional well-being: What is being done in the UK, and does it work? A systematic review. *British Medical Journal Open*, 8, e022560.

Mezirow, J. (2009). An overview on transformative learning. In K. Illeris (Ed.), *Contemporary Theories of Learning: Learning Theorists- In Their Own Words*. (pp. 90–105). London: Routledge.

Nagoski, E. (2019). *Burnout: The secret to unlocking the stress cycle*. New York: Ballantine Books.

Pelegrina, S., Justicia-Galiano, M. J., Martín-Puga, M. E., and Linares, R. (2020). Math anxiety and working memory updating: Difficulties in retrieving numerical information from working memory. *Frontiers in Psychology*, 11, Article 669. 10.3389/fpsyg.2020.00669

Schaeffer, M. W., Rozek, C. S., Maloney, E. A., Berkowitz, T., Levine, S. C., and Beilock, S. L. (2021). Elementary school teachers' math anxiety and students' math learning: A large-scale replication. *Developmental Science*, 24(4), e13080. 10.1111/desc.13080

Waller, R., Hodge, S., Holford, J., Webb, S., and Marcella, M. (2018). Adult education, mental health and mental wellbeing. *International Journal of Lifelong Education*, 37(4), 397–400. 10.1080/02601370.2019.1533064

Woolfolk, A. (2007). *Educational psychology* (10th ed.). Boston: Allyn and Bacon.

Wood, E. F. (1988). Math anxiety and elementary teachers: What does research tell us? *For the Learning of Mathematics*, 8(1), 8–13.

9 Teacher-led mathematical resilience research

Ben Sinclair, Telma Silveira Pará, Masha Apostolidu, and Aïcha Hadji-Sonni

Introduction

Teachers play an integral role in developing learners' mathematical resilience. They can cultivate a supportive and accepting classroom environment, where learners are willing to try, struggle and grow. They can select tasks and plan learning sequences which foster challenge and promote the value of mathematics. They can also draw on theoretical and evidence-based work shared by others to develop their practice further. However, teachers are not merely consumers of such professional knowledge; they are experts in their own learners and context, and often are researchers in their own right. As reflective practitioners, teachers are constantly interpreting, challenging, and adapting their teaching, to better serve their learners and develop as educators.

Many teachers have found it valuable to use a systematic approach such as *action research* or *educational design research* to guide the development of their practice. These two commonly used models of practitioner-led enquiry are helpful in organising and enhancing teachers' intuitive, day-to-day reflection. Action research is usually defined as an inquiry conducted by practitioners in their own educational settings to advance their practice and improve learners' learning (Efron and Ravid, 2013). Its emphasis on the role of practitioners steers teachers to reflect upon, embrace and challenge the privileges and prejudices their position affords them (Altrichter and Posch, 1989). Educational design research shares this same pragmatic, iterative and solution-focused nature, but also seeks to abstract broader design principles from the process (McKenney and Reeves, 2018). The understanding that is gained from a research project is used not only to shape an immediate solution to a problem in one teacher's practice, but also to inform others outside of that particular classroom.

This chapter presents four studies that used these methodologies to research mathematical resilience in a variety of contexts. Telma and Masha each employed participatory action research to reduce anxiety and increase empathy and motivation in mathematics classrooms. Ben and Aïcha both used a design-based research approach to develop tools to teach learners resilience and coping skills in secondary mathematics. Together we hope that these studies not only share some ideas for promoting mathematical resilience, but also encourage you to consider how – and why - you might research your own practice.

Telma: Participatory action research

Mathematics anxiety is a widespread problem in Brazil. One report (OECD, 2013) found that 49% of learners experience school-work related mathematics anxiety. In 2022, I conducted an action research study intended to address mathematics anxiety and build resilience in a public high school in Brazil. The study involved 32 learners (9 boys and 23 girls) aged between 15 and 17 years old and was a school-based intervention using tools such as the Growth Zone Model, the Hand Model of the Brain, and the Relaxation Response (Pará and Johnston-Wilder, 2023). The study aimed to address mathematics anxiety and build mathematical resilience in a public high school in Brazil by concentrating on the learners' subjective experiences in lessons.

The study used two main approaches. The first was a version of action research called *participatory action research* (Savin-Baden and Wimpenny, 2007). This involved the participation of the learners in order to generate new knowledge to address a specific issue or problem. I chose this methodology as it is a process which develops practical knowledge by combining theory, practice, action, and reflection to pursue worthwhile purposes. The participation of learners and teachers in the research process allowed them to articulate their concerns and ideas for the flourishing of those involved and their communities (Jacobs, 2016).

The second approach I used was *anxiety-informed mathematics teaching*. This model raises awareness of anxiety in the teaching of mathematics. It recognises mathematics anxiety in learners and encourages teachers to respond appropriately and to resist causing further harm. This approach was inspired by a recent report from Scotland which investigated teachers and other professionals working with traumatised people (Homes and Grandison, 2021) (Table 9.1).

The intervention was divided into four cyclic phases as depicted in Figure 9.1.

Findings

In line with the mathematical resilience framework, we looked for evidence of anxiety, feelings of psychological safety, growth mindset, motivation and value, struggle, and relatedness. A technique called *deductive latent thematic analysis* (Clarke and Braun, 2017) was used to analyse the interview data collected.

Table 9.1 The phases of the Participatory Action Research Cycle (Pará and Johnston-Wilder, 2023)

Reflecting	In phase one, we analysed the school context carefully and used that analysis to decide on what tools should be used in the intervention. An initial survey found out that most of the 32 learners had performance score levels lower in 2019 than they had in 2018 (63% of learners). We also found out that these learners suffered from low self-esteem, lack of confidence and school-related anxiety.
Planning	In phase two, we measured their level of mathematics anxiety – this gave us additional data to inform our intervention. We developed approaches to deal with different learner experiences, including which activities and exercises would best support the learners if they started suffering from anxiety outside of lesson time.
Acting and Observing	Moving to phase three, we introduced the growth zone model, the hand model of the brain and the relaxation response to the learners. This was followed by a discussion with the participation of learners and teacher to set up a new dynamic within the classroom. These activities were performed in two groups in mathematics lesson time.
Evaluating	Finally, we measured the level of anxiety of the learners and discussed the results with them. This is a key activity in participatory action research as it reinforces the involvement of the learners and gives the researcher feedback for the next the cycle of enquiry.

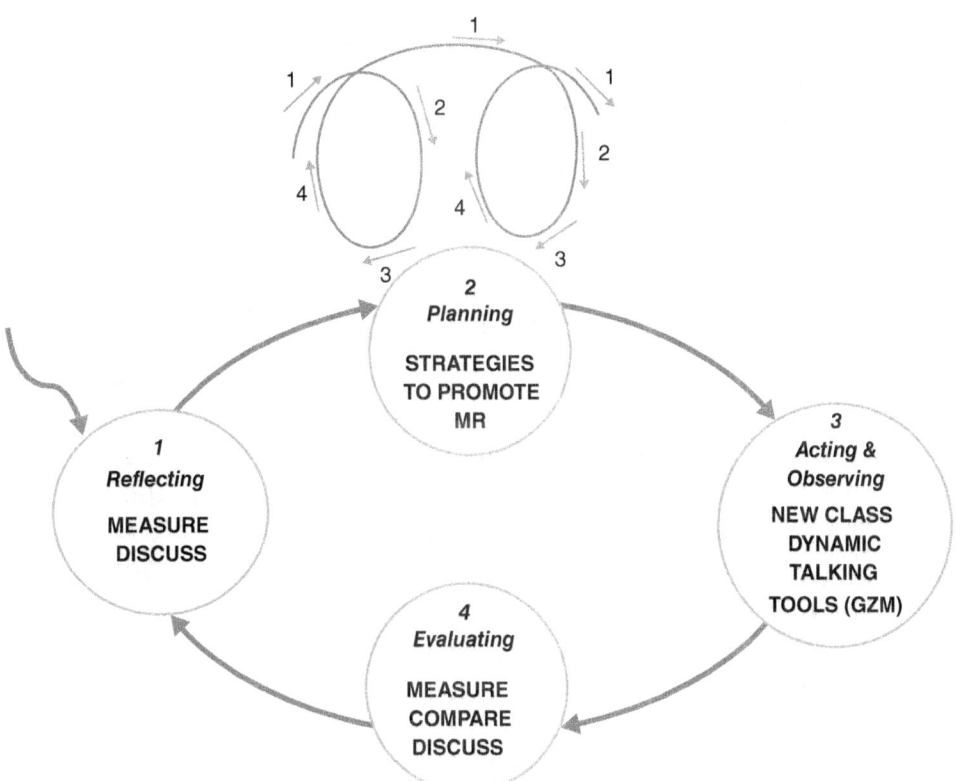

Figure 9.1 Methodology schema – Participatory Action Research Cycle (Source: Pará and Johnston-Wilder, 2023).

Mathematics anxiety

Only four learners (13%) had a low level of mathematics anxiety at the start of the study according to the measure we used, this increased to seven learners (22%) after the intervention. Evidence of anxiety and of underlying experiences that could be causal, was found when interviewing the learners. One learner, Vitoria, described feeling *"lost and confused in maths classes"*, saying: *"I always do maths tests as fast as I can to get rid of it"*. Another learner, Pedro, described the state of anxiety very vividly: *"I always had a good relationship with maths, and I had some teachers who made jokes in class until my 9th grade. When I began high school, I felt great difficulty as I was used to another methodology. Now, the same subject that I used to like because I thought it was easy and funny makes me terrified and in a crisis of anxiety, panicking and panting. I have tried to improve but it has been difficult."* He also declared that: *"I have an anxiety crisis"*.

Psychological safety

The learners were observed by their teacher to begin to feel safer as the study went on and, as a result, they asked more questions; and sometimes asked for a certain topic to be explained again or to be repeated. As the changes began to result in an environment designed to help their psychological safety, many learners expressed positive comments about the 'rhythm' of the lessons. One learner, Luana, said *"I feel myself capable of doing these classes, however I am not capable in normal maths classes"*.

Growth mindset

In their prior experience of learning mathematics, learners reported that they demonstrated a fixed mindset. Learner Beatriz said that: *"I was not born to deal with numbers"*. Learner Maria revealed her reaction by saying: *"… before I thought the subject of maths was not made for me"*. After the study, some learners recognised experiencing growth in their study of mathematics: *"I feel less dumb in your class"*; *"… today I have less difficulty in the subject"*; *"I am still confused however there is some progress"*; *"classes helped me to progress in maths"*; *"classes helped me to improve my performance in maths"*. Finally, one learner, Ester, proudly remarked: *"I feel relieved for being able to solve!"*.

Motivation and value

Before the intervention, most learners did not find mathematics motivating or see the value of it: *"I do not see the reason for studying maths"*. During and after the intervention, learner Pedro declared that *"time passes so quickly here"*, which could indicate changes in motivation. Another learner said that: *"after classes,*

I got more interested in the subject"; "I feel classes are more interesting and more useful". These quotes indicate changes in how learners now perceived value, interest and usefulness in mathematics.

Relatedness and struggle

Relatedness was built into the intervention design in that learners were encouraged to collaborate in groups and communicate if they experienced being in their red, or anxiety zone as indicated by the Growth Zone Model (see Chapter 3). After four classes of the new way of working, the teacher observed the learners forming informal discussion groups to find solutions to tasks they were struggling with, having previously waited passively for the teacher to help. After six classes, learners were more participative and asked more questions: *"I like your classes, sometimes I feel it is a bit complicated, but I find it interesting"*. Quotes like these indicated that changes in how learners perceived value, interest and usefulness in mathematics, were starting to change.

The changes in ways of working developed an environment of reduced anxiety, and increased empathy and motivation in class. Initially, 38% of the learners disagreed with the statement: 'I find math interesting'. This reduced significantly to 9% after the intervention indicating a more positive attitude towards mathematics. The activities we developed used tools to reinforce safety, relatedness and motivation towards mathematics classes. The evidence gathered suggests that mathematical resilience was developed in many learners, and they were helped in overcoming adverse prior experiences with learning mathematics.

Masha: Collaborative action research with mature learners

As a teacher of adult learners in a Further Education College in London, I saw a great need for simple, comprehensible, practical approaches to overcome the emotional barriers to learning mathematics I saw in my learners. With Sue Johnston-Wilder, I designed a small-scale action research project to introduce 'The Tools' (Johnston-Wilder et al., 2020) in Chapter 3 to the learners whilst also allowing me to explore the notion of psychological safety within the mathematical resilience model. The tools are part of "The Toolkit" (Johnston-Wilder et al., 2020). We used three key components to analyse and evaluate the effectiveness of the project, listening to prior experience, psychological education and the toolkit.

Listening to prior experience

In the first session, after introductions and procedure agreements, I invited each learner to share their personal mathematics story, including both good and bad experiences, and their coping strategies. All of my learners identified negative past experiences when learning mathematics; some described painful experiences, *"In class all other learners understood the topic, and I would still not understand. It made me very sad and discouraged"*, and others physiological elements *'heart beating very fast', 'feeling dizzy'*. Two learners remembered being physically punished by teachers for not understanding and would miss their classes to avoid punishment and the shame they felt as a result.

Psychological education

In the sessions that followed, activities were designed to help the learners understand mathematics anxiety and the role of the brain and emotions in aiding or preventing learning. The learners were presented with examples of connections between negative feelings about mathematics and low confidence and explored the effect of safety on learning. At the end of the sessions the learners were asked for their reactions. All of the learners confirmed that mathematics anxiety was a new concept to them. They were not aware that a fear of mathematics could disrupt their learning by preventing the effective use of their working memory and causing mathematics-avoiding behaviours (Dowker et al., 2016). Many of the learners expressed established beliefs in their inherent inability to learn mathematics, which Dweck (2017) describes as a fixed mindset.

The mathematical resilience toolkit

The Toolkit which comprises the four tools (Figure 9.2) was incorporated into my mathematics lessons. I encouraged my learners to use their preferred tool to practise managing anxiety and developing persistence outside the classroom.

The learners made use of the tools and they became part of the conversations that I overheard in my classroom.

Analysing the data

I had recorded and made notes on the learners' reactions to the ideas as presented in the Toolkit. Once transcribed, this formed my data. Analysis of that data resulted in four key themes and subthemes as summarised in Figure 9.3. These subthemes were grouped into those expressed ideas that seemed to indicate increased resilience and those which seemed to indicate decreased resilience. I also placed the subthemes that were apparent in the data on a spectrum from internal to external. More details of each theme follow.

Theme 1: Learners

Most of the learners spoke about the positive impact that the Growth Zone and Ladder Models had on their beliefs about the role of learners in the learning process, and specifically how these ideas helped them in developing self-efficacy. *"Now, I*

Teacher-led mathematical resilience research 133

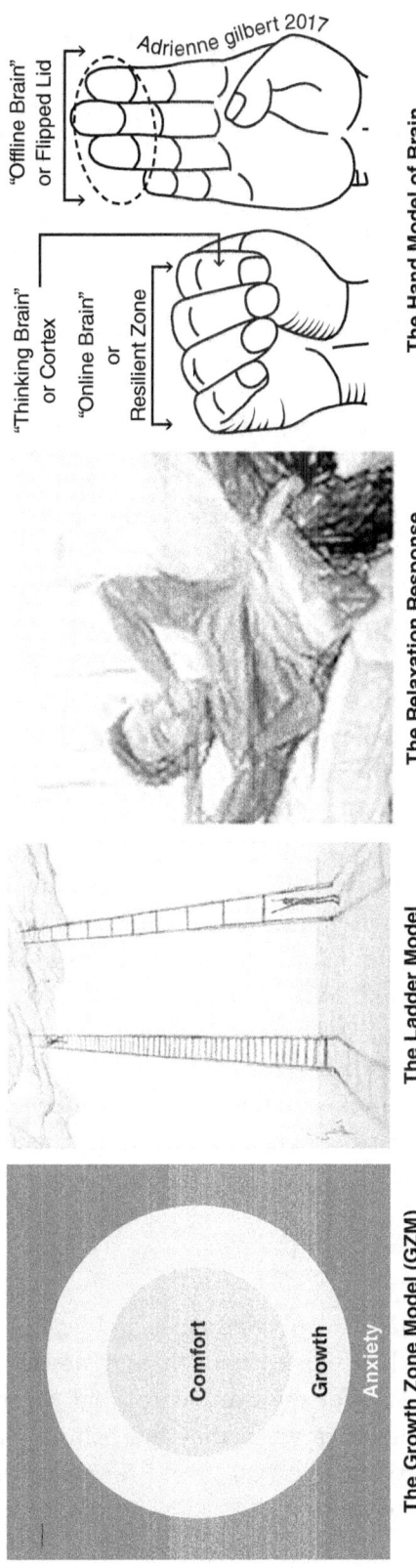

Figure 9.2 The Toolkit (Source: Johnston-Wilder et al., 2020).

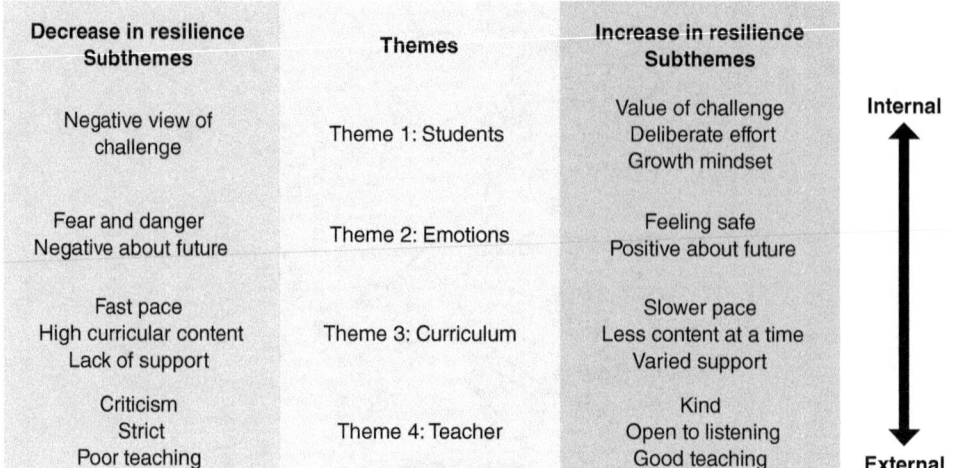

Figure 9.3 A summary of the four key themes and subthemes.

use Growth Model at work, it helps me to come out of the comfort zone and strive for more with confidence. I even share it with my colleagues when they go for promotions" (Mathilde). The learners concluded that having the tools gave them a set of "go to" strategies in the face of difficulties of mathematics, which had a positive impact on their confidence to have an active say in their mathematics learning.

Theme 2: Emotions

The learners told us that repetitive failure had convinced them that they lacked the ability to understand mathematics and made them feel helpless and anxious. This meant in turn that they opted for lower-paid and lower-skilled jobs. The learners valued the ideas in the Toolkit as practical approaches for addressing fear and hopelessness. *"I used to see mixed up numbers in maths problems and felt really stupid. I am not stupid, my brain can't learn when I am scared" (Anthony).* When asked about their view on the Relaxation Response, the learners commented that it was a refreshing break *'like a reset'* when they felt overwhelmed by mathematical rules and numbers.

Theme 3: Curriculum

Many learners associated their lack of confidence and feeling unsafe with the fast pace of previous lessons when the teacher felt they needed to cover the curriculum, rather than giving time for the learners to reach an understanding. They expressed feelings of hopelessness when they fell behind: *"it feels like you are running fast, but teacher is moving even faster. So that your fast is never fast enough in maths".* The learners also expressed feelings of being *'left behind'* when everyone else moved on.

Theme 4: Teacher

The learners noted that the teacher's role during lessons was very important. The learners found it valuable that the teacher displayed empathy rather than being critical and judgemental. They appreciated the teacher's sensitivity to their feelings, fears and worries, and how this was considered in lesson-planning. *"I always was scared of being behind and not catching up with other students. It made it only worse. This year was different because I knew I can always reach out for help"* (Ariana).

Overall, the data showed a positive shift in learners' attitudes towards challenge and struggle. The learners seemed to feel that the tools were easy to use outside of class and reported applying the ideas in work-related contexts. They felt that positive affirmations, relaxation and distraction strategies helped them manage their anxiety and fear of failure during independent work and assessments. All learners in the class attempted and completed homework and actively sought support from the classroom community and from online resources. They told us about their feelings of empowerment and confidence in overcoming the barriers and failures they encountered. One of the important aspects highlighted by the learners seems to have been that the teacher displayed empathy and had the teaching skills necessary to help them.

This action research project demonstrated how the principles and tools of mathematical resilience can positively influence FE learners' attitudes and efforts in mathematics. We are now looking to see how it might be expanded to larger cohorts and different contexts. Mathematics anxiety and avoidance are powerful barriers affecting many learners' vocational results in FE (Tobias, 1991), so further research is warranted on effective interventions that are accessible to learners and educators (Lyons and Beilock, 2012).

Aïcha: Educational design-based research with groups of learners

As a teacher of mathematics in disadvantaged areas in France, I frequently observed the symptoms of mathematics anxiety and low self-efficacy in learners. International assessments have likewise shown that French learners are significantly affected by mathematics anxiety (OECD, 2013). French learners are also revealed as showing reduced mathematics performance (TIMSS, 2019), indiscipline in classrooms (OECD 2019) and social inequality (OECD 2016). I set out to research how these issues could be mitigated with secondary school learners in a disadvantaged area of Paris.

To determine the extent of the problem, I used a paper-based survey to measure the mathematics anxiety (Betz, 1978) of 1902 learners aged 10–15 years old. Overall, 25% of learners reported having high mathematics anxiety and the proportion of learners reporting high mathematics anxiety increased significantly with age. However, there was no significant gender difference. I also conducted an online survey with 185 mathematics teachers in France to ascertain their perceptions of learners' attitudes toward mathematics. The survey showed that most of the teachers had observed avoidance and passivity in their learners and that their learners were easily discouraged from working towards a solution to a mathematics problem. All of these are signs of mathematics anxiety and demonstrate a lack of mathematical resilience.

To address these issues, I used a design-based research approach to plan a whole-class intervention. The tools developed were based on three of the existing mathematical resilience tools (Chapter 3) but were adapted to teach explicitly both problem-focused and emotion-focused mathematics coping skills. Coping skills are the cognitive processes used to respond effectively to a situation perceived as challenging or threatening (Lazarus and Folkman, 1984) to well-being in a mathematical context.

Resource 1: Mathematical resilience card

This card is a square divided in four parts (Figure 9.4), one for each zone of the *Growth Zone Model* which I extended to include a blue zone, labelled boredom. The boredom zone was used to help these learners recognise the difference between developing fluency and consolidating learned ideas in the green zone and the feeling engendered by doing nothing useful, just the same thing over and over again. When folded by the learner, the card presents the four zones along with the thoughts, feelings and behaviours associated with each. As each zone is unfolded, the card reveals a list of constructive coping skills for the chosen zone.

Resource 2: Mathematical resilience grid

The grid illustrated in Table 9.2, is to be completed by the learners. This can be done as a collective or individual task, in the presence of a teacher, carer, coach or alone. It is intended that the learner will complete it in a way that each learner will reflect on their personal situation. The grid invites learners to do this by structuring the reflection in three parts: past, present, and future.

The grid is inspired by the writing intervention described by Stogsdill (2013). If the learner chooses to reflect on a past and unpleasant situation, writing can help the learner to approach the past while managing their emotions. Writing about potential mathematics trauma can be a therapeutic exercise itself. An additional underpinning of the grid is Whitmore's (2017) GROW model. This coaching tool invites users to clarify their *Goals* and to describe their *Reality* by writing about the current thoughts, emotions and behaviours which they are experiencing. They then go onto considering possible *Options* before committing *Willingly* to their choice one of the options considered. Finally, the grid's analysis of the mathematics situation was inspired by the cognitive analysis framework developed by Finkel (2022), which asks the learner to identify the facts, emotions and actions they perceive if they are in a context strongly charged with emotion, and to recognise and store these ideas in their memories rather than allowing overly negative and destructive effects.

Teacher-led mathematical resilience research **137**

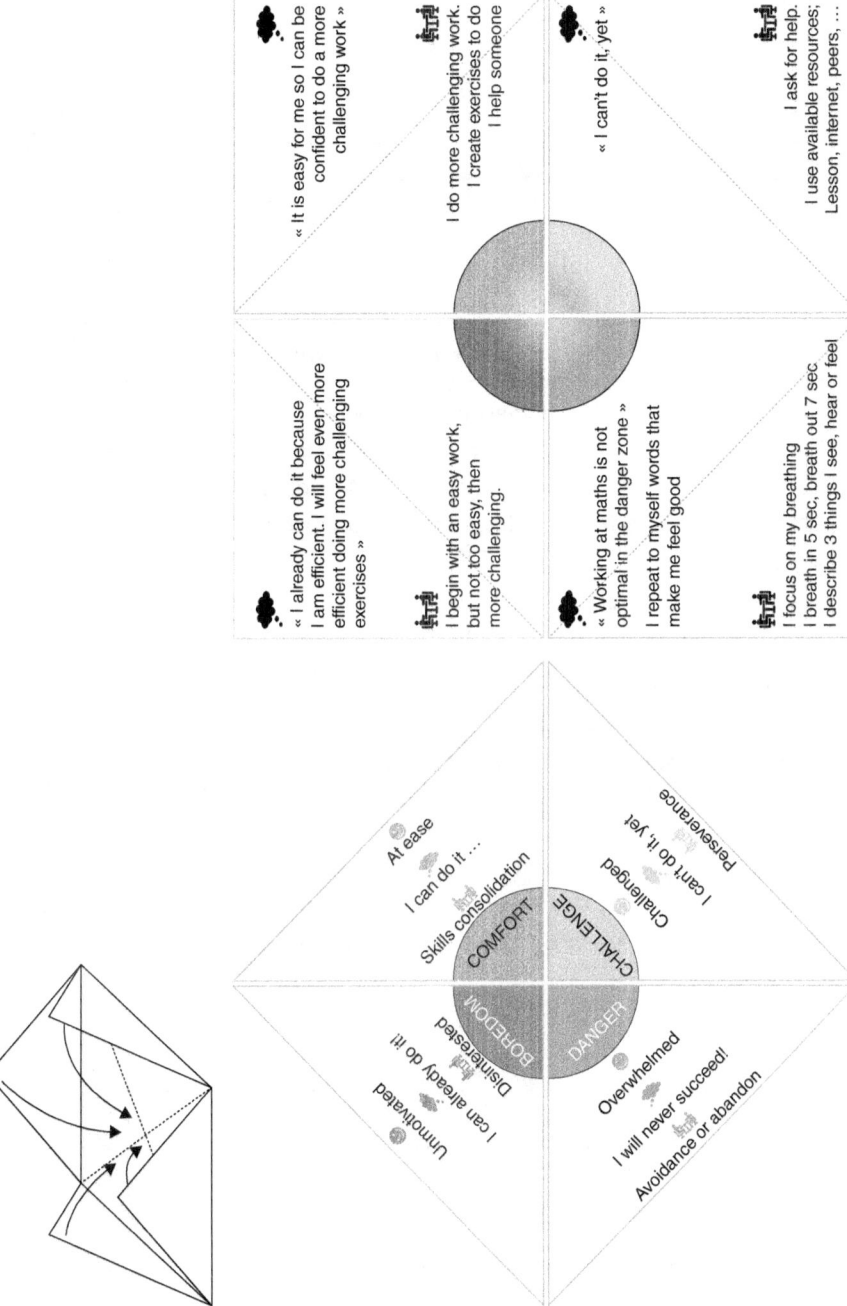

Figure 9.4 The mathematics resilience card folded (left) and unfolded (right).

Table 9.2 The mathematical resilience grid

I think about a maths moment, in the present, the past or the future ...			
When is it? Where?	**What happens?**		
What am I thinking?	**What am I doing?**		
What am I feeling?	**What is my emotion?**		
☐ A little ☐ Moderately ☐ A lot	☐ Sadness ☐ Fear ☐ Anger ☐ Disgust	☐ Joy ☐ Trust ☐ Anticipation ☐ Surprise	☐ Something else
What zone am I in?			
I decide I prefer the learning zone, I chose to think ...			
To reach (or stay in) the learning zone I list 3 things I can do ...			
I circle the action I choose			

Resource 3: The mathematical resilience posters

The posters (see Figure 9.5) are designed to be displayed in the classroom and used as a visual reference when learners feel stuck. The use of posters is inspired by the stuck poster developed by Chisholm (2017) as part of his action research project which explored several mathematical resilience strategies used with a class of

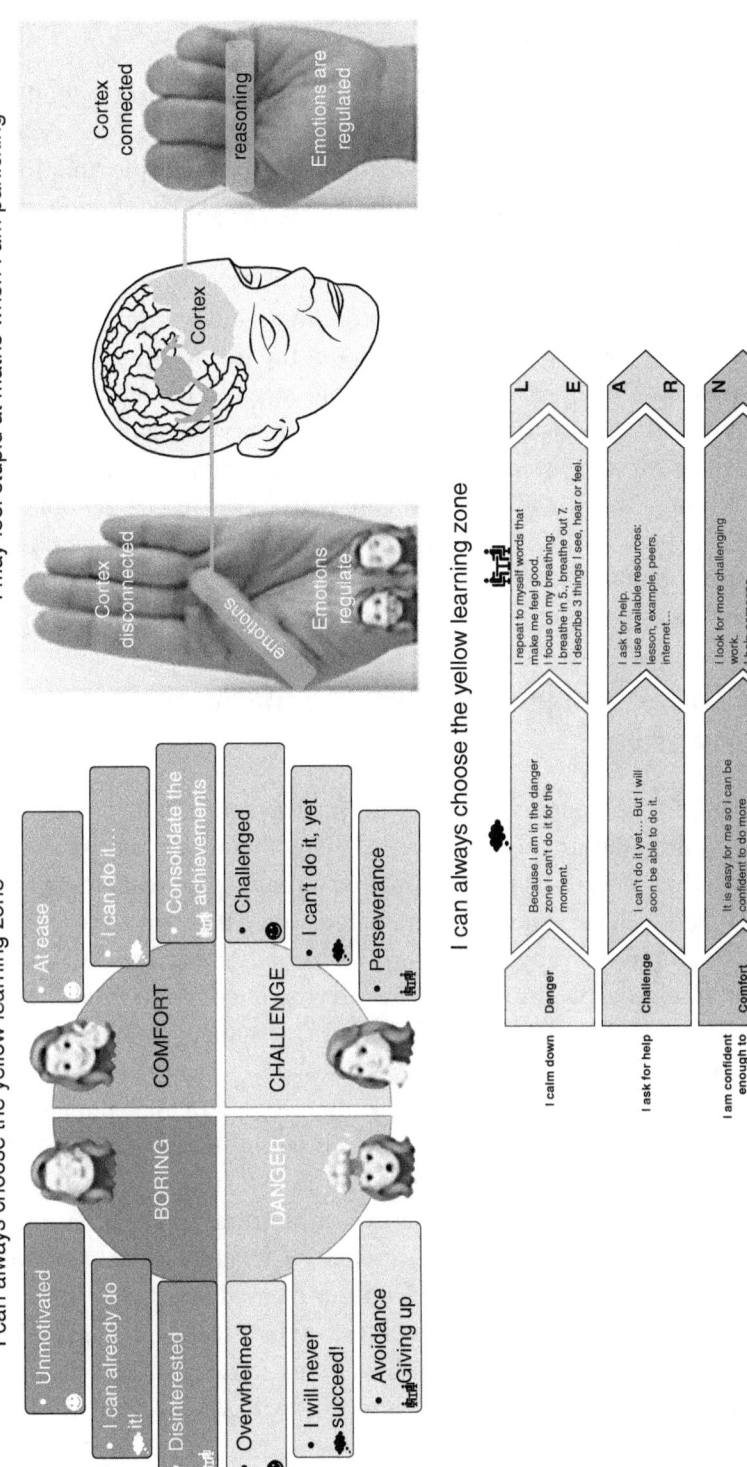

Figure 9.5 The mathematics resilience poster, tool 1 to 3 (from left to right).

learners. An explanation in French of the three tools can be found here: www.tinyurl.com/MRtoolsFR.

Resource 4: The relaxation video

This short video aims to induce the *Relaxation Response* (Benson, 2000) by helping learners use a breathing technique to ease from the danger zone. As it is silent, this video can be projected in a loop during a test, for example, to allow a learner to watch it and use it if experiencing the danger zone. The video can be accessed here: www.tinyurl.com/MRrelax

I have now improved the intervention design over four cycles of research. I have collected data from observing lessons, teacher interviews, and learners' work and analysed that data during each cycle. Cycles 1 to 3 of the intervention have been conducted in secondary schools in the disadvantaged areas in France where pre- and post-intervention mathematics anxiety questionnaires were conducted. The results so far are encouraging, especially with highly-mathematics-anxious learners and younger learners in their first and second years of secondary school. The data shows promise for the use of interventions that employ constructive coping skills, with problem- and emotion-foci. The next question to investigate is how the tools can be improved so that older learners can also take advantage of these ideas.

Ben: Maths ACTive – Designing web-based mathematical resilience interventions

Mathematically resilient learners struggle with mathematics but not against themselves. They do not take unhelpful thoughts like "I can't do maths" or "I'll never understand this" as facts or let them dictate who they are, or what to do next. When stuck or feeling a little anxious, they move towards, rather than away from, these uncomfortable sensations. They can bring their attention back to the work at hand instead of dwelling on past mistakes or worrying about future ones. Most importantly, they know what matters to them, and this guides their next steps when doing mathematics.

Fostering these behaviours in a wider context is the goal of *Acceptance and Commitment Therapy* (ACT) (Hayes et al., 2012). Pivoting towards these behaviours has been shown to improve everything from depression to sports performance, from psychosis to coping with tinnitus and from procrastination to parenting (Hayes, 2019). Using this powerful model for change, I created a web-based intervention, *Maths ACTive,* to help learners of mathematics to develop mathematical resilience and address mathematics anxiety (Sinclair, 2022). The self-guided course helps learners to explore how their minds work as they get 'stuck' and 'unstuck', before teaching them how to be *BOLD* (Ciarrochi et al., 2012):

- *Breathing deeply and slowing down;*
- *Observing what is happening;*
- *Listening to what matters;*
- *Deciding what to do and doing it.*

This technique not only helps learners who face difficulties in mathematics, but also helps them in situations that require clear, values-based decision-making. To help support mathematical learning, I incorporated many psychoeducational lessons, experiential exercises, metaphors, quizzes, homework, case studies, and interactive features into the course. In this section, I will consider in detail the purpose and value of three of the evidence-based principles used, so that others can apply them when conducting educational design research in similar contexts.

Experiential exercises

ACT (said as one word 'act') is so named because *action* must be taken to make progress. Developing mathematical resilience skills requires learners to undertake experiential practice in order to understand ideas that are often difficult to explain through language and rationale alone (Levin et al., 2012). *Maths ACTive* includes over thirty short exercises which learners are invited to both engage with and reflect on. For example, in order to help learners to *breathe deeply and slow down*, they are taught a technique called *balloon breathing*. This technique gives learners a way of 'dropping anchor' if they should feel overwhelmed when they are working on understanding mathematics. Crucially, the structure of the course means that the learners revisit the exercises to prompt transfer to different contexts. For instance, during the unit on *observing*, learners are encouraged to take a few balloon breaths before noting what thoughts show up when they plan how to achieve a mathematical goal. Exercises also provide a brief introduction into the ACT perspective of a meaningful life. Making mistakes, sometimes feeling a bit 'stupid' or frustrated, and wanting to give up, are all presented to the learners as being part of the learning process. The course suggests these actions and feelings are to be expected when stepping outside of the individual's comfort zone. The learner comes to understand that being willing to experience somewhat unpleasant difficulties in the service of a valued life is an important skill to cultivate. Each exercise affords an opportunity for learners to practise each idea, whilst also helping them realise that these experiences, whilst not comfortable, indicate that they are moving in a direction that will ultimately be of benefit to them (Figure 9.6).

Metaphors

Metaphors can link what you already know to domains where you are unsure what to do (Stoddard and Afari, 2014). Metaphors ground difficult or abstract concepts

```
Imagine that you are standing on one side of a deep dark pit.
On the other side is a huge monster - the negative feelings
monster, Badsadmad.
```

```
It stares at you with its red glaring eyes, showing rows of fierce
sharp teeth. A rope dangles across the pit. The monster is
gripping one end with its large claws and you are holding onto the
other end. The only way to win against the monster is to drag it
into the pit.
    You are sweating and exhausted. The monster only seems to be
    getting stronger!
        It's on the edge now...
            Just one more big pull...
                The monster is nearly gone forever...
```

Figure 9.6 The monster metaphor (Source: Maths ACTive, 2023).

in the familiar and can help learners come to an understanding and to generate their own conclusions and meaning. This perspective results in learners developing an ability to retain information and skills and transfer them to situations beyond the course (Otto, 2000). Maths ACTive uses metaphors extensively, adding many to the existing ones of the mathematical resilience toolkit. For instance, the story of *Badsadmad* (Figure 9.5) is repeatedly revisited to systematically explore how the characteristics of the monster are similar to that of unwanted thoughts and feelings. Learners recognise the exhaustion of a relentless tug-of-war; the demanding attention of the monster; and the way our minds readily evaluate it as 'bad'. These tangible features of the monster are equally applicable to feelings of worry, thoughts like 'I cannot do maths', and the uncomfortable sensations of anxiety. However, the purpose of a metaphor is not merely to illustrate a point, but to change behaviour. There is a powerful alternative to fighting – dropping the rope! Learners can choose not to engage in the constant, futile struggle with the monster, and use this gained space to listen to what matters to them. The goal is not to make the monster disappear, but to change the learner's relationship with it. It is possible to feel bad, sad, or mad without letting these emotions dictate the actions they take. The course helps the learner to understand this powerful message through the use of metaphor.

Homework

Outside of the Maths ACTive course, learners are likely to no longer receive guidance and feedback on their responses. Misconceptions may arise as they work, or they may return to using potentially unhelpful strategies when doing mathematics. These are two of the many reasons why homework is considered an essential aspect of ACT interventions (Twohig et al., 2007). When doing homework learners can consolidate the skills that they are introduced to in the course. By practising independently, they can develop confidence and fluency in a relatively controlled manner. Using homework, the skills introduced are transferred to contexts outside of the course where behaving with mathematical resilience may be challenging. To *decide what matters and do it*, learners are carefully helped to develop actions which are specific, meaningful, allow anxiety, realistic and trackable (SMART). Finally, homework reinforces the idea that understanding is achieved through intelligent practice over time. The review task for each homework sets the expectation of action by asking them to reflect on what they learnt in doing the homework, and to consider any barriers that prevented them from engaging with it fully.

Mastering mathematics is regarded in this course as being about learners using mathematics to do what matters to them. Some may value challenging themselves, learning something new, or always trying their best; others may care about getting the grades needed for a dream course or job, working with others, or being curious. Supporting learners to enact their values in mathematics is why *Maths ACTive* was created. However, the evidence-based strategies of ACT can be used by all educators to ensure mathematics is taught with willingness, flexibility and resilience.

Conclusion

These four snapshots of mathematical resilience research highlight the variety, richness and power of a teacher-led approach to testing how the principles of developing mathematical resilience play out in practice. Being able to step back and systematically reflect on one's practice can be challenging (Lee, 2006), not least because of the complex and pressured context in which this work is often undertaken. Conducting action research and educational design research requires teachers to balance the needs and interests of learners, colleagues, parents, schools, and researchers, all whilst managing their other professional commitments. As each of the authors found, this can be a demanding but valuable experience.

Every teacher's interpretation of the developing mathematical resilience principles provides further insight into the nuances of the construct. Engaging with these important perspectives undoubtedly is, and will continue to be, central in the development of mathematical resilience research.

References

Altrichter, H., and Posch, P. (1989). Does the "grounded theory" approach offer a guiding paradigm for teacher research? *Cambridge Journal of Education*, 19(1), 21–31. 10.1080/0305764890190104

Benson, H. (2000). *The relaxation response*. New York: Avon Books.

Betz, N. (1978). Prevalence, distribution, and correlates of math anxiety in college students. *Journal of Counseling Psychology*, 25(5), 441–448.

Chisholm, C. (2017). *The development of mathematical resilience in KS4 learners*. Unpublished PhD thesis. University of Warwick. Centre for Education Studies.

Ciarrochi, J. V., Hayes, L. L., and Bailey, A. (2012). *Get out of your mind and into your life for teens*. New Harbinger Publications.

Clarke, V., and Braun, V. (2017). Thematic analysis. *The Journal of Positive Psychology*, 12(3), 297–298.

Dowker, A., Sarkar, A., and Looi, C. Y. (2016). Mathematics anxiety: What have we learned in 60 years? *Frontiers in Psychology*, 7, 508.

Dweck, C. S. (2017). *Mindset: Changing the way you think to fulfil your potential*. London: Robinson.

Efron, S. E., and Ravid, R. (2013). *Action research in education: A practical guide*. New York, USA: Guilford Publications.

Finkel, A. (2022). *Manuel d'analyse cognitive des emotions*. Paris: Dunod.

Hayes, S. C. (2019). *A liberated mind: The essential guide to ACT*. London, UK: Vermilion.

Hayes, S. C., Strosahl, K. D., and Wilson, K. G. (2012). *Acceptance and commitment therapy: The process and practice of mindful change* (2 ed.). New York, USA: The Guilford Press.

Homes, A., and Grandison, G. (2021). *Trauma-informed practice: A toolkit for Scotland*. The Scottish Government. https://www.gov.scot/publications/trauma-informed-practice-toolkit-scotland/

Jacobs, S. (2016). The use of participatory action research within education-benefits to stakeholders. *World Journal of Education*, 6(3).

Johnston-Wilder, S., Baker, J., McCracken, A., and Msimanga, A. (2020). A toolkit for teachers and learners, parents, carers and support staff: Improving mathematical safeguarding and building resilience to increase effectiveness of teaching and learning mathematics. *Creative Education*, 11(8), 1418–1441.

Johnston-Wilder, S., Kilpatrick-Baker, J., McCracken, A., and Msimanga, A. (2020). A toolkit for teachers and learners, parents, carers and support staff. *Creative Education*, 11(8).

Lazarus, R., and Folkman, S. (1984). *Stress, appraisal, and coping*. New York: Springer Publishing.

Lee, C. (2006). *Language for learning mathematics – Assessment for learning in practice*. Buckingham: Open University Press.

Levin, M. E., Hildebrandt, M. J., Lillis, J., and Hayes, S. C. (2012). The impact of treatment components suggested by the psychological flexibility model: a meta-analysis of laboratory-based component studies. *Behavioural Therapy*, 43(4), 741–756. 10.1016/j.beth.2012.05.003

Lyons, I., and Beilock, S. (2012). Mathematics anxiety: Separating the math from the anxiety. *Cerebral Cortex (New York, N.Y. 1991)*, 22(9), 2102–2110.

McKenney, S., and Reeves, T. (2018). *Conducting educational design research*. Abingdon, UK: Routledge.

OECD (2013). *PISA 2012 results: Ready to learn: Students' engagement, drive and self-beliefs (Vol. III)*. Paris: PISA, OECD Publishing. 10.1787/9789264201170-en

OECD (2016). *PISA 2015 results (volume I): Excellence and equity in education*. Paris: PISA, OECD Publishing. 10.1787/9789264266490-en.

OECD (2019). *PISA 2018 results (volume I): What students know and can do*. Paris: PISA, OECD Publishing. https://doi.org/10.1787/5f07c754-en.

Otto, M. W. (2000). Stories and metaphors in cognitive-behavior therapy. *Cognitive and Behavioral Practice*, 7(2), 166–172. 10.1016/S1077-7229(00)80027-9

Pará, T., and Johnston-Wilder, S. (2023). Addressing mathematics anxiety: A case study in a high school in Brazil. *Creative Education*, 14(2), 377–399.

Savin-Baden, M., and Wimpenny, K. (2007). Exploring and implementing participatory action synthesis. *Qualitative Inquiry*, 18, 689–698. 10.1177/107780041245285

Siegel, D. (2010). *Mindsight: Transform your brain with the new science of kindness*. London: Oneworld Publications.

Sinclair, B. (2022). Maths ACTive: Mastering mathematics with psychological flexibility. *Proceedings of the Third International Conference on Developing Mathematical Resilience*.

Stoddard, J. A., and Afari, N. (2014). *The big book of ACT metaphors*. Oakland California USA: Context Press.

Stogsdill, G. (2013). A math therapy exercise. *Journal of Humanistic Mathematics*, 3(2), 121–126. athsac

TIMSS (2019). TIMSS 2019 international results in mathematics and science, IEA TIMSS and PIRLS. https://timss2019.org/reports/

Tobias, S. (1991). Math mental health. *College Teaching*, 39(3), 91–93.

Twohig, M. P., Pierson, H. M., and Hayes, S. C. (2007). Acceptance and commitment therapy. In N. Kazantzis, & L. L'Abate (Eds.), *Handbook of homework assignments in psychotherapy: Research, practice, and prevention* (pp. 113–132). Springer US. 10.1007/978-0-387-29681-4_8

Whitmore, J. (2017). *Coaching for performance: The principles and practice of coaching and leadership*. London: Murray Press.

PART 3
Working within the wider community

Communicating ideas about mathematical resilience to parents

Rosemary Russell and Donna Wright

Introduction

Parental engagement can have a significant and positive impact in general. Care is needed though, as it has been found that for mathematics, in some cases, parents' own mathematics anxiety has had a negative impact on progress when they helped their children at home.

This chapter will show how parents can overcome their mathematics anxiety, leading to a positive experience for parent and child. It will also report on an ongoing project about communicating ideas about mathematical resilience to parents. In this chapter, the term 'parents' includes parents, carers and guardians.

Donna's story

The Covid lockdown in March 2020 brought Donna some interesting challenges with regards to supporting pupils' learning at home. Working as a senior education improvement adviser at the time, Donna was asked to oversee remote education and to work with schools to enable high quality learning for pupils in her area during lockdown periods.

Lucas, Nelson and Sims (2020) reported that, during this time, around 55% of pupils' parents were engaged with their children's learning at home. This engagement was slightly lower for parents of secondary aged pupils as parents were more inclined to believe that their children were able to manage their own learning. Even though parents were spending more time with their children whilst they were learning, Andrew et al. (2020) reported that over half of parents found it difficult to support their child's learning due to time constraints; their understanding of the content that their child was learning; their child's ability to work independently and the quality of the resources provided by the school.

Teachers began reporting to Donna that parents were increasingly contacting schools requesting support and help with teaching at home, particularly with mathematics as approaches to teaching were different from parents' own experiences. Some parents reported that their children were becoming upset and not wanting to participate in the mathematical activities set by their school.

At this time, Donna was also involved in a mathematics research work group, led by Sue Johnston-Wilder, with a small group of primary teachers. The group were focusing on how to support children and young people with mathematics anxiety. Mathematics anxiety is a fear of mathematics to the extent that it can stop the brain effectively processing mathematics (Johnston-Wilder and Marshall, 2017). This fear is usually developed due to poor teaching methods, previous embarrassment related to learning mathematics, negative life experiences, social pressures and expectations, or desires for perfection and stereotyping (Arem, 1993). Pupils experiencing mathematics anxiety tend to perform worse in examination situations and the anxiety also affects their ability to learn new concepts and procedures. The anxious thoughts negatively affect the recall of mathematical facts and procedures (O'Connor, 2020).

In addition to assisting schools and teachers with remote learning, Donna decided to explore ways in which to support parents with children who were experiencing mathematical anxiety at home and who were not able to demonstrate resilience. To investigate this further Donna asked all schools in the borough to send a communication to their families, asking concerned parents to make contact so that a working group could be established. Using the information gained from parents, Donna intended to trial some strategies before sharing them more widely across the community. The parents who became part of this working group had children who ranged from those in nursery (aged three to four years) through to key stage 3 (aged 11–14 years).

To gather an understanding about what parents were experiencing, Donna collated some baseline information through a series of short multiple choice style surveys. Around 25 parents initially showed an interest in taking part in the project, but this reduced to 10 families of primary aged pupils who responded to the surveys and engaged with the creation and use of the materials throughout. This was a small representative group, not large enough to draw any conclusive themes, but the responses provided a starting point for creating supporting documents.

The first survey concentrated on "How does maths anxiety present in your child?" Questions focused on observed triggers for anxiety and how they presented and explored parents' current understanding of the causes. The information shared revealed that many of the children had always had some form of mathematical anxiety, but for some it had been exacerbated through lockdown and working at home. Some parents reported that their child did not like mathematics and became anxious when asked to engage with a mathematical problem, often becoming teary or angry and refusing to take part. Three key emotions shown by the children in the group became evident: anger, low self-esteem (a feeling of worthlessness) and frustration leading to crying and upset.

Most of the children's anxieties seemed to occur after reading the problem they were given and then realising that they did not know where to start. Others refused to even attempt the problem as they said they perceived that they were not good enough at mathematics and had decided that they would not be able to do it. Some children felt that their brains 'stopped working' when they were forced to answer questions quickly or when they felt under pressure to provide a correct answer. They said they saw themselves as not good at mathematics as they believed good mathematicians were people who answered questions quickly and accurately. Anderson, Boaler and Dieckmann (2018) suggest that if a teacher prioritises speed over depth of understanding, then pupils begin to believe that speed is an important part of learning mathematics even though the teacher may not have intended to portray this. Quite often this may be communicated to pupils implicitly as teachers aim to keep up the pace of a lesson or to cover certain prescribed lesson content.

Parents reported trying to help their children at home by providing reassurance and attempting, where they could, to break the problem down into smaller steps but often ended up doing much of the problem solving themselves. Some of the parents shared their own experiences and it was interesting to see that many of the group also recognised that they either experience mathematics anxiety themselves and/ or did not enjoy mathematics at school. A small proportion of families reported that all of their children experienced mathematics anxiety.

Donna decided to provide support for parents which would help them to understand what their children were experiencing and to help their children develop resilience through managing their anxiety. Part of this support included introducing parents to the Growth Zone Model (Johnston-Wilder et al., 2016), described in detail in Chapter 3 of this book. The Growth Zone Model structure has three zones: the comfort zone (green); the growth zone (amber/orange) and the anxiety zone (red) (Figure 10.1).

The comfort zone (or green zone) represents everything the learner can already do. When a learner metaphorically stays in this zone, they are likely not to be learning anything new, but may be building confidence and automaticity.

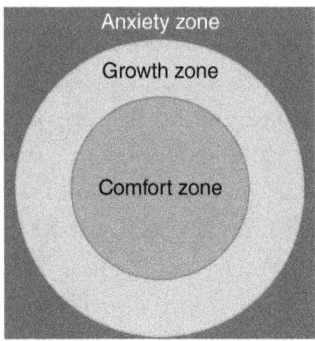

Figure 10.1 The Growth Zone Model.

The growth zone (or amber/orange zone) is the space where the learner is learning. Here what they are doing pushes them slightly beyond what they currently know. This zone is where new learning happens and although the learner may feel some levels of anxiety because they are being challenged, they manage this well and persevere with their learning. To stay in this zone, they must be comfortable to make mistakes or to get stuck and be motivated to continue learning.

The anxiety zone (or red zone) is where the learner is asked to attempt something that is mathematically beyond their reach for various reasons. In this zone, the learner will feel high levels of stress and anxiety and less, if any, useful learning will take place. The brain is in 'fight, flight or freeze' mode. It is when the learner is in this zone that they may choose to avoid tasks, become overly dependent on support or be unable to engage. From this zone, the pupil should be encouraged back to the comfort zone where they can recover.

To share this information with parents, Donna created some help sheets focusing on:

- ***What is maths anxiety?***
  A document providing an overview of what maths anxiety is, how to identify a child experiencing maths anxiety, what causes maths anxiety and why a child may feel like this.

- ***Helping my child using the Growth Zone Model***
  A document providing an explanation of the 'Growth Zone Model' and how it can be used to help children understand their feelings and begin to manage their anxiety

- ***Helping my child to have a 'can do' attitude***
  A document providing an explanation of how to help children develop a more positive attitude towards maths and an overview of what a 'good' mathematician could look like.

- ***Finding out more about maths anxiety***
  Suggested websites for parents who wish to explore maths anxiety further.

These materials were shared with parents via a council website focused on Maths Anxiety so that all parents could be signposted to this area and could access these materials. The parents, who were working with Donna directly, were asked to read the documents and to try out some of the suggestions. Some parents printed out an outline growth zone model chart and joined their children in colouring in the sections. They asked their children to write down what they might feel when they were in each zone. The parents reported that this enabled them and their parents to identify which zone they were working in and any early signs of anxiety. Lee and Johnston-Wilder (2018) suggest that one of the characteristics of a resilient learner in mathematics is that they have the

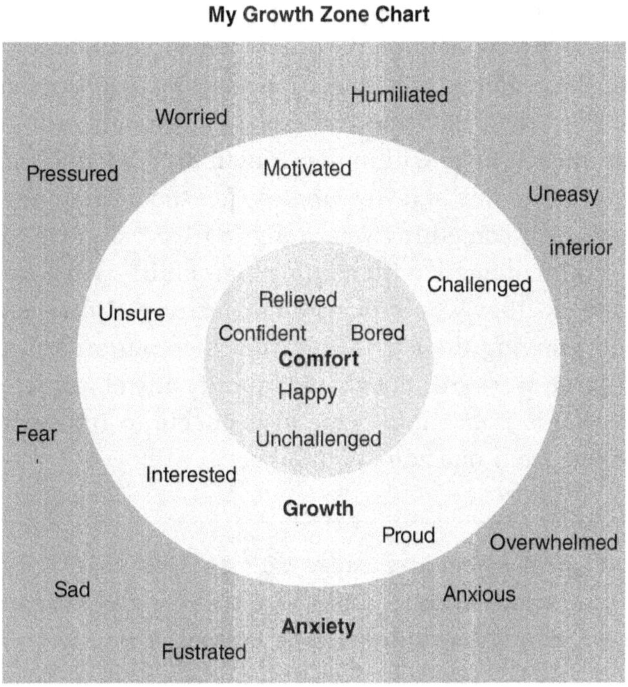

My Growth Zone Chart

(Chart showing three concentric zones with emotion labels)

- **Comfort** (inner zone): Relieved, Confident, Bored, Happy, Unchallenged
- **Growth** (middle zone): Motivated, Unsure, Challenged, Interested, Proud
- **Anxiety** (outer zone): Worried, Humiliated, Pressured, Uneasy, inferior, Fear, Sad, Overwhelmed, Anxious, Fustrated

Figure 10.2 My Growth Zone Chart, created by Key Stage 3 pupil.

appropriate language to express their feelings as they start to feel out of control. If they are able to recognise these feelings, they can request the support they need that will enable them to stay in their growth zone longer. Figure 10.2 shows a chart completed by a key stage 3 learner who recognised that whilst the comfort zone was a safe place to be, if in this zone for too long, periods of boredom can set in. The learner further noted that in the growth zone, they felt challenged and motivated, but they also acknowledged feeling unsure. They recognised that feeling unsure was part of experiencing some level of challenge which is necessary for new learning. Interestingly, the parent fed back that when completing the chart, the pupil quite readily found words to describe how they felt in the anxiety zone but had to think more deeply and reflect more on their emotions in the other zones.

> " ... the majority of parents reported feeling more confident with how they could help their children and felt they were able to understand their child's emotions at each stage within the growth zone model."

The second survey explored "How is the Growth Zone Model supporting my child?" These questions focused on to what extent the Growth Zone Model, and the supporting documents, were supporting the young person to manage their own anxiety and how they were helping the parents to assist their child.

At this point, the majority of parents reported feeling more confident about how they could help their children and felt they were able to understand their child's emotions at each stage within the Growth Zone Model. However, most of the children were still not able to identify for themselves where they were within the model, or if something was going to trigger a move to another zone. Parents said they felt confident with helping their child escape from the red zone to the comfort zone as they would just stop the activity. However, they said that they needed help moving their children from the comfort zone to the growth zone. As the children were still not able to identify any changes to their working emotional state at this point, they were also unable to independently support themselves to move from one zone to another.

> "It also became apparent from parental feedback that their children also did not see themselves as good mathematicians as they were not always able to answer questions quickly or because they did not always answer questions correctly."

The next set of survey questions focused on "Does your child have a positive attitude towards maths?". This survey explored the parent's and young person's attitude towards mathematics and their perceptions of what skills and attributes a good mathematician has. It also explored whether the parents focused discussions with their children on speed and accuracy or effective approaches to solve the problems. It became apparent from parental feedback that the children did not see themselves as good mathematicians as they were not always able to answer questions quickly or because they did not always answer questions correctly.

LaMar, Lesin and Boaler (2020) also found that high proportions of pupils believed that successful mathematicians worked at speed and answered questions correctly with 80% of pupils surveyed agreeing that those who really understand mathematics will get an answer quickly. This is a limiting belief which can damage pupils' progress and engagement with the subject. Through developing a more positive mindset, working with pupils to study mathematics at depth, not speed and working more collaboratively, pupils can change their ideas and attitudes towards mathematics (Boaler et al., 2021).

A poster (Figure 10.3) was designed identifying some of the different attributes of a good mathematician. Speed and accuracy did not feature in the poster. The poster was designed to allow the children to identify something they had done well in their mathematics activity which showed that they were acting as mathematicians, even if they had not achieved a correct response. The attributes include:

Figure 10.3 I am a good mathematician poster.

- finding patterns;
- seeing connections;
- taking risks;
- finding different ways to solve problems;
- communicating mathematics clearly;
- persevering with challenges;
- learning from mistakes;
- checking results;
- asking good questions;
- being resilient when challenged;
- making links in the real world.

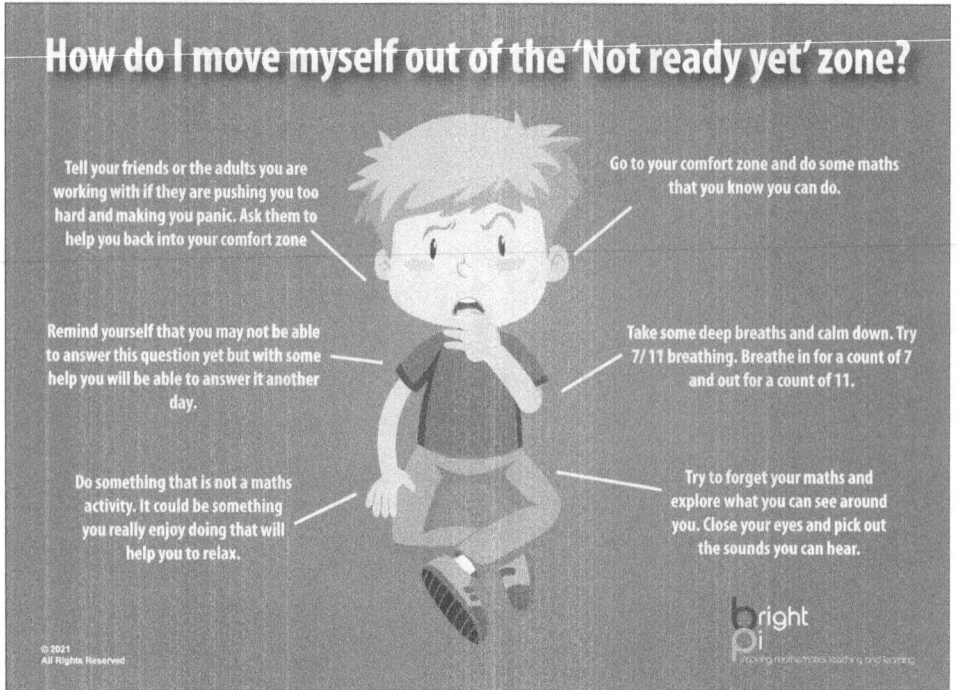

Figure 10.4 Moving out of the "not ready yet" zone poster.

A further three posters were designed for parents to use at home with primary aged pupils with the red zone renamed as 'the not ready yet' zone to further instil the idea that the child has the capacity to learn the mathematics. The 'not ready yet' label emphasises that they will be ready when whatever barriers are preventing them accessing the mathematics are discovered and dealt with. One of the most important strategies is to identify how to move out of the red or 'not ready yet' zone (Figure 10.4).

Lee and Johnston-Wilder (2018) suggest that *"learning and doing mathematics can feel risky"*. Pupils need to understand that learning new things may make them feel anxious but if they are aware of this and have their own strategies to move themselves back to their comfort or growth zones, they are more likely to accept a challenge willingly. If learners understand why they experience the emotions and behaviours that present at moments of anxiety, they may be better equipped to manage these feelings and become more resilient.

Previous negative experiences can lead the brain to view mathematics as a threat. The role of the amygdala within the brain is to enable survival and recognise threatening situations. When a threat is identified the brain often initiates the 'fight or flight' response before the actual threat can be interpreted more fully. The brain's reaction is to avoid the situation that is triggering the response. When the learner experiences mathematical learning as a threat too often and these feelings are is not addressed, avoidance will result and anxiety will develop (Johnston-Wilder et al., 2017).

> "Breathing exercises and mindfulness activities such as being in the moment and noticing sights and sounds and smells may help the child to relax and distract them from their negative emotions."

When using this poster, children are encouraged to address the threat they are feeling by talking about how they feel at each stage of their learning. They can develop the language to tell the adult they are working with if they start to feel any of their identified anxiety triggers. Parents can help their child maintain a positive mindset by assisting their child to see that they may not yet be ready for this piece of mathematics. The mathematics may feel tricky now but will become easier over time with perseverance and support. Sometimes, the child may need to move their thinking away from the mathematics and do something else they enjoy in order to relax the brain and come back to the learning later. Breathing exercises and mindfulness activities such as being in the moment and noticing sights and sounds and smells may help the child to relax and allow them to control their negative emotions.

Parents suggested that their children were happy in the comfort zone and began to show signs of anxiety at any introduction of challenge. As children move into the growth zone, they will need support to manage their feelings of risk taking and potential failure (Figure 10.5). This support is best thought out before the child experiences these feelings (Lee and Johnston-Wilder, 2017).

Figure 10.5 Moving out of the comfort zone poster.

Figure 10.6 Keeping in the growth zone poster.

At this stage, children should be encouraged to take very small steps in learning and avoid big conceptual leaps which could send them into a state of anxiety. Encouraging a child to focus on the other, good mathematician (Figure 10.3) attributes they could demonstrate as they engage with a problem may also take away the child's perceived pressure to achieve a correct response to a mathematical question. Parents could also point out other challenges their child has taken on and succeeded with outside of mathematics and how this has made them feel.

The balance comes with supporting the child to remain in their growth zone (Figure 10.6) and develop resilience through being able to self-regulate their emotions and develop a positive mindset. With a positive mindset to learning, pupils will develop a comfort with struggle and come to know that struggle is part of learning (Boaler et al., 2021) (Figures 10.6 and 10.7).

This can be supported through many strategies including:

- keeping jottings of thought processes and different stages of problem solving;
- maintaining a positive mindset and a 'can-do' attitude;
- being aware of useful resources;
- talking strategies and ideas through with peers or supporting adults;
- taking risks without fear of mistakes;

Figure 10.7 Learning ladder poster.

- knowing what to do when stuck;
- asking questions.

This poster (Figure 10.6) also refers to the learning ladder poster (Figure 10.7) which is an analogy to understanding learning as a climb up a long ladder. Sometimes, learning seems to require a large conceptual leap but, where this is recognised, 'rungs' on the ladder can be put in which require a smaller stretch for the child. Many children are able to stretch and reach the next rung as they have long legs and strong arms but for some the reach is too much and they cannot get to the next rung. They watch as their peers climb ahead to the top of the ladder and often feel discouraged. After many attempts to reach the next rung, they give up. For these children, the missing rungs need to be replaced – the curriculum needs to be reconsidered and smaller, or replacements for missed, steps of learning planned so that each step becomes successful. Children can use this analogy to explain when they feel the stretch in understanding is too much and ask for another rung for their ladder.

In addition, to help build resilience, pupils can produce their own 'what to do if I am stuck' posters. Creating the poster themselves gives them ownership of the strategies noting what works for them as opposed to generic strategies that may not be useful to all individuals. These strategies could include:

- asking a friend to explain in their words (without giving the answer);
- looking back in your book, on the board or at a working wall in the classroom;

- using some mathematical resources;
- drawing a representation of the problem;
- rewording the problem in your own words;
- asking an adult.

The final questionnaire sent to parents was designed to capture progress, changes in thinking and attitudes and changes to how the young people were managing their anxiety and was shared after six months of working on the project.

> "… to help build resilience, pupils can produce their own 'what to do if I am stuck' posters. Creating the poster themselves gives them ownership of the strategies noting what works for them as opposed to generic strategies that may not be useful to all individuals."

Parents reported that the posters had become a useful point of reference for them. Some of the children pinned their growth zone models on notice boards or put them on display at home and are using these to begin to understand their own emotions more and how to take control of how they manage their mathematics.

Parents also shared that, when helping their children, they now focus on strategies and reasoning rather than getting a correct answer and speed of response. The parents felt more knowledgeable about how to support their children with mathematics anxiety and to become more resilient using the growth zone model. At this point, some of the key stage 1 children in the group, were able to identify when they were in the 'not ready yet' zone but could not distinguish between the 'comfort' and the growth zone. The focus here remained on supporting these children to develop strategies to get themselves out of the 'not ready yet' zone when they became anxious.

Donna also received feedback from a parent stating that they felt that working on the strategies at home would be even more helpful if they were used in their child's school as well thus highlighting the importance of partnership working between schools.

Rosemary's story

Rosemary's doctoral research was at the University of Bristol (Russell, 2002). Her expertise is parental engagement with mathematics. She is currently focusing on helping parents develop and nurture mathematical resilience in their children.

There are many important key messages that parents may benefit from knowing which would help them nurture mathematical resilience in their children. The challenge is how to communicate these to parents in a meaningful and accessible way. Inviting parents for a curriculum meeting at school is one solution, but many parents have practical difficulties and are busy and have difficulty arranging childcare (Harris and Goodall, 2007). There are also those who may themselves have negative attitudes towards school (Skyrme, Gay and Ratcheva, 2014) and may not wish to attend.

One very successful way Rosemary has found of communicating these messages and overcoming the practical difficulties is through her author talks and more recently her online author talks, based on her latest book, *Help Your Child Do Maths Even If You Don't* (Russell, 2020). In the book, Rosemary looks at key messages regarding parental engagement and developing mathematical resilience that parents could benefit from knowing. These messages are:

Parental engagement

Parental engagement is to be welcomed as a good thing in general (Desforges and Abouchaar, 2003) and there is much literature to support this. But what is it that parents do that is important? Harris and Goodall (2007) found it is not academic ability, but what the parent does with their child that has the most impact. It is engagement that really helps. Taking an interest in their child, having conversations.

However, if parents with mathematics anxiety helped their children at home with mathematics, Maloney et al. (2015) found this can have a negative impact on progress. But Goodall, Johnston-Wilder and Russell (2017) report that parental mathematics anxiety can be overcome, leading to a positive experience for parent and child.

Parental engagement and mathematics

Considering parental engagement and mathematics, it is important for parents to listen to their child and not ignore or dismiss their child's answers but instead to get into their child's thinking. One way to do this is to ask questions such as, 'How did you get this answer?' In summary, start from where they are (Hughes, 1986).

Parents need to be made aware that there is often more than one valid way to calculate an answer. Rosemary found an instance where the parent said school was teaching subtraction wrongly simply because the parent did not recognise the method of subtraction the child was using. The school was teaching subtraction by decomposition, and the parent had been taught subtraction by equal addition and was completely unaware that there is more than one valid way to calculate a subtraction (Russell, 2002). It is also important to pass on a positive attitude and not speak negatively about mathematics (Eccles and Jacobs, 1986).

Developing mathematical resilience

Sharing the Growth Zone Model, developed by Lee and Johnston-Wilder (2017), which Donna has discussed previously, with parents and children can be very helpful (Maths on Toast, 2019). It can help children express their feelings as they are learning mathematics and parents to understand what their child is feeling and so empathise and give support.

If, in the course of learning, a child finds themselves in the Red Zone, the Not Ready Yet Zone, it is important to realise what has happened and for the child to find a positive way out of the Red Zone. Donna has discussed techniques and strategies parents can use to help.

In order to develop mathematical resilience, there are four areas that need to be considered: growth mindset, valuing mathematics, understanding how to work at mathematics and learning how to recruit help. Here these areas will be looked at in more depth.

Growth Mindset

Firstly, having a Growth Mindset. This is based on the work of Dweck (2000). Ability is not fixed. It is not a case of either you can or cannot do mathematics. We can all improve. A failure should be regarded as an opportunity to learn. The adage, 'It is not that you can't do it, it's that you can't do it yet', is very helpful.

Valuing Mathematics

Secondly, Valuing Mathematics. Mathematics is part of our lives, from a toddler learning which shoe goes on the left and which on the right foot (Spatial Reasoning), to making budgets. It is an important part of society. Also, the learners' ideas are to be valued and they have an important contribution to make.

An understanding of how to work at mathematics

Thirdly, an understanding of how to work at mathematics. It is important to realise that at times, all learners may struggle to grasp concepts. They will need both perseverance and persistence. Mathematics also has structure; it is built on previous work and new ideas need to be fitted in to what is already known (Skemp, 1971). Thus, working for understanding is important at each stage. There is also no one fixed way of doing things, learners may need to experiment to make connections and reach an understanding. And in all of this, learners need to manage their emotions. The Growth Zone Model helps learners express how they feel. Being in the Red Zone is not pleasant. They may know many things, but it can all seem so far away. The learner needs to find a positive way out of the Red Zone.

Knowing how to recruit help

Fourthly, an understanding of how to recruit appropriate help when needed. This is particularly helpful if a learner finds themselves in the Red Zone. There are many strategies available, such as working in collaboration which has been discussed.

Rosemary puts all these messages over in an accessible and meaningful way for parents in her book, and her online author talks have been based on the book. It was these online author talks that led to Rosemary to be introduced to Donna and the mathematics research work group and their wider contacts. It was through this network that Rosemary gave her author talk to school leaders in Coventry, Warwickshire, and Solihull. They wanted to hear the messages she was giving and how she communicated these messages with parents.

Invitations to attend Rosemary's online author talk were then sent out to parents from these schools. Rosemary gave her online author talk and Donna co-hosted the meeting. Those attending were able to ask questions and participate using the Chat facility and also (at the end) by unmuting their microphone.

Quotes from Chat

> *"I just love the 'I can't do it yet' attitude instead of 'I can't do it'. Can apply it to my professional life too. Thank you so much for that."*
>
> *"I definitely have issues with my own maths and am mindful I don't want to pass this onto my child. The points about how to get out of the red zone make total sense and have made me now think how I can deal with problems and make it better for me and my child."*
>
> *"I think it's untraining my brain in how I was taught and learning to think about maths in a different way, a more positive way."*
>
> *"I would also be interested is this talk given to schools and teachers?"*

Many found the talk very helpful, and the event has been very successfully repeated. Other author talks are now being planned. In this way, teachers and parents were able to hear the same messages. By communicating these ideas about mathematical resilience, many parents were helped and encouraged.

Conclusion

In this chapter, two different, successful projects have been described that have enabled the ideas involved in developing mathematical resilience to be communicated to parents. Each project has offered support to parents in supporting their children to develop a positive stance towards mathematics, knowing that they can understand and 'do maths' even if they cannot do it 'yet'. The have each shown parents how to help their child understand and express their feelings as they try to progress their mathematical understanding. This includes helping children know that speed and accuracy are not the be all and end all of mathematics. If children understand that everyone has to struggle to learn something new and challenging for them in mathematics then they can stop seeing needing time to think, making mistakes and taking wrong turns as unacceptable but rather a characteristic of being in the growth zone and learning something new.

Another common feature of these two projects is the wish of both parents and children for the school to adopt these ideas so that their child's mathematical learning can be a true and fruitful partnership between school and home.

References

Andrew, A., Cattan, S., Costa-Dias, M., Farquharson, C., Kraftman, L., Krutikova, S., Phimister, A., and Sevilla, A. (2020). *Family time use and home learning during the COVID-19 lockdown*. London: Institute for Fiscal Studies.

Anderson, R., Boaler, J., and Dieckmann, J. (2018). Achieving elusive teacher change through challenging myths about learning: A blended approach. *Education Sciences*, [online] 8(3), 98. 10.3390/educsci8030098

Arem, C. A. (1993). *Conquering math anxiety*. Brooks Cole.

Boaler, J., Dieckmann, J. A., LaMar, T., Leshin, M., Selbach-Allen, M., and Pérez-Núñez, G. (2021). The transformative impact of a mathematical mindset experience taught at scale. *Frontiers in Education*, 6. 10.3389/feduc.2021.784393

Desforges, C., and Abouchaar, A. (2003). *The impact of parental involvement, parental support and family education on pupil achievement and adjustment: A literature review*. London: Department of Education and Skills.

Dweck, C. (2000). *Self-theories: Their role in motivation, personality and development*. Philadelphia, PA: Taylor and Francis.

Eccles, J., and Jacobs, J. (1986). Social forces shape math attitudes and performance. *Signs*, 11(2), 367–380, cited in *Parents' Beliefs about Math Change Their Children's Achievement*. Stanford Graduate School of Education: You Cubed Retrieved from Parents' Beliefs about Math Change Their Children's Achievement – YouCubed (Accessed 17 June 2022).

Goodall, J., Johnston-Wilder, S., and Russell, R. (2017). The emotions experienced whilst learning mathematics at home. In U. Xolocotzin Eligio (Ed.), *Understanding emotions in mathematical thinking and learning* (pp. 295–313). London: Academic Press.

Harris, A., and Goodall, J. (2007). *Engaging parents in raising achievement: Do parents know they matter?* London: Department for Children, Schools and Families.

Hughes, M. (1986). *Children and number*. Oxford: Blackwell.

Johnston-Wilder, S., Pardoe, S., Almehrz, H., Evans, B., Marsh, J., and Richards, S. (2016). Developing teaching for mathematical resilience in further education. In: 9th International Conference of Education, Research and Innovation, ICERI2016, Seville (SPAIN), 14–16 of Nov 2016. Published in: *ICERI2016 Proceedings* (pp. 3019–3028).

Johnston-Wilder, S., and Marshall, E. (2017). Overcoming affective barriers to mathematical learning in practice. In *IMA Conference (online)*. London: Institute of Mathematics and its Applications. https://cdn.ima.org.uk/wp/wp-content/uploads/2016/07/Overcoming-affective-barriers-to-mathematical-learning-in-practice-Marshall-Johnston-Wilder-paper.docx

LaMar, T., Leshin, M., and Boaler, J. (2020). The derailing impact of content standards–an equity focused district held back by narrow mathematics. *International Journal of Educational Research Open*, 1, 100015. 10.1016/j.ijedro.2020.100015

Lee, C., and Johnston-Wilder, S. (2017). The construct of mathematical resilience. In U. Xolocotzin Eligio (Ed.), *Understanding emotions in mathematical thinking and learning* (pp. 269–291). London: Academic Press.

Lee, C., and Johnston-Wilder, S. (2018). *Getting into and staying in the growth zone.* [online] Available at: https://nrich.maths.org/13491

Lucas, M., Nelson, J., and Sims, D. (2020). *Pupil engagement in remote learning.* Slough, UK: National Foundation for Educational Research Report.

Maloney, E. A., Raminez, G., Gunderson, E. A., Levine, S. C., and Beilock, S. L. (2015). Intergenerational effects of parents' math anxiety on children's math achievement and anxiety. *Psychological Science.* 10.1177/0956797615592630.

Maths on Toast (2019). How talking about maths suddenly became easier – The toast model from a parent's perspective – Maths on toast.

O'Connor, P. (2020). *What a fear of maths does to children – New research.* [online] Queen's University Belfast. Available at: https://www.qub.ac.uk/Research/feature/fear-maths-children/#:~:text=Maths%20anxiety%20is%20the%20feeling,stomach%20and%20a%20racing%20heart.

Russell, R. (2002). *Parents helping their children with mathematics.* Bristol: University of Bristol.

Russell, R. (2020). *Help your child do maths even if you don't.* Poole: AR and RR Education. http://www.amazon.co.uk/dp/1838112812?ref_=pe_3052080_397514860

Skemp, R. (1971). *The psychology of learning mathematics.* London: Penguin.

Skyrme, S., Gay, S.-J., and Ratcheva, V. (2014). "It's a massive confidence boost having your mum or your dad there": Discovering attitudes and barriers to parental engagement in mathematics with school age students. In G. Adams (Ed.), *Proceedings of the British Society for Research into Learning Mathematics 34(3)* (pp. 61–66). https://bsrlm.org.uk/wp-content/uploads/2016/02/BSRLM-IP-34-3-11.pdf

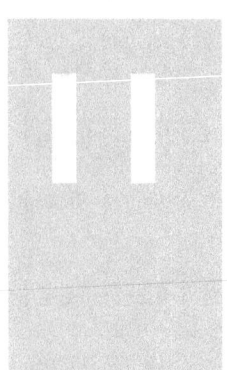

Working in Further Education with adult learners

Holly Heshmati and John Morgan

Introduction

For many, if not most, learners, mathematics anxiety does not simply disappear when they leave school. Mathematics anxiety continues to cause feelings of unease when facing mathematical tasks even after leaving formal schooling (Ashcraft, 2002). Mathematically anxious adults typically continue to feel uneasy when facing mathematical tasks such as managing finances or interpreting statistics, and this uneasiness has the potential to trigger panic and lead to avoidance (Ashcraft, 2002). This can in turn lead to problems and disadvantages at home and at work. There is therefore a need to address mathematics anxiety and promote mathematical resilience specifically in adults, using strategies and approaches which reflect the experiences and orientations of adult learners.

This chapter will initially discuss the differences between teaching adult learners compared to children and the different ways they seem to learn. It will then explore how mathematics anxiety affects adult learners in more detail, including the extent of that mathematics anxiety. It will also report on a successful targeted intervention where adult learners were supported and provided with the necessary toolkit to overcome or prevent mathematics anxiety through building mathematical resilience. Finally, both authors will discuss barriers to the development of mathematical resilience among adult learners, exploring implications for Further Education (FE) and Initial Teacher Education (ITE) providers working with adult learners.

What is the same and what is different?

Pedagogy is the act of educating children (Conner, 2008). Although teachers can deploy a range of pedagogic strategies and tools, a lot of the pedagogy in contemporary mathematics classrooms is traditional, transmission based, teacher led and passive on the part of the learner. Ramnarain (2010) indicated that, in traditional transmission pedagogy, learning is teacher directed with "much emphasis on the transmission of scientific knowledge, [and] was teacher-centred, and portrayed the learner in a passive role" (cited in Lehesvuori, Ramnarain, and Viiri, 2018, p.1050).

By contrast, the idea of andragogy, which has a long history, refers to the teaching of adults. Knowles (1998) maintains andragogy is different to teaching children. Smith (2002) considers many of Knowles' (1998) ideas remain relevant today. Smith (2002) following on from Knowles (1998) considers that adult learners have the following characteristics: they

- are self-directed, so can be involved in the planning and evaluation of what they learn;
- have life experience, so can build on what they already know, make mistakes and increase their learning resource;
- are ready to learn, as the person matures so does their willingness to learn;
- have an orientation to learning, here the learner's perspective changes from what was once one of postponement to one of more immediacy of application;
- are motivated to learn: as the person matures the motivation to learn increases.

Having taught younger learners as well as adults, we both see that the ideas developed by Knowles and Smith relate clearly to our experiences in adult education. In John's experience in FE, returning adult learners need to have all five characteristics to learn effectively and be able to take the opportunities to learn that are presented to them. In Holly's experience in ITE the adult learners she teaches are motivated to learn in order to take up the career to which they aspire.

Learning mathematics in an FE college

Over the years John has taught in an FE college; he has seen that many, if not most, adults have difficulties with the mathematics that is usually a mandated part of their course of study. As will be shown, their difficulties seem to be a result of the extent of the mathematics anxiety which they have developed before returning to study. Mathematics anxiety is known to cause avoidance (Ashcraft and Krause, 2007) and present a barrier to participation in mathematical problem solving (Beilock and Maloney, 2015). Relatives, friends, employees or teachers report hearing statements such as 'I hate maths', 'I will never understand maths', 'Where am I likely to use this stuff anyway?' from adult learners when they are told that mathematics will be part of their course. As their teacher John has heard many learners using emotive terms to describe how they feel about mathematics, often with great negativity. He found that many learners have negative attitudes towards mathematics, which in some people may be described as *"a general fear of contact with mathematics"* (Hembree, 1990, p.45).

This 'difficulty with mathematics' led to John's PhD research in this area. He saw in many of his mature learners that fear can inhibit effective learning and block future pathways for the learners (Ashcraft and Krause, 2007). It became apparent that this fear was present in the learners when they were in school and that it has 'travelled with them' into adulthood. For new adult learners, all is not negative. John's adult learners brought with them a variety of life experiences and enough general resilience to allow them to start studying again having left school behind many years ago. However, most of John's adult learners also brought the negative experience of "hating maths" into his classroom.

Adult learners are likely to have more discipline and motivation for learning when they arrive at FE college because their life experience has told them that learning and qualifications are important. Adults bring a variety of work and social experiences, which can contribute to effective in-class discussions and be exploited as learning opportunities. Younger learners may need more help to becoming engaged with mathematics; they may not see the point and may need more experience in understanding the value of mathematics. As a teacher, it is likely that more time will be spent on active learning with adults and less time on discipline, classroom management and addressing pupil behaviour. Learners usually attended the courses that John taught because they did not achieve the qualifications in school which they now need for the career they want to pursue. John's learners along with the great majority of the population still find engaging with mathematical tasks difficult, to the point that they exhibit phobia or anxiety, or at least avoidance from engaging in any endeavour that could require mathematical reasoning (Johnston-Wilder and Lee, 2010). Thus, many of his learners over the years have been unable to progress in their chosen field as they could not face mathematics.

The extent of mathematics anxiety in John's adult learners will be considered along with how that anxiety may be addressed. He will also talk about how mathematical resilience can be developed to allow adult learners to engage and succeed with mathematics.

John's research

As part of a PhD research project between 2019 and 2022, I surveyed and interviewed many learners who have mathematics anxiety. My target group was adult or mature "Access to Higher Education (HE)" Health learners, who were returning to education. As part of the research, I measured mathematical anxiety and its debilitating consequences and instigated a programme of interventions to help adult learners more easily engage with mathematical learning.

The extent of mathematics anxiety in adults

Much of the literature in the field of mathematics anxiety indicates that 'unpleasant outcomes' are a common theme. Learners often report feeling sick when faced with mathematics problems and various other associated feelings such as fear and even dread (see, for example, Hembree,1990; Ashcraft, 2002; Lewis 1970). If learners perceive a threat to their well-being brought on by their mathematics anxiety, their progress will be impeded (Morgan, 2022). Even within traditionally technical vocations, mathematics anxiety may still be present. Dowker et al. (2016) cites Johnston-Wilder et al. (2013) who found that "about 30% of a group of apprentices showed high mathematics anxiety, with a further 18% affected to a lesser degree" (Dowker et al., 2016, p3). I measured anxiety levels in 561 engineering learners and found that 48% reported feeling anxiety in mathematics, supporting the idea that mathematics anxiety is widespread even amongst learners in more technically based courses.

Why is mathematics anxiety in adults important?

Much published research on mathematics anxiety tends to involve school-age learners. However, Dowker et al. (2016) found that negative attitudes towards mathematics increase as children reach secondary school age and that this negativity continues to increase into post-secondary education and then through to adulthood. Research by Carey et al. (2019) confirms that the UK could be going through a 'mathematics crisis':

> functional literacy skills amongst working-age adults are steadily increasing but the proportion of adults with functional mathematics skills equivalent to a GCSE grade C has dropped from 26% in 2003 to only 22% in 2011. (Carey et al., 2019, p.2).

If more people are to progress to the professions where functional numeracy is required, then the findings by Carey et al. (2019) indicate that the low level of mathematics skills in working age adults must be addressed. This is likely to involve, among other aspects, addressing mathematics anxiety in adult learners going back into education.

Data collection

The first step in my research was to discover whether (or not) mathematics anxiety was present in my target groups. I used the Mathematics Anxiety Scale, MAS; devised by Betz (1978) which had already been used successfully in the UK to measure mathematics anxiety in apprentices (Johnston-Wilder et al., 2013). It was known to have test/retest reliability, making it suitable for duplication across all groups (Pajares and Urdan, 1996). My pilot cycle for data collection commenced in 2019–2020, where 77 learners participated in the MAS surveys. I found that a total of 81% of all learners (Figure 11.1) experienced some form of anxiety towards learning mathematics. These learners were either 'visibly anxious' or 'anxious but not visibly'.

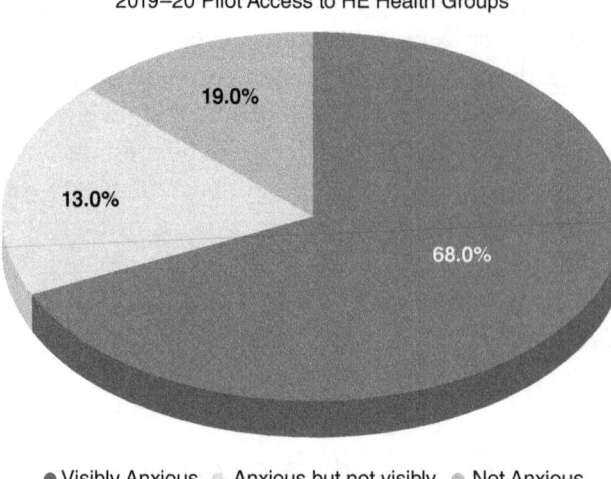

Figure 11.1 MAS scores for October 2019.

I also asked learners for one word that described their feelings towards mathematics. Only 13% of learners (10) used terms like 'ok', 'fine', 'like' and 'comfortable'. Most learners, 74% (57) were using emotive terms such as: 'terrified', 'dreadful', 'scared', 'cripplingly anxious', 'hate' and 'dislike'. Cates and Rhymer (2003) confirm that mathematics anxiety would cause these emotive reactions to the idea of studying mathematics and is also likely to be the cause of the learners' previous underachievement at mathematics.

An intervention to support the learners

Having established that many learners were mathematically anxious, an intervention was developed to help learners understand their feelings and to introduce strategies to progress their mathematics learning. Uusimaki and Kidman (2004) stated that, when trying to reduce mathematics anxiety, learners must know that anxiety exists and that there are ways to overcome its debilitating effects. Dweck (2007) also asserted that if learners realise that mathematics anxiety is 'normal' and widely accepted as a debilitating state, then they will be that much more able to overcome their negative feelings in pursuit of mathematics proficiency. In response, I developed 'The Maths Anxiety Presentation', showing that mathematics anxiety exists, is real and can be minimised or overcome.

Intervention : the maths anxiety presentation

My intervention was based on research by Johnston-Wilder and Lee (2017) where they developed five characteristics of a learning environment that is known to build mathematical resilience; these are:

1. helping students know that brain capacity can be grown;

2. enabling everyone to feel included and supported;

3. helping students to see mathematics as relevant in the world in which they live;

4. asking students to struggle, but not too much;

5. modelling ways to work at mathematics, showing how to get support.

The presentation consisted of several slides where I discussed briefly:

- Mathematics anxiety exists, is commonplace and can be reduced.
- You are not alone with mathematics anxiety.
- Mathematics is relevant. Many jobs require mathematics proficiency.
- Mathematical resilience can be improved.
- We can all develop a growth mindset assisted with good teaching and support.
- Growing mathematics ability requires struggle and perseverance.
- It is fine to learn collaboratively.
- Challenging tasks grow our minds and makes us smarter.

Morgan (2022)

The before and after results of the presentation

A before and after survey was used to measure the effects of the presentation, helping learners to recognise and overcome negative feelings towards mathematics. The results are shown in the bar chart in Figure 11.2. It shows there was a high level of anxiety in the room before the presentation, but that changed considerably following the presentation.

The learners' reactions

I took the opportunity to ask the learners how they felt about the intervention. They made several statements which included:

> "I didn't realise that there were so many people in the world who felt the same as me about mathematics".
> "I thought it was just me".
> "Now that I know I'm not on my own, I feel that I will be able to give mathematics a much better try than I did in school".
> "When I attended school, I felt isolated and alone, now I feel much better placed to do this and win".

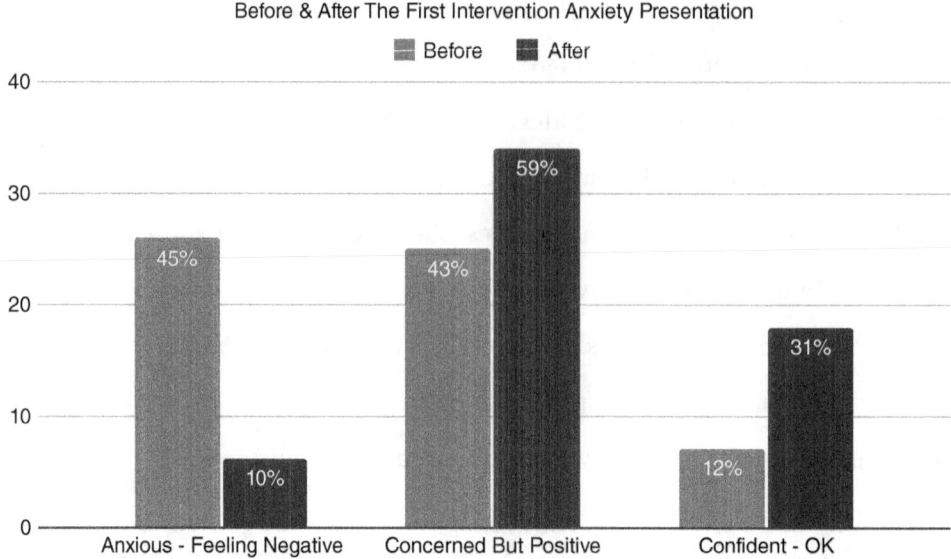

Figure 11.2 Learners' feelings before and after the first intervention.

"I felt brushed aside when I was in school, I couldn't do mathematics, my teacher knew it and he just ignored me and concentrated on the students he thought could do mathematics".

(Morgan, 2022)

When I reviewed the survey data along with the informal 'after class discussion', I felt that this certainly looked very positive for the success of the intervention and that this intervention was well worth using again.

In most cases, the learners were very positive about my research and the various ideas such as the Growth Zone Model in Chapter 3. Throughout my research, I tried to support them in studying mathematics despite their anxiety. They all willingly participated. Whilst some continued to assert that they do not particularly like mathematics, they also indicate that it does not cause them as much stress now and their anxiety levels are considerably reduced. In fact, my research has consistently shown, over the three-year period, that with appropriate interventions, learners can feel a reduction in anxiety and feel able to progress towards their desired qualifications and career goals.

Barriers and affordances for adult learners

The mindset barrier

The phrase that "either you're good at it or not" is used by many adults in our experience when referring to mathematics. We see this phrase as a reflection of the prevalent fixed mindset beliefs in society (Dweck, 2000). This is the widespread belief that individuals are either born with the capacity for mathematical thinking or they are not. This belief means that making any effort in mathematics is not worthwhile as the individual will not be able to 'do' the mathematics and may end up looking stupid. Therefore, positioning oneself as being mathematically less capable than others, and avoiding opportunities for meaningful engagement with mathematics (Dowker et al., 2016), may well be a reasoned choice for an adult learner, who has this belief. These fixed mindset beliefs will also shape their self-efficacy, their deeply held beliefs about the possibility of success when engaging in mathematics, which Bandura (1997) suggests will in turn affect their agency, which is their feeling of being in control of their environment and actions. Given their mathematical outcomes are likely to confirm that thinking, this can lead to 'anti-resilient' thinking in the adult learner.

Dweck (2007) has proven that where learners are open to struggle and accept that there are problems that they do not currently know how to or cannot yet solve, then it is possible to start to combat the limitations that a fixed mindset place on them. However, in the case of mathematics, Schoenfeld (2022) has shown that adult learners' school experience is the major factor that shapes individuals' beliefs about the likelihood of their success in subjects with a high mathematical content in their futures. The adult learner with lower previous school attainment in mathematics is likely to believe that these results confirm that they are less capable in mathematics, and as a result, they stop working to remedy the situation. If they find they cannot figure out a solution to a mathematical problem in a short period of time, they will usually give up. A fixed mindset prevents the possibility of perseverance or trying a different way. If you are sure you 'can't do maths', why would you spend your time trying? In addition, the adult learner who believes that school mathematics did not prepare them for solving any real-life problems may fail to notice how mathematics applies to real-life contexts, as they will not consider it worth looking. An adult learner who does not expect mathematics to make sense, will see effort and struggle as futile and may not engage with developing problem-solving or mathematical thinking skills which, in turn, can lead to lack of experiences that can help them develop mathematical resilience.

The issue of adults having a fixed mindset about learning mathematics as a result of previous educational experiences is one of the main barriers (Dowker et al., 2016) to feeling able to learn, use and understand mathematics as an adult learner. Helping learners to develop a positive attitude and a growth mindset could be considered a first step in an effective andragogy for educators working with adult learners. Working as John has described earlier, to challenge widespread theories of fixed intelligence and show the possibility of growing mathematical capability with support, will help many learners begin again with learning mathematics. Developing mathematical resilience will take time as it requires struggle and perseverance (Lee and Johnston-Wilder, 2017) which, as has been discussed, many learners associate with an inability to understand mathematics. Helping adult learners safely experience the link between experiences of struggle and succeeding in mathematics is an important aspect of a mathematics education relevant to adult learners.

The need for support

Embedding many of the approaches known to develop mathematical resilience (Lee and Johnston-Wilder, 2013) in younger learners will also help adult learners although they may need some modification. Creating opportunities for learners to explore an idea with the support of others and struggle together towards an answer will help them understand the vital role of struggle in succeeding in learning mathematics. As has been seen, adult learners are apt, due to fixed mindset beliefs and previous experiences when in school, to interpret struggle as an indication of weakness or incapability. Supporters of adults learning should instead, help them to attribute the causes of struggle to the context or gaps in their understanding that can be addressed. For example, difficulties around learning mathematics are more likely to be due to a misunderstanding of the abstract nature of the discipline, or a misconception about some underlying concepts rather than to any limitations in their capability to learn mathematics. It will help if the supporter of their learning explains to the adult learner that it will be a struggle to discover what is preventing understanding, but if the adult learner is prepared to put in the effort and accept support from their peers and teachers, understanding is possible. The realisation that struggle is a vital process in learning mathematics can help adult learners overcome the feelings of shame and guilt for 'not being good at mathematics' that they have experienced previously and to seek help and recruit support in their 'brave' struggle to understand. Reluctance to seek help and recruit support is often the result of the social and emotional challenges many adult learners have faced previously, where it may have been interpreted by others as a sign of personal and professional weakness. An important step is to acknowledge that all mathematicians struggle, even those that study mathematics to the highest level. In fact, if the learners are not struggling then they are likely not to be learning and should be seeking greater challenges. Being able to seek appropriate help and recruit the right support for each individual when learning mathematics is an essential attribute of mathematical resilience.

Drury (2018) and Schoenfeld (2022) regard teaching and learning mathematics as particularly challenging compared to other subjects. From this perspective, learning mathematics requires both the resilience and the investment of time from the learners. As a result, time constraints and competing priorities can be other barriers to developing mathematical learning for adult learners. Many adult learners have busy lives, with work, family and other responsibilities, which can make it difficult to find time to engage with opportunities for meaningful learning in mathematics. Mathematical qualifications often act as gatekeepers, that is not having the right mathematics grade can prevent many adult learners accessing the training they need for the career to which they aspire. The time and struggle they must put in to make up for not having been able to learn mathematics when at school can create a sense of being overwhelmed and avoidance behaviour may be the result. Support and encouragement from teachers, peers and family members, therefore, can make it easier for adult learners to persevere in the face of challenges. These learners might first be encouraged to demonstrate their growing resilience by asking others, such as family members to support them by, for example, temporarily taking on household tasks to give them the time they need to study.

Making learning more authentic

Mathematics problems do not have to be approached in one right way even if there is one right answer. The idea that different methods can be used, and some prefer this way but others that way, can be a revelation to adult learners. Listening to learners' unique mathematical approaches is very important in the process of teaching in ways that develop their mathematical resilience. This is often ignored when teaching of mathematics especially when the focus is mainly on the teacher's actions rather than what learners are learning. Working with adult learners involves teachers using patience and their own resilience, actively listening to the learners and discussing mathematical ideas with them. Adult learners may well find this way of working is very different to those used when they were younger but is necessary to develop resilient adult learners.

While the teaching of mathematics in schools may have been mainly focused on developing the fluency skills of learners so that they can pass examinations, efforts exerted by the teacher to help learners to make connections between mathematical ideas and concepts would have been more helpful to adult learners. They may have tried and failed to 'remember how to do it'. A challenge in mathematics learning for adult learners can be to help them develop problem-solving and self-directed learning skills, underpinned by principles of andragogy (Knowles, 1984; Kearsley, 2010), while addressing the common misconception that the subject solely revolves around memorising procedures. In school, mathematics curricula tend to be heavily centred on systematised teaching, practising methods and memorising rules and formulas, the value of such ideas in the real world is not always evident to learners. However, adult learners can benefit from thinking about the mathematics they will come across in real life and employment and are more likely to see the value in their learning. Thus, adult learners can be helped to learn effectively if the supporters of their learning seek out mathematics based on the personal and professional interests of their learners. They can use approaches such as problem-based learning and opportunities for authentic problem-solving based firmly in their learners' experiences and interests, which can make for more meaningful learning.

Autonomy and agency

Another challenge for those who support adults learning mathematics is to create learning environments that support every learner in developing not only the knowledge, skills and experiences that underpin effective mathematical learning, but also help them develop a sense of autonomy and agency in their mathematical work. Developing autonomy and agency will mean they can go beyond merely reproducing the processes and procedures modelled by teachers, to engaging in sense making (Schoenfeld, 2022). This means using alternative productive approaches when teaching mathematics to adult learners, which allow them to develop another view of mathematics. Such alternatives are generally based on socio-constructivist (Vygotsky, 1978) approaches to learning, stressing the role of collaborative learning and problem solving in teaching mathematics (Swan, 2006). Such problems and ways of working can be used to motivate learners or to enable them to make connections and apply concepts through discussions and sensemaking. The social dimension of such learning, through interactions among adult learners and between teachers and learners is heavily emphasised in mathematics andragogy (Smith, 2002). These ways of working also encourage learners to reconstruct their attitude and beliefs towards mathematical learning by making their ideas explicit and explaining their reasoning (Swan, 2006), providing opportunities for them to build their mathematical resilience. Unlike children, adult learners can recognise why they are learning, so their motivation to learn can be high compared to younger learners (Smith, 2002). Working with highly motivated learners, teachers may adapt their practices to allow socio-constructivist discourse to predominate. They may, for example, set authentic and meaningful problems for learners to engage in collaborative and discovery-based practices following the principles of andragogy.

However, sharing the mathematics learning responsibilities between teachers and adult learners implied by this andragogy is far from easy. It requires the teachers to find suitable tasks which are meaningful for the learners and to give sufficient guidance but not too much. It requires teachers who are capable of coping with the unexpected issues and ideas which may arise and of identifying, and dealing with, any mathematical ideas that have yet been understood by individuals in the group. Successful implementation of this way of working, requires teaching expertise much greater than that required in traditional teaching practices.

Adult learning and teacher training

In ITE programmes, university courses try their best to affect teachers' beliefs about effective pedagogy in mathematics (Boz, 2008), but new teachers are too often presented with models of practices at university and FE colleges that are different from the accepted principles of androgyny. An example of this has been noted by student teachers completing their ITE programmes at FE colleges. They state that there is a lack of attention to wider issues and barriers to developing their learners' mathematical resilience because of a focus on developing mastery of the subject. Perhaps, while suitable professional training at ITE can assist new teachers in understanding the knowledge, skills and experience of mathematical andragogy, for adult learners to develop the skills required for mathematical resilience, within the constraints of the classroom, changes to traditional practices within FE colleges must be made. Professional training will be required by existing teachers in the colleges which must be supported by appropriate resources. The training should include discussion around approaches which can lead to meaningful and productive mathematical learning for adult learners.

The socio-constructivist model may inspire educational activities suitable for adult learners, however, this model can take very different forms, depending on the social and cultural context. Moreover, it is not the only possible model. While successful implementation of such approaches requires teachers with good subject knowledge and an understanding of andragogy approaches to teaching mathematics, the high percentage of non-specialist teachers of mathematics (Allen and Sims, 2018) in FE colleges will be a limiting factor. One of the particular issues is that non-specialist mathematics teachers are very likely to suffer from mathematics anxiety themselves. This can make the process of implementation of meaningful and authentic ways of learning mathematics challenging and out of reach in the short term. A direct link between teacher quality and their ability to enable their learners to develop mathematical resilience can be established. Thus, addressing issues with teachers and teaching quality continues to limit effective teaching of adult learners. Perhaps, one way to address the issue could be collaborations between schools, FE colleges and ITE providers to discuss barriers to building mathematical resilience for both teachers and adult learners which could identify the learning needs of both teachers and adult learners. Such collaborators could discuss and explore ways of working which can help build mathematical learning and mathematical resilience within the context of adult education.

Conclusion

Mathematics-anxious individuals end up with lower mathematics proficiency and achievement (Ashcraft, 2002; Ashcraft and Krause, 2007). As a mathematics qualification is vital to taking up many career choices, suffering from mathematics anxiety, and the resulting lower achievement, in school will mean that adults have to return to learn mathematics in FE colleges if they are to take up a career to which they aspire. This chapter has shown the extent of mathematics anxiety among adult learners and discussed the kinds of ways of working with adults, andragogy, that will build their mathematical resilience and allow them to overcome the barriers they face when learning mathematics.

A nurturing environment and the possibility to learn at a rate suitable for each of the adult learners will help many to succeed with mathematics. Helping learners understand that everyone has to struggle when they are learning mathematics can enable them to understand that having to struggle does not mean they 'can't do it'; it means they 'can't do it YET'. Helping adult learners to understand the process of learning mathematics is rather steep and not smooth, but that they have a right to all the support they need to clamber their way to understanding, will promote mathematical learning with resilience.

This chapter highlights only some of the more prominent issues and solutions surrounding mathematics anxiety and mathematical resilience in teaching adult learners; there is still more to do.

References

Allen, R., and Sims, S. (2018). How do shortages of maths teachers affect the within-school allocation of maths teachers to pupils? [online] Available at: <www.nuffieldfoundation.org>.

Ashcraft, M. (2002). Math anxiety: Personal, educational and cognitive consequences. *Current Directions in Psychological Science*, 11, 181–185. 10.1111/1467-8721.00196

Ashcraft, M., and Krause, J. (2007). Working memory, math performance, and math anxiety. *Psychon Bull Rev.*, 14(2), 243–248. 10.3758/bf03194059

Bandura, A. (1997). *Self-efficacy: The exercise of control.* New York: Freeman.

Beilock, S., and Maloney, E. (2015). Math anxiety: A factor in math achievement not to be ignored. *Policy Insights from the Behavioral and Brain Sciences*, 2(1), 4–12.

Betz, N. (1978). Prevalence, distribution, and correlates of math anxiety in college students. *Journal of Counselling and Psychology*, 25, 441–548. 10.1037/0022-0167.25.5.441

Boz, N., (2008). Turkish pre-service mathematics teachers' beliefs about mathematics teaching. *Australian Journal of Teacher Education*, 33(5), 66–80.

Carey, E., Devine, A., Hill, F., Dowker, A., McLellan, R., and Szucs, D. (2019). *Understanding mathematics anxiety: Investigating the experiences of UK primary and secondary school students.* University of Cambridge.

Cates, G., and Rhymer, K. (2003). Examining the relationship between mathematics anxiety and mathematics performance: An instructional hierarchy perspective. *Journal of Behavioural Education*, 12(1), 23–34.

Conner, M. (2008). Andragogy and pedagogy. *Ageless Learner, 1997–2004.* http://agelesslearner.com/intros/andragogy.html, Accessed 09/09/08.

Dowker, A., Sarkar, A., and Looi, C. (2016). Mathematics anxiety: What have we learned in 60 years? *Frontiers in Psychology*, 7, 508.

Drury, H. (2018). *Teaching mathematics for mastery.* Oxford: Oxford University Press.

Dweck, C. (2000). *Self-theories: Their role in motivation, personality, and development.* Philadelphia PA, USA: Psychology Press.

Dweck, C. (2007). Child development. *The Society for Research in Child Development*, 78(1), 246–263.

Hembree, R. (1990). The nature, effects, and relief of mathematics anxiety. *Journal for Research in Mathematics Education*, 21(1), 33–34.

Johnston-Wilder, S., and Lee, C. (2010). Developing mathematical resilience. In *BERA Annual Conference 2010*, 1–4 Sep 2010, University of Warwick.

Johnston-Wilder, S., and Lee, C. (2017). Addressing the affective domain to increase effective-ness of mathematical thinking and problem solving. In: *IMA and CETL-MSOR 2017: Mathematics Education beyond 16: Pathways and Transitions,10–12 Jul 2017*, University of Birmingham.

Johnston-Wilder, S., Brindley, J., and Dent, P. (2013). *A survey of mathematics anxiety and mathematical resilience among existing apprentices.* London, UK: The Gatsby Charitable Foundation. Grant Reference: GAT3358/DSS.

Kearsley, G. (2010). Andragogy in action. The theory Into practice database. https://www.instructionaldesign.org/theories/

Knowles, M. (1984). *Andragogy in action. Applying modern principles of adult education.* San Francisco: Jossey Bass.

Knowles, M. (1998). *The making of an adult educator.* San Francisco: Jossey-Bass.

Lee, C., and Johnston-Wilder, S. (2013). Learning mathematics – Letting the pupils have their say. *Educational Studies in Mathematics*, 83(2), 163–180.

Lee, C., and Johnston-Wilder, S. (2017). The construct of mathematical resilience. In U. Xolocotzin Eligio (Ed.), *Understanding emotions in mathematical thinking and learning* (pp. 269–291). San Diego CA, USA: Elsevier.

Lehesvuori, S., Ramnarain, U., and Viiri, J. (2018). Challenging transmission modes of teaching in science classrooms: Enhancing learner-centredness through dialogicity. *Research in Science Education*, 48, 1049–1069. https://doi-org.libezproxy.open.ac.uk/10.1007/s11165-016-9598-7

Lewis, A. (1970). The ambiguous word "anxiety". *International Journal of Psychiatry*, 9, 62–79.

Morgan, J. (2022). Recognising and coping with mathematics anxiety in adult learners. *Proceedings of the 3rd international conference of Developing Mathematical Resilience.* Open University.

Pajares, F., and Urdan, T. (1996). An exploratory factor analysis of the mathematics anxiety scale. *Measurement and Evaluation in Counselling and Development*, 29, 35–47.

Ramnarain, U. (2010). Grade 9 science teachers' and learners' appreciation of the benefits of autonomous science investigations. *Education as Change*, 14(2), 187–200.

Schoenfeld, A. (2022). Why are learning and teaching mathematics so difficult? In M. Danesi (Ed.), *Handbook of cognitive mathematics.* Cham: Springer. 10.1007/978-3-031-03945-4_10

Smith, M. (2002). Malcolm Knowles, informal adult education, self-direction and andragogy. *The encyclopedia of pedagogy and informal education.* [https://infed.org/mobi/malcolm-knowles-informal-adult-education-self-direction-and-andragogy/. Retrieved: 10/02/2023].

Swan, M. (2006). *Collaborative learning in mathematics: A challenge to our beliefs and practices.* Leicester UK: National Institute of Adult Continuing Education.

Uusimaki, L., and Kidman, G. (2004). Reducing maths-anxiety: Results from an online anxiety survey. *Presented at the AARE Annual Conference*, Melbourne, Australia: Australian Association for Research in Education.

Vygotsky, L. S. (1978). *Mind in society: The development of higher psychological processes.* M. Cole, V. John-Steiner, S. Scribner, & E. Souberman (Eds.), Cambridge: Harvard University Press.

Mathematical resilience for lifelong learning

Clare Lee

Introduction

Mathematical resilience matters, not because it enables people to pass examinations, although it does, but because if a learner develops mathematical resilience, they know they can, with appropriate support, learn and use whatever mathematics they need throughout their lives personally and in their career. There is an imperative to continue to learn and use new mathematical ideas beyond formal education, but the support used by each learner must change. This chapter is about helping learners develop approaches to support their learning that will continue to be useful throughout their lives.

Here I also discuss how learning might be supported and continued mathematical growth facilitated when the learner is no longer sat in front of a teacher for hours every week. Where learners leave formal education understanding that learning mathematics is likely to involve some struggle, but that appropriate support is available which they can and should access to lead to a successful outcome, they may be said to have developed a mathematically resilient identity. With such an identity, they are more likely to continue to learn mathematics. With such an identity, they will know that they can learn, understand and use mathematical ideas, and that with perseverance from them, and support from appropriate others, they will be able to understand and use any mathematics they need throughout their lives. The ways they might continue to proactively seek the support that everyone needs when they attempt to learn mathematics will be discussed later in this chapter.

To begin with I will discuss what is meant by lifelong learning and how having a mathematically resilient identity helps learners become lifelong learners. Then I will go onto discuss how an identity can be built whilst in school that will help someone cope with the mathematical learning they need after school. In the future they will need to use mathematics in ways we know about and in ways that, as yet, we do not know about, hence, a resilient identity will be important.

Lifelong learning

Lifelong mathematical learning is often thought of as learning the mathematics that is needed to be successful in a chosen career or workplace. However, mathematics has a wider role to play in everyone's life, although currently elitism (Nardi and Steward, 2003) and mathematics anxiety (Ashcraft and Moore, 2009) mean that role is often curtailed. If someone has developed mathematical resilience, they can use mathematics in their personal life, perhaps just for pleasure, and they can be more democratically active through increased understanding of what might really be behind any bluster, which may help increase social justice.

The European Commission argues that *"lifelong learning is not only about employment and adaptability ... [but] is also a means to personal fulfilment, active citizenship and social inclusion"* (Van der Pas, 2001, p. 16). Biesta (2006) defined lifelong learning as a triangle of dimensions; these dimensions are economic, personal, and democratic and social justice. OECD Education Ministers *"are all convinced of the crucial importance of learning throughout life for enriching personal lives, fostering economic growth and maintaining social cohesion"* (OECD, 1997, p. 21) and that *"[f]uture economic prosperity, social and political cohesion, and the achievement of genuinely democratic societies with full participation – all depend on a well-educated population"* (ibid., p. 24). Philosophies of education vary worldwide, and therefore the meaning of lifelong learning is situated and is often politicised.

Most people leaving formal education are looking forward, often with trepidation, to earning a living in a position where they will need to learn some new mathematics and many people return to learning mathematics as part of working towards qualifications for better paid employment, so perhaps the economic aspects of lifelong learning take precedence.

Dimension I – economic

Economic lifelong learning can be thought of as acquiring the new skills and knowledge that are required to feel skilled and confident in a particular position in the world of work. Such learning is important for a person's employability, for their financial well-being and the well-being of the economy as a whole. Trying to envisage what mathematical skills and understandings will be needed in ten years is impossible, thinking even two years into the future feels difficult because of the rate of change that is currently a feature of society. For example, the current rise and changes in the uses of data science and machine learning mean that the mathematics that is part of most jobs has changed and will continue to change at a rapid rate. The willingness, capability and flexibility to learn new mathematics will be important to a person's economic viability. Another example would be how exams change over time; a teenager who passes their GCSE now will need to cope with many changes in content and analysis by the time they seek to help their children succeed in their examinations. We do know that where people are able to learn new mathematical ideas and put those ideas to use, that is they have mathematical resilience, they are more likely to prosper. Having more people with sufficient mathematical resilience to continue to learn and grow mathematically will mean society will, for example, be better able to deal with current sustainability issues and continue to thrive.

Learning beyond the classroom

Those learners that have become mathematically resilient learners in the classroom will be able to go on to make good choices about learning once they leave formal education. Having been given an understanding of how to learn mathematics when in formal education, they will know that they can continue to grow their understanding of new mathematical ideas if they wish to do so. Importantly they will, as adults, understand the value of the mathematical ideas they are seeking to learn. They are likely to continue to use their understanding that learning mathematics requires struggle and perseverance, to seek ways to gain appropriate support and to collaborate with others who are also learning mathematics.

Those that have used discussion to learn mathematics in school will realise that they are likely not to be alone in needing to discuss mathematical ideas in order to learn effectively. They may seek to collaborate with others on their course or in their workplace offering support and gaining support at the same time. Adults at work often have occasion to use mathematics and they may well find that they experience success and display a proficiency that those who have not developed mathematical resilience are less well placed to do. They may find the tables turned and that they are the ones who are expert. If they have engaged in collaborative discussions as part of learning mathematics, they will know to respect everyone's contribution as something to learn from. They will also understand that sometimes they know more than others, but at other times their peers may have more to offer. In this way, they will be better placed to develop good resilient working and learning relationships. An understanding of what it means to be mathematically resilient, and that others may not have had that opportunity and may well display mathematics anxiety, will also help learners handle finding themselves the expert and be able to model and pass on resilience.

Depending on their context, learners beyond formal education may need to 'go online' to establish a supportive community for learning more mathematics. Knowing that needing support is an integral part of growing mathematical knowledge, and not a weakness, will make it more possible for them to set up collaborative groups with others who are seeking to progress their mathematical learning. There are many on-line support sites that can offer help with mathematical learning. Again, learners must look to their mathematical resilience, as they are seeking *appropriate* support. This means they should not settle on the first site they encounter but consider if the way the site is configured really works for them. Many sites will offer support in the form of short cuts and quick tricks or procedures to memorise, but a mathematically resilient learner will want to fully understand so will look for a site that will help them do that. A short session in a few lessons at school where the learners search for mathematics sites and assess if they work for them, will emphasise that everyone needs support sometimes and what is appropriate for one learner is not always appropriate for another. Such sessions will also help them after leaving school to continue to be discerning and look for appropriate support. Learners who have depended solely on a teacher to support their understanding can be lost and helpless after school (Ward-Penny et al., 2011).

Accessing adult learning courses will often be the first step taken by those who need to extend their mathematical learning after school. Many of these courses will be facilitated by teachers who are empathetic and seeking to enable their learners to gain the qualifications they need. Where the teachers of adult courses continue to allow their students to build their mathematical resilience, those learners who already have a mathematically resilient identity will thrive. Where the difficulties that colleges encounter in recruiting the right staff to meet demand result in learners encountering a less-than-ideal teacher, mathematically resilient learners will seek to make the best of what is on offer and support others to do the same. They are more likely to be the ones who know how to seek ways to succeed and get the support they need. Making sure that learners know they are well equipped to continue to learn mathematics later, even if they choose to leave mathematical learning at the end of formal schooling, seems a valuable gift to give them.

Dimension 2 – personal

Mathematical resilience also helps when someone engages in the kinds of lifelong learning that have to do with personal development and fulfilment. Someone with mathematical resilience is much better situated to learn from the encounters and experiences with mathematics that make up their life, than someone who experiences mathematical anxiety and cannot manage those emotions. Mathematics has a role to play in enabling people to reason things out and maybe find pleasure in solving logic puzzles or succeeding in many on-line games. Being able to reason mathematically helps in making decisions about what money must be used for food and housing and whether some can be saved for other pursuits. Those who suffer from mathematics anxiety are likely to avoid thinking about personal finances because of the negative feelings it invokes.

There is a sense of personal satisfaction to be gained from achieving and succeeding in learning as an adult in order to take on a role that may have seemed out of reach (Daker et al., 2023) but which with the resilience to continue to learn mathematics becomes achievable. Mathematical resilience enables people to assess risk and make good choices because they can engage in the learning needed in order to use whatever mathematics or statistics they need to understand situations. Thus, mathematical resilience can help to enable the personal dimension of lifelong learning.

A mathematically resilient identity

The term identity is often discussed as if the meaning of the term can be taken for granted, just as most people understand what is meant by the term 'table'. However, looking more deeply at what is meant by identity can lead to the idea that the term is nebulous and difficult to define. In this section, I will unpick what a mathematically resilient identity means before moving onto how such an identity might be developed at school and maintained so that mathematics can be learned throughout someone's life.

The dictionary definition of the term identity is 'the distinguishing character or personality of an individual', but authors such as Buckingham (2008) consider identity to be a slippery and ambiguous term that is overused in contemporary society. Buckingham (2008) states that individuals strive to find and establish their identity as they grow towards adulthood. All too often, adults establish an identity that is in opposition to one which is mathematically resilient. Too many people see themselves as people who do not engage with mathematical ideas or reasoning and that identity or stance is seen as acceptable within much of wider society (https://www.nationalnumeracy.org.uk/).

Lave and Wenger (1991) consider that people develop identities within a community of practice or a community of people with a common aim or purpose, which could be a school but could be, for example, a family or a gospel choir. The community of practice shapes and develops the way that someone thinks of themselves and what they can and cannot do as they move from peripheral to full participation in the practice of the community. That an identity can be built through interaction in a certain situation seems useful in thinking about how learners can be enabled to consider themselves mathematically resilient. Acknowledging that someone has the right, for example, to be supported in understanding mathematics can change the way they see themselves and help them build an identity as someone who will be able to understand mathematics given time and appropriate support. When someone has built over time an identity as someone who cannot 'do' mathematics, they will need considerable time and support to realign that identity and take on a mathematically resilient identity that enables lifelong learning. Coaching support as described in Chapter 5 of this book is one of the most promising ways to help someone change the way they think and take on that mathematically resilient identity that allows for lifelong learning.

It is in thinking about changing identities where the term can become "slippery and ambiguous". Identity is seen as a socio-cultural construct built through engagement of some sort in community and seen this way, identity is relatively stable. For example, Lave and Wenger (1991) investigated the learning that was needed to move from peripheral participation in a community of tailors to taking on the more stable identity of a master tailor. Once the learner had attained the status or identity as a master tailor, they continue to learn about new technology or fabrics and continue to see themselves as a master tailor. However, we also know identity is open to change, for example, at one time I could not knit but now I am a constant knitter. People want to feel they have 'found themselves' or they have begun, as my old school motto has it, to "know thyself", and that they have an established identity (Buckingham, 2008). However, given the right conditions they can learn and take on a new identity. Most classrooms are places where identities are being formed bit by bit; the older the learner the more difficult change can be, but it is not impossible as many of the examples in this book testify.

So, learners throughout their lives are seeking to establish their identities and to feel that they "know" themselves; hence, if they are enabled to establish a mathematically resilient identity as they grow through school, they are more likely to build on that identity in their lives. There is an important caveat that if someone has not developed a mathematically resilient identity at school, all is not lost! The ideas and tools that feature in many chapters of this book have convinced many that the idea they had of themselves as someone who 'can't do maths' is untrue and self-limiting. With appropriate help and support, they can not only 'do maths', but they can also learn mathematics independently and support others in doing so.

An environment that supports building a mathematically resilient identity

An environment that supports establishing a mathematically resilient identity will both help learners understand how to work at mathematics when in a formal learning environment and to be able to continue to use this knowledge when they move beyond that environment. Many of the characteristics of a mathematically resilient learner can be framed as 'knowing their rights'. A resilient learner will claim their right to progress their own mathematical thinking from where they are rather than where someone else would like them to be. They will know they have the right to use their existing knowledge, skills, understanding and strategies to develop new understandings. Someone with a mathematically resilient identity knows that they have the right to:

- be allowed and encouraged to actively work at understanding mathematical ideas, not just passively accept them. Which means they have the right to discuss and question the ideas they are learning;
- work collaboratively and feel part of an extended mathematical community.
- make mistakes and see them as an inevitable part of making progress in mathematics;

- seek help in ways that are appropriate to them, whether from their peers or a more knowledgeable other;
- help others who are also learning because that confirms and extends their own understanding;
- take responsibility for their own understanding;
- self-assess and build a realistic understanding of their own strengths and limitations and know how to work to improve;
- keep themself safe and challenged when learning mathematics.

A classroom that enables learners to take on a mathematically resilient identity which will help them to continue to learn throughout their lives will therefore look different from many traditional classrooms. The learners will be active. They will be collaborating with others, both helping and seeking help. They may be making use of books and technology to progress their understanding of mathematics. They will be exploring ideas and framing questions. They will use peer-assessment (Black et al., 2003) regularly to explore how others have approached a particular task, what ideas they have brought to provide a solution. Through that process of peer-assessment, they will have developed their own facility to self-assess, so that they know what gaps to fill or what they need to understand next.

Of course, having a classful of questioning, active learners who are seeking to take responsibility for their own learning demands a teacher who is prepared and able to provide an environment where these rights can be exercised. Hence that teacher must have also explored their own mathematical resilience and know what it means to safeguard their own and others' wellbeing. Mathematics anxiety is contagious (Dowker et al., 2016); you get it from your teacher, your parent or carer or your peers. If you are worried that your mathematics anxiety might 'infect' the learners who you are responsible for, you will need to seek to address this in ways discussed in Chapter 3.

Mathematical resilience is developed in classrooms where teachers deliberately act to allow their learners to do so. When learners have developed mathematical resilience, they will be much better placed to learn and to adapt their thinking to any given situation.

Dimension 3 – democratic and social justice

The third aspect of lifelong learning where mathematical resilience has a part to play is the learning required to exercise democracy and work for social justice. This learning empowers and emancipates individuals, enabling them to live their lives alongside others in more democratic, just and inclusive ways. In doing so, they ensure their own, and other's, well-being and increase the quality of democracy within society. Faure (1972) presents a vision of lifelong learning in which democratisation is the main driver; the aim of education is 'to enable man to be himself', yet learning-to-be always must be understood in democratic terms, that is, as learning-to-be-with-others. Faure (1972) concludes that the knowing human being and the producing human being are not enough. What is needed is the human being "in harmony with himself and others" (ibid., p. xxxix). The recent pandemic, when the Covid-19 virus threatened society world-wide, is an example of how important this aspect of mathematical lifelong learning can be. Those with a mathematically resilient identity would be able to find out how to read the graphs presented to them using a logarithmic scale if they chose to. Then they could reason out what they showed and assess the associated risks. When only 40% of 2000 US citizens could respond correctly to a basic question about the information on the graphs (Romano et al., 2020), the use of such a complex idea to present vital information should have been questioned. However just asking whether this is the best way to convey information requires mathematical resilience, and those who are not comfortable thinking mathematically were unlikely to ask the needful questions. Would a mathematically resilient society have made better decisions? There is no way of knowing, but those with mathematically resilient identities are better placed to engage with and understand the statistics for themselves and therefore make better decisions. Thus, mathematical resilience is implicated in democratic lifelong learning.

Conclusion

It is clear from the previous discussions that the three dimensions of lifelong learning should not be seen as separate, rather as interrelated. Having people who are able to continue to learn and use the diverse mathematical ideas that are needed in a changing society is important from an economic viewpoint. The changes that are needed to make our society sustainable and economically stable are and will continue to be underpinned by mathematics. Being able to learn to use and understand new mathematical ideas is important from a personal standpoint as well, as a sustainable and stable economy also allows individuals to thrive. Mathematical reasoning is part of an understanding needed for a socially just and inclusive democratic society. Furthermore, the fact that so many in current society are mathematically anxious and unable to engage with mathematics could be undermining everyone's wellbeing.

Many people have considered that learning mathematics is about 'learning for earning' and there is no doubt that those with a qualification in mathematics will earn much more over their lifetime that those without, (see the introduction chapter for details). However, although those with a mathematically resilient identity are more likely to remain employable and productive in the face of the demands of the new, global economy, they are also more likely to be able to protect their own mathematical wellbeing and that of those around them. They are more likely to be the ones able to learn and use the mathematical skills and understanding needed, as things change to establish a socially just and truly democratic society.

Mathematical resilience is an important component in allowing learners to access the lifelong learning that will allow them to thrive economically and personally and is part of an education that will allow the growth of democratic participation and social justice. A great deal can be done in school to help learners be ready and able to learn any new mathematical ideas that they want to as they continue their lives and careers after formal education, by working in ways that help them develop a mathematically resilient identity.

This chapter has laid out what learners require from formal education in order to accomplish the lifelong learning that will allow them to flourish in a world that is changing and will change in ways that we cannot predict. Sustaining our world through the inevitable crises to come will require people that work with mathematical ideas creatively and flexibly. People who have been taught that mathematics is a set of rules and procedures to memorise will not be well placed to offer the ideas based in modelling patterns and risk that are likely to be needed. This will be partly because they may well have developed mathematics anxiety and wish to avoid anything to do with mathematics.

Other chapters in this book have made clear that if learners have not developed mathematical resilience at school and even if mathematics anxiety has been the result of their school experiences, all is not lost. It is never too late. Adults can be helped to recognise that the resilience they have shown in their daily lives can be used to learn mathematics and mathematics anxiety can be treated and diminished. But how much better would it be for learners to develop a mathematically resilient identity at school and be prepared to take on the challenges they may encounter?

It would be disingenuous not to echo the UNESCO (2023) Social Contract for Education here, in asking for changes in assessment practices to allow and encourage teachers to teach the skills that will be needed to sustain our world. They say:

> Assessment should be ... meaningful for student growth and learning. Exams, tests, and other assessment instruments should harmonize with educational purposes and intents. A great deal of important learning cannot be easily measured or counted. Teacher-driven formative assessments that promote student learning should be prioritized. We must reduce the importance of competitive, high-stakes standardized assessment. (UNESCO, 2023: Principles)

I concur with these sentiments and would add that change in assessment procedures are needed because of the harm they cause to the mathematical wellbeing of learners, the hold they have on the way that teachers feel forced to teach and the harm they cause to the willingness of learners to engage in life-long learning of mathematics. If learners are to be encouraged to see learning as a lifelong enterprise, which becomes more urgent as the world changes around them, then changes must be made to the pedagogies used in formal education, which requires changes to the assessment system and the way that schools are held accountable. Whilst mathematically resilient learners can and do succeed at standard examinations, and a great deal else besides, many schools feel that changing pedagogic practice from the norm is too risky when so much is at stake. This in turn leads to learners unprepared to engage in the lifelong mathematical learning that will help them, their communities and the world to thrive.

References

Ashcraft, M. H., and Moore, A. M. (2009). Mathematics anxiety and the affective drop in performance. *Journal of Psychoeducational Assessment*, 27(3), 197–205. 10.1177/0734282908330580

Biesta, G. (2006). What's the point of lifelong learning if lifelong learning has no point? On the democratic deficit of policies for lifelong learning. *European Educational Research Journal*, 5(3–4), 169–180. 10.2304/eerj.2006.5.3.169

Black, P., Harrison, C., Lee, C., Marshall, B. and William, D. (2003). *Assessment for learning - Putting it into practice*. Maidenhead, U.K.: Open university Press. http://www.mcgraw-hill.co.uk/html/0335212972.html

Buckingham, D. (2008). Introducing identity. Youth, identity, and digital media. In *The John D. and Catherine T. MacArthur foundation series on digital media and learning* (pp. 1–24). Cambridge, MA: The MIT Press. 10.1162/dmal.9780262524834.00

Daker, R., Gattas, S., Necka, E., Green, A., and Lyons, I. (2023). Does anxiety explain why math-anxious people underperform in math? *npj Science of Learning*, 8. 10.1038/s41539-023-00156-z

Dowker, A., Sarkar, A., and Looi, C. Y. (2016). Mathematics anxiety: What have we learned in 60 years? *Front Psychology*, 7 article 508. 10.3389/fpsyg.2016.00508

Faure, E. (1972). *Learning to be: The world of education today and tomorrow*. Paris: UNESCO.

Lave, J., and Wenger, E. (1991). *Situated learning: Legitimate peripheral participation*. Cambridge: University of Cambridge Press.

Nardi, E., and Steward, S. (2003). Is mathematics T.I.R.E.D? A profile of quiet disaffection in the secondary mathematics classroom. *British Educational Research Journal*, 29(3), 345–367. 10.1080/01411920301852

OECD (1997). *Labour Market Policies: New Challenges Lifelong Learning to Maintain Employability* OCDE/GD(97)162.

Romano, A,. Sotis, C., Dominioni, G., and Guidi, S. (2020). *The public do not understand logarithmic graphs used to portray COVID-19*. https://blogs.lse.ac.uk/covid19/2020/05/19/the-public-doesnt-understand-logarithmic-graphs-often-used-to-portray-covid-19/#:~:text=Many%20media%20outlets%20portray%20information,exponential%20nature%20of%20the%20contagion

UNESCO (2023). *Renewing the social contract for education*. https://www.unesco.org/en/futures-education/new-social-contract?hub=81942

Van der Pas, N. (2001). Address by the European commission. In *Adult Lifelong Learning in a Europe of Knowledge. Conference Report* (pp. 11–18). Sweden.

Ward-Penny, R., Johnston-Wilder, S., and Lee, C. (2011). Exit interviews: Undergraduates who leave mathematics behind. *For the Learning of Mathematics*, 31(2), 21–26.

PART 4
International considerations

Mathematics anxiety as a global problem

International authors

Introduction

In this chapter, the many ways that mathematics is seen as a key area in the curriculum for different nations is reviewed by authentic voices from those countries. In each contribution, mathematics is seen as enabling the understanding and appropriation of scientific knowledge which is vital for developing the understanding needed to use and expand new technologies strategic for the economic growth of each country. The extent and effects of mathematics anxiety in each country are explored, through personal views from each of the international authors. The authors in this chapter are all working within their own contexts in some way to address the problem of mathematics anxiety and to build resilience in their learners. The authors are based in: Argentina, Brazil, Colombia, Ireland, Kenya, South Africa and Zambia, Turkey and USA. Their contexts are on a wide continuum which stretches from academia, with the potential for wide-ranging reach and research, to working with individuals to make a difference. All show facets of how mathematics anxiety is disabling individuals to the detriment of whole countries, and all add valued contributions to our knowledge. In Chapter 14, the same authors present their work related to mathematical resilience; here they set the scene in their country.

The way in which mathematics is traditionally taught crosses international boundaries and these traditional pedagogies seem to cause mathematics-specific harm, and even trauma, worldwide. There appears to be a need to think globally if the problems caused by mathematics anxiety and avoidance are to be overcome. Each author has their own writing style.

Brazil

Telma Silveira Pará, João dos Santos Carmo, Karina Lumena de Freitas Alves

Although in the last two decades, Brazil has tried to enhance education and ensure that children take part, it is far from reaching OECD levels of participation. One of the problems linked to this issue is the prevalence of mathematics anxiety.

In Brazil, an operational definition of mathematics anxiety was proposed by Carmo, Gris and Palombarini (2019). This operational definition consists of the description of three sets of simultaneous reactions, in situations in which mathematical problem-solving skills are required. Carmo, Gris and Palombarini (2019) suggest that the first set of reactions involves physiological responses common to all manifestations of anxiety. The most common of these reactions are excessive sweating, increased heart rate, changes in blood pressure, cold body extremities, changes in sleep and stomach pains. The second set of manifestations is composed of cognitive reactions, such as forgetting algorithms and problem-solving steps which are probably related to failures in working memory, as suggested by Ramirez et al. (2013) and Ashcraft and Kirk (2001). This second set of reactions also includes the development of unhelpful beliefs related to mathematics, for example, mathematics is a very difficult subject; boys are better than girls at mathematics; mathematics is for geniuses. In addition, it is quite common for students with mathematics anxiety to develop negative self-attributions in relation to their performance in mathematics, which evidences low self-efficacy. The third set of reactions concerns escape, avoidance, and freeze behaviour. Figure 13.1 summarises this operational definition.

It is noteworthy that presenting the reactions previously described is not enough to conclude that a student has mathematics anxiety, rather Carmo, Gris and Palombarini (2019) state it is necessary to identify the high frequency with which reactions occur in specific contexts of mathematics learning; the high intensity with which the student evaluates how aversive it is to study and do mathematics; and the feeling that they cannot control (change) the aversive situation that is the study of mathematics.

In summary, faced with study situations or situations that require the use of mathematics, students can present those three sets of reactions. The higher the frequency with which such reactions occur, and the greater the feeling of aversion and lack of control, the greater the probability that the student is experiencing mathematics anxiety. In these cases, the resulting increase in errors in attempts to learn mathematics generates more insecurity and aversion, characterising a cycle that repeats itself incessantly.

Regarding reading, writing and mathematics, Brazilian students perform poorly in large-scale assessment programs, both national and international, such as the Basic Education Assessment System (SAEB); the National Exam for Secondary Education (ENEM), and the Programme for International Student Assessment (PISA). In the SAEB in 2021, data reveals that Covid-19 had a very negative impact on students' performance, with negative percentages compared to SAEB in 2019 (INEP, 2021). The PISA assessment measures how countries are preparing learners to use mathematics in their personal and professional lives. Although PISA shows that there have been some improvements in learning outcomes in Brazil, student performance remains behind OECD and comparable countries (OECD, 2019). In

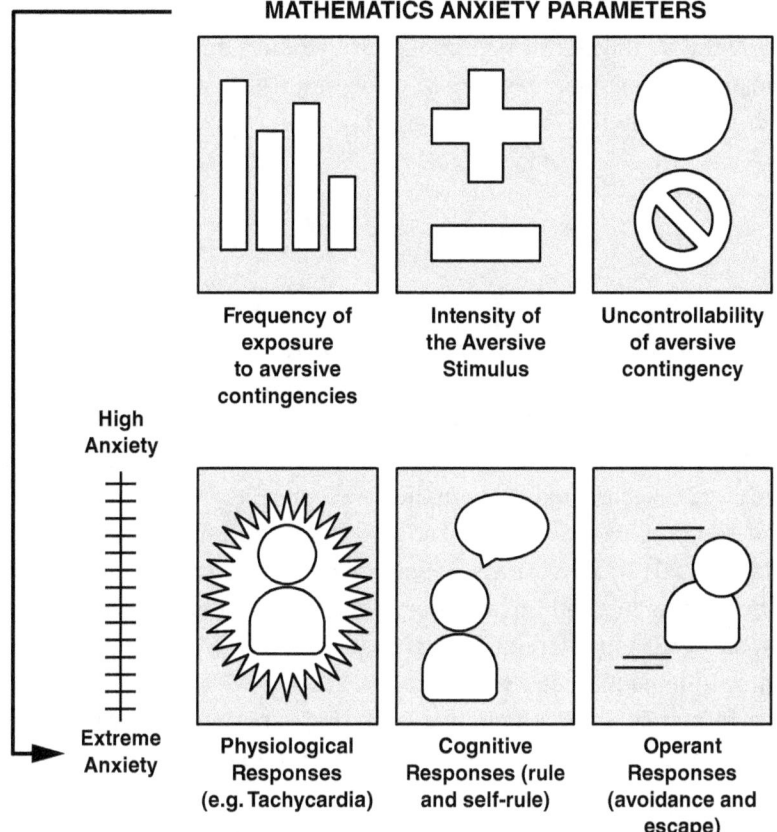

Figure 13.1. Mathematics anxiety parameters (adapted from Carmo and Henklain, 2022).

PISA 2018, students in Brazil scored 384 points on average in mathematics, while the OECD average is 489 points.

How have we faced mathematical anxiety in Brazil?

According to the panorama presented, it is necessary and urgent to search for strategies that evidence the reversal or, at least, the reduction of mathematical anxiety in our country. As mathematics anxiety is still little known by parents, teachers and educational authorities, some dissemination and research strategies have been developed with relative success. Among the dissemination strategies, the theme of mathematical anxiety has been discussed in scientific events, reports and science dissemination events. Familiarity with the theme is increasing, although there is a need for more dissemination due to the large territorial extent of our country. With an area of 8,510,000 km², Brazil is the fifth largest country by area on the planet. The challenge of mathematics anxiety becomes even greater when considering the size of our country.

In Brazil, there are at least three research groups investigating mathematical anxiety. Two groups are from the Southeast region and are associated with the

Federal University of São Carlos and the Federal University of Minas Gerais. One group is located in the South region, at the Federal University of Rio Grande do Sul. These groups have made efforts to understand different factors present in the development of mathematics anxiety and how they may be mitigated.

Ireland

Maria Ryan

In Ireland, mathematics is lauded as both an important and difficult subject (Conway and Sloane, 2005), an opinion which perpetuates the viewpoint that being good at mathematics equates to high intelligence and having a 'math brain' (Boaler, 2016). The topic of school mathematics attracts much negative discourse both as an academic subject and in its relevance for practical application on a day-to-day basis. This negativity is exacerbated through the gatekeeper status afforded to mathematics in terms of progressing through educational levels and facilitating career progression (Finlayson, 2014). There is also a perceived difficulty in being able to apply the mathematical skills learned in secondary school to problem solving for real world situations, as well as an evident lack of mathematical knowledge among new undergraduate students (Gill and O'Donoghue, 2008; O'Meara et al., 2017).

Poor performance in mathematics and the experience of repeatedly failing mathematics examinations can be distressing for many students, giving rise to mathematics anxiety (Dowker et al., 2016). However, mathematics anxiety is still an under-researched topic in the Irish context, and there is a paucity of normative data pertaining to the prevalence of mathematics anxiety among Irish students across all levels of education (Ryan et al., 2023). There is much anecdotal evidence that mathematics anxiety exists, although this is typically expressed through different terminology characterising a dislike or fear of mathematics with varying levels of intensity (Ryan, 2019).

Much media attention is given to reports of deficient student performance in mathematics, particularly in Irish secondary school state mathematics examinations, as well as performance by Irish students in international assessments of mathematics, which was below average in the successive Programme for International Student Assessments (PISA) with a mathematics focus in 2003 and 2012 (Moran et al., 2013). Results of the PISA 2012 survey showed that Irish 15-year-olds demonstrated significantly higher levels of mathematics anxiety than the OECD average, with a noted increase since the 2003 survey (Perkins et al., 2013); further, mathematics anxiety was reported as being more prevalent among female students than males in both years (OECD, 2004; OECD, 2013).

An insight into the mathematics education landscape of Ireland provides a context for understanding the evolution of the subject and reforms in mathematics education in Ireland (O'Reilly et al., 2017). The motivation for education reform in Ireland had progressed since the 1960s, driven by the need for the sustainability of the Irish economy through the development of its human capital (OECD, 2013). Graduates now require a varied skillset underpinned by competence in mathematics which is beneficial to every aspect of life. Consequently, the need for students to have quaifications in science, engineering and technology disciplines became the stimulus for the formulation of successive government policies in education (Lynch et al., 2019).

The absolutist tradition of mathematics had for decades shaped the competitive character of Ireland's secondary school mathematics curriculum. Preparation for the State Leaving Certificate examination generally dictated what materialised in the mathematics classroom (Oldham, 2006). This focus determined the curriculum for mathematics and seems to have impacted negatively the desire for secondary school students to take higher levels of mathematics (Ní Ríordáin and Hannigan, 2009). For decades, there was a lack of constructive discussion on alternative approaches to teaching mathematics (Lyons et al., 2003). The need to address such negativity and rigidity around mathematics at national level, as well as the impact of successive, deficient PISA test results, provided a worthwhile impetus for policy discussion and reforms in school mathematics education.

The Literacy and Numeracy for Learning and Life (LNLL) Strategy was launched in 2011. The intention behind the proposed reforms of the LLNL Strategy was to make mathematics relevant to real world contexts, and not simply a curriculum with linear strands of topics that are not interconnected (Atweh and Goos, 2011; NCCA, 2016). A concerted effort to tackle these issues resulted in a number of significant reforms, including changes to the primary level mathematics curriculum in 2002 and 2023 (NCCA, 2020; NCCA, 2023); and at second level, changes to the Junior Cycle curriculum introduced in 2003, as well as the implementation of 'Project Maths' across both Junior and Senior cycles since 2010 which adheres to the investigative and practical philosophy of Realistic Mathematics Education (Ní Shúilleabháin, 2014). More recently, the national Science, Technology, Engineering and Mathematics (STEM) Education policy was launched to 'enhance STEM learning for learners of all backgrounds, abilities and gender' (DES, 2017).

The introduction of the bonus points initiative (BPI) in 2012 for achievement in state mathematics examinations gave mathematics a 'special status' among subjects (Prendergast et al., 2020) and aimed to address the deficit in numbers taking higher level mathematics, which stood at 16% in 2011. The BPI awards 25 bonus points to students who achieve 40% or more in their Leaving Certificate examination and incentivises senior cycle students to take higher level mathematics, by increasing their chances of accumulating sufficient points to apply for higher education courses (DES, 2011; Prendergast et al., 2020). However, criticism of the initiative refers to the challenge for teachers of mixed ability classes, and variations in student proficiency within higher level mathematics classes, as well as continued mixed motivation for doing the subject (Prendergast et al., 2020).

Research with mathematics teachers in Ireland showed that 48% of teachers surveyed in 2008/2009 were out-of-field mathematics teachers (Ní Ríordáin and Hannigan, 2009). This finding provided the impetus for the upskilling of many out-of-field mathematics teachers through the Postgraduate Diploma in Mathematics for

Teachers (PDMT) programme which was launched nationally to address the professional development needs of these teachers (Ní Ríordáin and Hannigan, 2009). A review of the PDMT described how those who engaged with the programme reported a positive impact on teachers' mathematics teaching self-efficacy, with a shift in emphasis from procedural to conceptual understanding, and changes in teaching styles and skills (O'Meara and Faulkner, 2021).

Inadequacies in students' understanding of mathematics, and difficulties in the application of mathematical knowledge outside of the school context (Cosgrove et al., 2012) have accompanied students into the higher education context, materialising as low achievement and a deficit in basic mathematical skills, known as 'the mathematics problem' (Hunt and Lawson, 1996). Many undergraduate programmes have service mathematics modules, where mathematics is a component, but not the main discipline of study (Gill and O'Donoghue, 2008). This may adversely impact the transition to higher education for some students (O'Reilly et al., 2017), thereby posing challenges for mathematics lecturers and resulting in mathematics anxiety among students (Finlayson, 2014; Ryan, 2019). Ryan's (2019) study on mathematics anxiety among undergraduate mature students substantiated research that students with high levels of mathematics anxiety tend not to engage effectively with their mathematics coursework, preferring to learn procedurally rather than conceptually, aiming to pass their examinations, just to get by, and not engaging with mathematics support in order not 'to look stupid' (Ryan, 2019).

The rollout of dedicated mathematics support facilities in individual HEIs aimed to address the shortfall in suitable mathematical skills among undergraduate students, and particularly among first year students (Mac an Bhaird et al., 2013). However, research on undergraduate students shows that teaching methods used in lectures in Ireland are reflective of more traditional teaching methods with an emphasis on very large lecture cohorts and much 'teacher talk' (Mulryan-Kyne, 2010; O'Donoghue, 1999). On the other hand, mathematics support facilities are positively received by students, as help is available all year round in most HEIs, and in some cases preparatory mathematics courses are available before students start their programme (Mac an Bhaird et al., 2013; Ryan, 2019).

While these initiatives and reforms have aimed to address the shortcomings pinpointed in the teaching and learning of mathematics in Ireland, there is a dearth of evidence to show that these reforms have counteracted the existence of mathematics anxiety among students at all levels of education in Ireland. In this regard, creating and spreading awareness of the challenges posed by the presence of mathematics anxiety is necessary and urgent within the Irish context.

Colombia

Japcy Margarita Quiceno and Stefano Vinaccia

Mathematics anxiety is a phenomenon that has gradually gained relevance in research conducted in Colombia, due in part to the alarming scores obtained and the country's position in national and international tests. The State Examination of Middle Education, *ICFES Saber 11°*, which students take to obtain official results for admission to higher education, has revealed in recent years that public schools, low socioeconomic strata, and female students obtain lower scores in mathematics (LEE, 2022). Notably, in the 2012 PISA test, Colombian girls were adjudged to be more affected by mathematics anxiety than boys.

In general, studies conducted in Colombia with school students have demonstrated a negative relationship between mathematics anxiety and mathematics performance. In some cases, girls have exhibited better average performance in mathematics than boys, but their levels of mathematics anxiety are more prevalent, particularly in mathematics problem solving and mathematics tests (Reali et al., 2016; Villamizar et al., 2020). It has been discovered that the lack of cognitive clarity (reduced ability to identify weaknesses in learning and to correct them), along with increased test anxiety, are variables that can predict anxiety towards statistics. This anxiety may also generalise to the entire course because of test anxiety (Mercado et al., 2018, p. 93).

In relation to mathematics professionals, Fernández-Cézar et al. (2020) conducted a study to analyse the beliefs and attitudes towards mathematics anxiety of Spanish and Colombian basic education teachers. The authors found that Colombian teachers tend to have Euclidean and constructivist epistemic postures. Additionally, teachers with job stability and higher levels of education, which indicates greater preparation, exhibit lower levels of anxiety towards mathematics, instilling confidence in their work. Similarly, Ávila-Toscano et al. (2020) concluded that a greater cognitive understanding of the discipline, achieved through preparation and experience, improves the affective profile towards it. They found that in undergraduate mathematics students, three typological profiles emerged in the interaction between anxiety and attitude towards mathematics: neophyte students with a negative approach, intermediate students with an ambivalent approach, and self-efficacious and mathematically competent students.

On the other hand, although there are successful programmes in Colombia designed to reinforce mathematical competencies among elementary and middle school students, there are very few programmes that address the emotional competencies necessary for learning mathematics. Additionally, there is a significant lack of programmes that specifically target mathematics anxiety management.

In this regard, Durán-Sánchez (2022) developed a proposal that proved to be effective in reducing mathematics anxiety among Colombian sixth and seventh grade students. She found that students had several negative attitudes towards mathematical learning, such as relying on memorisation rather than analysis, giving up easily on solving mathematical problems due to a perceived lack of concentration, persistence, and patience, and a lack of understood usefulness in mathematics, all of which led to a lack of understanding. On an emotional level, students felt frustrated or not interested in learning mathematics, and on a belief level, they did not consider it their strength. Mathematics is perceived as difficult, torturous, or only for talented and smart people. Following these results, Durán-Sánchez designed three active learning guides with concrete material to awaken students' interest and strengthen their mathematical knowledge in three types of thinking: random and data systems, metric thinking and measurement systems, and variational thinking and algebraic systems using a constructivist approach. Activities included games, stories, problem-solving in practical and real-life situations, among others.

The results of the Colombian studies suggest research should be focussed on two lines of study: one oriented towards the measurement of mathematics anxiety and the other focused on intervention. In the area of measurement, the design and validation of scales for measuring mathematics anxiety need further research; to analyse the prevalence of mathematics anxiety in Colombia by region, in both student and teacher populations, with special emphasis on identifying causal factors such as: teaching methodologies used, whether these methods contribute to differential behaviour according to gender and type of school, teacher preparation, support and educational levels of parents, family dynamics, cultural stereotypes and beliefs, ethnicity, socioeconomic status, and the quality of nutrition. Similarly, emotional and psychological factors that could be involved in mathematics anxiety, such as: violence, bullying, learning problems, low self-schemas, depression, motivation, lack of purpose in life, social phobia, personality, study styles and habits, and burnout in students and teachers, should also be analysed.

The outcomes of the aforementioned proposals would provide key elements for designing specific programs in mitigating mathematics anxiety. However, our recommendation is to focus on motivation, especially intrinsic motivation. As students progress through school, they tend to perceive a decrease in motivation, which leads to increased emotional stress due to various associated factors when entering high school. These factors range from comparisons between students, instead of emphasising personal growth and achievement, to large groups that further limit the opportunity for relationships, and low academic performance in subjects such as mathematics, language, natural sciences, and social sciences being attributed to their complexity and new and different methodologies (Burchinal et al., 2008).

In Colombia, there have been programmes addressing motivation towards mathematics through contextualised teaching processes from a socio-environmental perspective. These programs have shown improvements in anxiety levels and in the perception of mathematics' usefulness, with an emphasis on activities that apply the concepts learned in class (Barrera et al., 2015). Other intervention proposals could aim to train teachers on self-regulated learning strategies. Colombian studies have found that these strategies favour intrinsic over extrinsic motivation and lead to a reduction in mathematics anxiety levels (Puentes, 2016). Additionally, working on small group leadership, where peers accompany and help each other in their mathematics learning experiences, has been found useful for the mathematical progress of students. However, this strategy did not significantly affect emotional regulation, as anxiety tends to increase more in the first sessions compared to the last ones in this type of program (González, 2018). However, more research is required to determine the long-term effectiveness of these interventions.

It should be noted that programmes designed with an emphasis on cognitive approaches to address emotional issues may increase anxiety at the beginning, but this effect tends to decrease over time (Quiceno et al., 2016). This pattern is also observed with expressive writing techniques used to address mathematics anxiety in students (Ganley et al., 2021). Therefore, we recommended that programs with these characteristics that aim to reduce anxiety, for example, for examinations or other activities requiring cognitive effort, should not be implemented near or during the peak of such activities, as the opposite effect may occur. Colombian studies also propose designing gender-differentiated programs that address affective factors and emotional responses to mathematics, particularly in girls (Villamizar et al., 2020).

Research in Colombia suggests that mathematics anxiety could be a public health issue. The negative attitudes associated with mathematics anxiety, such as impatience and a sense of defeat towards effortful tasks, can lead to a sense of indifference rather than frustration (Durán-Sánchez, 2022). This reinforces irrational beliefs of failure, ultimately becoming a part of the broader culture. The attitude of dislike towards mathematics seems to be repeated among most Colombian students. The studies of Colombian students and teachers indicate that as one advances in age and educational level, the perception that mathematics is useful tends to increase, although the dislike persists (Ávila-Toscano et al., 2020; Franco-Buriticá et al., 2019; Granados and Pinillos, 2008). As children become adults and take on adult responsibilities, they realise the practical importance of mathematics. Issues arise for individuals who do not acquire the necessary mathematical skills, through avoidance generated by mathematics anxiety, which could result in long-term disadvantage and limited opportunities.

For all these reasons, the development of intervention programs is urgent, especially in the area of resilience. Resilience programs are required not only to protect the mental health of students from the stress and anxiety generated by mathematical experiences, but also to increase the human strengths associated with the virtue of courage, such as bravery, perseverance/persistence and vitality/passion for learning. The aim is to develop positive attitudes towards mathematics and to enable students to enjoy the subject.

Finally, we suggest that a Colombian network of mathematical resilience be established. Why resilience? Resilience is a macro construct that can accommodate, support, and respond to all the proposals outlined. This would allow professionals working on mathematics anxiety and other associated problems to join efforts and carry out systematic studies. Mathematics anxiety requires a multidisciplinary and interdisciplinary approach, and the research available to date has either been developed in isolation or some proposals have not had continuity over time. This network could lead to a bank of successful experiences that would serve to promote and replicate the studies that have the greatest impact, in addition to providing training and support on the topic of mathematics anxiety to the academic community and other professionals involved and interested.

Turkey

Abdulvahap Yorgun

High quality in learning mathematics or a high score in nation-wide mathematics achievement is a desired outcome in Turkey. These outcomes may be evaluated using various national and international assessment resources. The Programme for International Student Assessments (PISA) reports, students' performances on mathematics subtests at high school and university entrance examinations and scientific research reports all provide invaluable data about the current situation in Turkey. They show that learning or doing mathematics is a nation-wide problem. In other words, it can be concluded that a mathematical pandemic is observed among students. The most crucial and concrete sign of the pandemic may be considered to be mathematics anxiety. For example, the results from PISA conducted in 2018 by the OECD showed Turkish students achieved lower points from the tests related to problem solving skills compared with participants from other member countries (Schleicher, 2018).

Similarly, in the Turkish educational system, to attend some high schools students are required to take the High School Entrance Examination (HSEE) organised by the Ministry of Education. According to statistics released by the Ministry, the candidates answered correctly only 5.09 of a total of 20 mathematics questions in 2019; 4.89 in 2020 and 4.02 in 2021 on average (MEB, 2021). In Turkey, to attend higher education, students are required to take two examinations conducted by Measurement, Selection and Placement Center (MSPC). These tests are BFT- Basic Proficiency Test and FQT-Field Qualification Test. Each test contains a 40-question subtest of mathematics and the mean for correct answers was calculated as 5.1. In other words, approximately 2.5 million students could answer only 5 of the 40 mathematics questions correctly (OSYM, 2021). Therefore, the reasons for this situation deserve to be investigated in order to develop policies and interventions aiming to help students learn and do mathematics more effectively and in a psychologically healthy environment.

Non-healthy educational environments generate anxiety, stress and feelings of tension among students and teachers. From this point of view, mathematics anxiety is considered as one of the main predictors of loving or hating mathematics or being successful or unsuccessful at mathematics (Baloğlu, 2001). In other research, Baloğlu (2010) investigated the anxiety levels of 220 elementary students and compared the results with their mathematics achievements. The results showed that high levels of anxiety leads to poor performance among Turkish elementary school students.

Similarly, Yenilmez and Özbey (2006) investigated the mathematics anxiety levels of students who attend public and private schools and the relationship between certain variables such as type of school, gender, mathematics achievement and parental educational level and mathematics anxiety. They reported that mathematics anxiety hinders students' mathematics performances regardless of other factors. To sum up, as a specific type of anxiety, mathematics anxiety may have a high prevalence rate among Turkish students (Baylan, 2020; Yorgun and Mert, 2023) and mathematics anxiety has a strong relationship with mathematics performance or success.

The number of research studies on mathematics anxiety has increased gradually in recent years. More concretely, a chronological review of the theses and dissertations on mathematics anxiety in Turkey showed that six theses were written between 2005–2010; 7 between 2011–2015 and 24 between 2016–2021 (Turkish Council of Higher Education, 2023). Similarly, a significant increase in the number of research articles on mathematics anxiety between 2016–2022 years has been detected. Baylan (2020) investigated the scientific efforts on mathematics anxiety between 2010–2017 in Turkey and found that most of the studies were conducted between 2015 and 2017. Thus, in the last decade, the researchers have paid an increasing attention to mathematics anxiety. Turkish literature on mathematics anxiety focuses on certain aspects of mathematics anxiety: the level of mathematics anxiety, its relationship with various variables, the relationship between certain mathematics subjects and mathematics anxiety, and scale development (Baylan, 2020; Gökbulut and Şahin, 2022). Most of the studies used quantitative research methods. However, qualitative studies that aim to develop a theory by examining the problem in depth and experimental studies that investigate treating mathematics anxiety and enabling students to develop healthier responses are almost non-existent. Therefore, future research may focus on this weakness to fill the gap in the relevant literature.

Turkish literature on mathematics anxiety revealed that the reasons for the development of mathematics anxiety are the nature of mathematical concepts, teaching and learning styles, age, gender and socioeconomic status of students, attitudes towards mathematics and self-esteem (Baloğlu, 2001; Baloğlu, 2004; Nazlıçiçek, 2007; Alkan, 2011; Durmaz, 2012). Nazlıçiçek (2007) investigated the predictors of mathematics success of 10th grade students and she found that the previous mathematics success, mathematical academic self-concept and the students' beliefs about mathematics were the significant factors predicting their mathematics success. However, these predictors were identified in an individual context; social or pedagogical predictors which may trigger mathematics anxiety may also be significant factors.

Considering social problems from an ecological perspective may provide a wider range of findings. As an example of the researchers using a broader perspective, Alkan (2011) examined the reasons for mathematics anxiety through qualitative research and she found that the reasons for mathematics anxiety may be categorised under four dimensions: teachers, students, parents and peers. Teacher-related reasons included teachers' teaching style and methods, their behaviours perceived as discrimination between students doing well and not and their limited support. On the other hand, self-efficacy and self-confidence about mathematics were student-related reasons. Peers may be one of the contributors to mathematics anxiety. Namely, the sense of being humiliated as a result of being teased by peers may increase mathematics anxiety. Similar findings are achieved by Yorgun and Mert (in press). The researchers investigated the experiences of high school students with low mathematics grades during mathematics classes. They found that 95% of the participants reported experiencing a traumatic event which triggered their mathematics anxiety. Lastly, parents' coercive attitudes towards homework and unsupportive behaviours were reported as familial reasons for mathematics anxiety.

Kenya

Harrison Njaru Mbogo

The Programme for International Student Assessment (PISA) is an international assessment for comparing quality and efficiency in learning outcomes across all member countries. It is managed and coordinated by the Organisation for Economic Cooperation and Development (OECD) since its inception in 2000 when the first cycle was implemented. The main aim of PISA is to measure the ability of children in the use of science, reading and mathematics knowledge and skills which can prepare them to be able to solve real-life challenges. Unfortunately, Kenyan teachers and students have not been participating in the PISA activities. The Kenya National Examinations Council (KNEC) is mandated to improve Kenyan students' scores in the next PISA assessments on behalf of the Ministry of Education in line with the National Education Sector Strategic Plan (NESSP, 2018–2022), This commits the Ministry of Education to participate in a large-scale International Assessment as one of the strategies towards reforming learning assessment practices in Kenya. The Kenyan Government intend to participate in PISA 2025 assessment activities implemented under the Kenya Primary Education Equity in Learning (KPEEL) programme funded by the World Bank, Global Partnership for Education (GPE) and Lego education.

A systematic literature review of articles published from 1980–2020 (Kakoma and Sun Zuma, 2023) reveals that little has been done in terms of ascertaining the extent of Mathematics Anxiety (MA) in Sub-Sahara Africa. Of interest is that out of the nine studies reviewed, one was from South Africa, eight from Nigeria and none from Kenya. Such a small number of articles and distribution with Kenya having none is a cause of concern. Having student who want to study mathematics at higher levels is important especially in this era of technological advancement, as solving the huge climatic change and economic injustice challenges require mathematical thinking skills.

Children with low grade attainment in Kenya are forced to repeat classes. Sadly, the teachers and the stakeholders in Kenya do not appear to recognise any affective barriers in mathematics and the psychological harm which may be caused to children's well-being. The situation in Kenya has been further exacerbated by shortages of qualified mathematics teachers. In Kenya, for example from 2012–2016, about 61% of the young grade three children could not successfully compute grade two mathematics tasks (Uwezo, 2016). A significant body of research on interventions to improve attainment in Kenya (Kwayumba et al., 2017; Piper et al., 2018; Piper et al., 2018; Piper et al., 2016) hardly acknowledges the prevalence of mathematics anxiety in Kenyan schools. This is despite the fact that Kenyan children have a long tail of mathematics underachievement (Uwezo, 2016). The Kenyan government has responded to mathematics under-achievement in Kenya by attempting to build the capacity of mathematics teachers to teach, but without developing teachers' understanding of mathematical resilience.

Many children may have experienced repeated mathematics failure in the classroom. Therefore, their typical response regarding mathematics attainment expectations may include, among others, 'I will never get it', fear, hopelessness, anger, or in some instances outright refusal to come to school or to do mathematics. In my experience, some of these children are not able give any examples of good experiences in Kenyan mathematics classrooms. As a result, they exhibit self-preservation strategies such as avoidance and passive-noncompliance. Thus, their mathematics attendance is often low, also their mathematics attainment expectations is low while their mathematics anxiety level is usually high. Based on the long tail of mathematics underachievement and that research hardly acknowledges the prevalence of mathematics anxiety in mathematics classrooms, teachers in Kenya seem to lack awareness of the impact of affective attributes on progress in mathematics. Consequently, children in mathematics classrooms may unconsciously preserve their mathematical well-being by attempting work they have already mastered and not engaging in mathematics tasks which may pose challenges to them resulting in them making little progress beyond basic mathematics skills acquisitions in Kenya. For many children in lower strands, this has resulted in grades that are stuck at learning below age level while some children have deteriorated to very low mathematics attainment which represents mathematical under-achievement.

Hence, developing Mathematical Resilience seems very important given the consistent of poor mathematics attainment in Kenya and the evidence that mathematics anxiety can negatively affect children's mathematical performance (Gunderson et al., 2018) and their willingness to pursue careers involving mathematics (Johnston-Wilder et al., 2020).

South Africa and Zambia

Brighton Mudadigwa, Sakyiwaa Boateng, Nomzamo Xaba, Tawanda Chinengundu, Folake Adelabu and Royda Kampala

In the South African context, Sithole et al. (2017) highlight that there are pervasive anxieties surrounding the learning and understanding of STEM programs.

These anxieties have been identified as major obstacles to achieving educational excellence and hinder the country's progress in STEM programs (Mkhize, 2019; Mutodi and Ngirande, 2014). According to Sithole et al. (2017), the concept of mathematics and science anxiety refers to the fear, tension, and apprehension experienced by learners when engaging with these subjects. These anxieties often lead to reduced motivation, avoidance behaviours, and even lower academic achievement.

Howie (2003) and TIMSS (2019) have raised concerns about the low achievement of South African students in mathematics and science. South Africa's mathematics and science learners have come into the international spotlight since it started participating in the Trends in International Mathematics and Science Study (TIMSS). South Africa participated with grade 4 and 8 learners in 1999 and 2003 and with grade 4 and 9 learners from 2011 to 2019. The South African grade 9 learners are a year older (average age 15 years) than the rest of the participants in grade 8, who have an average age of 14.

Figure 13.2 exhibits the mathematics and science performance trends for South African grade 8/9 learners from 1999 to 2019 (Mullis et al., 2004; 2016; 2020). In 1999 and 2003, South African grade 8 learners performed poorly, decreasing their performance in the years under review. However, when the cohort of grade 9 learners started participating in 2011, their average scores gradually increased for mathematics and science subjects. Nonetheless, the performance is well below the TIMSS average benchmark score. The national average for South Africa's mathematics and science success scores increased from "extremely low" in 1995, 1999, and 2003 to "low" in 2011 and 2015 (Mullis et al., 2020). Mullis et al. (2020) assert that, in comparison to the other participating countries, South Africa has one of the worst mathematics and science performance rates.

In addition, Sithole et al. (2017) noticed that 45 per cent of first-year university students had severe difficulty with mathematics, which is crucial to competencies in STEM programs, and they blame the high dropout rate in STEM degrees on students' lack of understanding in mathematics. In the South African context, to

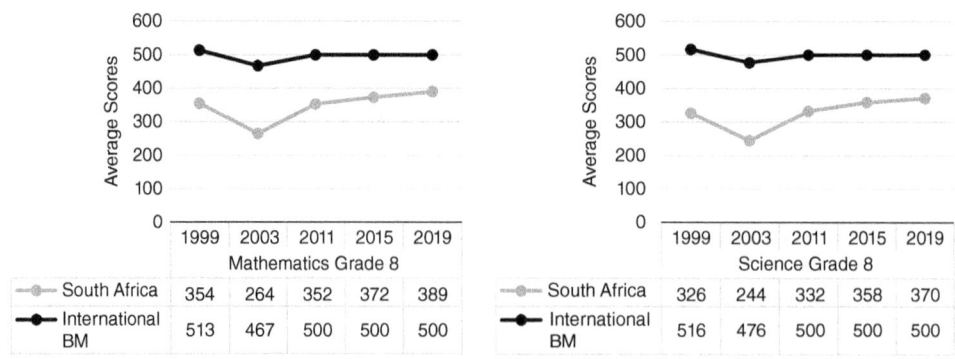

Figure 13.2 South African performance in mathematics and science (Mullis et al., 2020).

increase the retention rate of STEM students, we acknowledge the need to focus on the causes of high failure rates in mathematics and science subjects at secondary school level. Studies have shown that anxieties in mathematics and science are significant setbacks and cause low performance in teachers and learners (Mkhize, 2019; Mutodi and Ngirande, 2014).

Numerous studies have been conducted in South Africa to determine mathematics and science anxieties, and several interventions have been developed to reduce anxiety levels (Mkhize, 2019; Mutodi and Ngirande, 2014; Sithole et al., 2017). While there has been considerable discussion regarding the developmental trajectory of mathematics anxiety and negative mathematics attitudes, little emphasis has been paid to the effect of past mathematical experiences, poor teaching, and personal and intellectual aspects that cause mathematics anxiety among learners and pre-service teachers in South Africa. Studies have also established that little has been done on instructional intervention strategies that reduce mathematics anxiety in Sub-Saharan Africa (Luneta and Sunzuma, 2022; Spangenberg and van Putten, 2020). Despite an abundance of research investigating the underlying cognitive processes that contribute to mathematics anxiety (Samuel and Warner, 2021; Smith and Capuzzi, 2019; Spangenberg and van Putten, 2020), there is limited research leveraging this knowledge to develop embedded instructional intervention strategies to explicitly address this problem in South Africa. Blazer (2011) suggests there are techniques teachers, parents, and students could use to alleviate mathematics anxiety. Thus, awareness and measures to minimise the impact of mathematics anxiety are necessary among students in schools.

Therefore, in the South African context, there is a need to address these anxieties to improve learner performance in STEM subjects. Learners' attitudes towards mathematics and science should be determined and managed to achieve this goal. We suggest that improvements can be made by addressing the anxiety in teaching and learning mathematics and science among teachers, undergraduate students, and high school learners. Hence, addressing anxieties in individual learners by building mathematics resilience (Johnston-Wilder and Lee, 2019) is our priority in the mathematical resilience project in South Africa.

Argentina

Silvia Renata Figiacone

"Students from the 10 Latin American countries that participated {were} placed well below the world average. Their worst subject was math" (Programme for International Student Assessments results, 2018). Poverty is surely a source of inequity in education. In Argentina, and in many countries all over the world, being poor seems to lead to not attaining well in mathematics. Not attaining well in mathematics leads to a lack of opportunities to get out of poverty. It seems to be a vicious cycle. In 2022, in Argentina, we faced 100% inflation rate and poverty rose to 39.2% of the population (INDEC, 2022). We are not doing well in

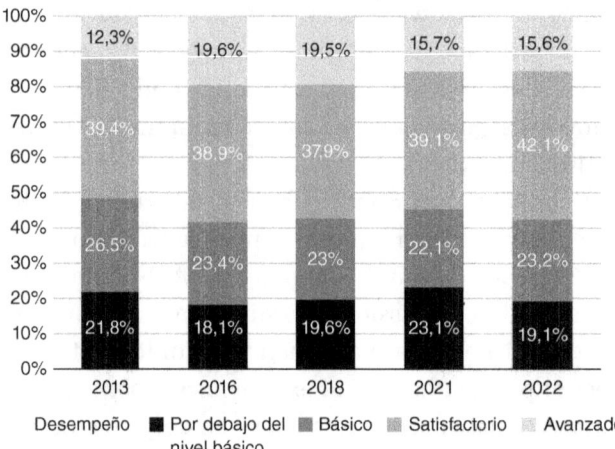

Figure 13.3 Students' distribution by level of performance in Mathematics since 2013 (Ministerio de Education, 2022).

the economy, and we are certainly not doing well in mathematics. In 2021, we had 11.6 million students all over the country (INDEC, 2022) attending schools in the public and private systems, an increase from 10.6 million in 2011. The most recent data we have about how our children are doing at school date from 2022, and in the report by *Argentinos por la Educación* (2022) it can be seen that we are doing worse in Mathematics than before the pandemic.

As Figure 13.3 shows, at least 32.3% of students did not reach satisfactory levels in mathematics performance in 2022, and 19.1% of them did very badly (Ministerio de Education, 2022). It seems in Argentina we are not good at teaching mathematics; we are not good at doing mathematics, and as a country that faces 100% inflation in one year, we surely need to be flexibly able to adjust our mental mathematics just to buy groceries.

As a clinical neuropsychologist, I work every day with children who have been diagnosed with a specific learning disorder. Many of them have dyscalculia. You do not need to be dyscalculic to have difficulties in learning mathematics. You can be either dyslexic or you can have an attention deficit disorder and you can also be bad at mathematics (Knight, 2021). And you can have no developmental disorder at all and not be good at mathematics. At NeuroEduca (www.neuroeduca.com) we know that having a specific learning disorder increases the possibility of having mathematics difficulties as well as mathematics anxiety. Much has been said already about mathematics anxiety in this book (see chapter 2) and I am not going to go deeper into definitions, but I can go deeper into what some children say and what we do to help them.

Marco is an 8-year-old boy who goes to weekly sessions to improve mathematics performance at school. He has dyscalculia and he knows it. Talking with his neuropsychologist about doing mathematics at school he says every time he is doing mental mathematics, he feels nervous and dizzy. He is afraid he is going to get a wrong result and he tries to avoid the teacher's eyes, praying not to be called. Ema is a 9-year-old girl who tells her therapist she does not want to go to mathematics classes because numbers "keep turning around" and her teacher "keeps asking me to write them well". She started with stomach aches every morning and her parents started to think she might be sick, but she does not have stomach aches on weekends and holidays.

Clearly, Marco and Ema have mathematics anxiety and use avoidance as a strategy to regulate this emotion. Avoidance as a regulation strategy has long been defined as useful but pervasive (Hattie and Donoghue, 2016). As soon as they tell us, mathematics anxiety and its regulation become a treatment objective. We know, as has been repeatedly told us by Hattie (2008), that making things visible will help children to learn everything. We also know that we have to address two objectives at the same time: regulating mathematics anxiety and improving mental mathematics.

USA

Allison Dillard

In this section, I explore my perceptions of mathematics and the state of mathematics anxiety in the United States. As a former mathematics teacher at a community college, I have first-hand experience of working with students to experiment with different strategies to help them overcome their mathematics anxiety. These experiments are discussed in my books, *Crush Mathematics Now* and *The Crush Mathematics Experiments*. Additionally, as the host of the *Allison Loves Mathematics* podcast and the *Empowering Mathematics Teacher Virtual Conference,* I have had the privilege of interviewing leaders in mathematics education about their perspectives on improving student success in mathematics. My goal here is to give a glimpse of the current state of mathematics, mathematics anxiety, and mathematics resilience in the United States from my personal perspective, to best identify areas of potential positive change. It is important to note that my experiences and perspectives are not representative of all teachers, as we all have different backgrounds and experiences.

How do people in the United States view mathematics? In the United States, many people view mathematics as a challenging, prestigious, and important subject. While many individuals may dislike mathematics or feel inadequate in their abilities, a significant percentage still value mathematics and view it as important. Despite the growth mindset movement in education gaining popularity, many students continue to approach mathematics with a fixed mindset (Dweck, 2000).

How do people in the United States view mathematics anxiety? Perceptions of mathematics anxiety vary greatly in the United States. On one hand, some students have described their experience with mathematics to me as being similar to PTSD. The term "mathematics PTSD" resonated so that other students felt it accurately reflected the emotional and physical reactions they had to mathematics.

On another hand, when I mentioned to someone that I was writing a book called *The Mathematics Anxiety Handbook*, they laughed and made a dismissive comment about there "being an anxiety for everything now". This reaction highlights the misunderstanding and stigma that still surrounds mathematics anxiety in the United States.

How widespread is mathematics anxiety in the United States? Surveys indicate that as high as 93% of Americans have experienced some level of mathematics anxiety (Blazer, 2011). It is a problem that can start early, as mathematics anxiety has been documented in students as young as five years old. Rates of mathematics anxiety also differ across populations, affecting girls and boys differently, low-income and high-income students differently, etc. For instance, mathematics anxiety affects college-aged students differently: approximately 25% of four-year college students and 80% of community college students suffer from a moderate to high degree of mathematics anxiety in the United States (Luttenberger et al., 2018).

As a former mathematics teacher at a community college, I can confirm that the statistic of 80% of community college students suffering from moderate to high mathematics anxiety is reflected in my experience. In one of my classes, I asked my students if anyone had mathematics anxiety and, to my surprise, every single student in my class raised their hand. This high rate of mathematics anxiety repeated itself in my other classes, regardless of whether I was teaching Prealgebra, College Algebra, or Statistics. In every case, 80–100% of students acknowledged experiencing mathematics anxiety.

What factors contribute to mathematics anxiety in the United States? Several factors seem to contribute to the prevalence of mathematics anxiety, including:

- **Many parents have mathematics anxiety.** Many parents and teachers grew up in a time when people were considered "mathematics people" or "not mathematics people," and skills and struggles in mathematics were considered innate. The emphasis on getting mathematics right easily– without trying hard– was prevalent and the idea of approaching mathematics with a growth mindset (Dweck, 2000) was not a common conversation in classrooms or households.

- **Many elementary school teachers have mathematics anxiety:** According to a study by Brown et al. (2011), a significant number of elementary school teachers in the United States experience mathematics anxiety, with 41% of elementary mathematics teachers reporting some level of mathematics anxiety. This is problematic as research suggests that a teacher's mathematics anxiety affects their students' mathematics achievement (Dowker, et al. 2016).

- **Teachers lack sufficient training and resources to address mathematics anxiety.** Helping students to develop a growth mindset about mathematics is crucial to overcoming mathematics anxiety. Unfortunately, many teachers in the United States do not feel equipped to teach growth mindset effectively. A national survey (Yeager et al., 2022) found that only 20% of teachers believe they can encourage a growth mindset through their teaching. Even more concerning, 85% of teachers stated they needed professional development around teaching growth mindset.

- **Students face high-stakes exams in school.** The pressure of high-stakes exams in schools and their emphasis on getting to the correct answer can also exacerbate mathematics anxiety. This emphasis is not just limited to high school students, as even elementary school children may face high-stakes exams. I spoke with a mother recently who shared that in her district, third-grade students must pass state exams in order to be promoted to the next grade level. Can you imagine the pressure of being a third grader with one test determining if she's ready to move on to fourth grade?

- **Disconnect between messages and experiences.** The messages that students hear about having a growth mindset and learning from mistakes often do not align with their actual experiences in mathematics classrooms. Although students may be told to "try hard," "learn from mistakes," and "have a growth mindset," they are not always given experiences, assignments, or environments where they are able to practice these things.

- **Teacher resources emphasise procedural mathematics.** In many classrooms in USA, mathematics is still more about following the rules than it is exploring and problem-solving. Teacher resources are often helpful for teaching to the test, getting the right answer, and replicating steps to get to the right answer. As a result, teachers often follow procedural-focused lessons and are judged based on their students' test scores, which leaves little room for conceptual, exploratory, problem-solving. This creates a disconnect between the skills students need for long-term success and what they are taught in the classroom.

Mathematics anxiety can result from a combination of many different factors, and one way to understand it is through the story of Priscilla. As a tutor I worked with Priscilla before I started teaching formally. She was taking Precalculus at a community college, had previously failed the class multiple times, and had mathematics anxiety so severe that she had panic attacks during her examinations. A combination of factors brought her to this point. She indicated that her mathematics anxiety started when she was in elementary school, and a teacher told her she was bad at mathematics. Her parents, trying to make her feel better, introduced her to me by explaining that she was "so smart, but just so bad at math". Throughout all her years of education, she had never heard about a growth mindset or realised that mathematics was just a skill, like playing the piano or reading. The high-stakes exam of her Precalculus class, which determined whether she would be accepted into a nursing program, further fuelled her mathematics anxiety.

While Priscilla's story is just one of many, it is a common one that I have seen in other students that I have taught. Their mathematics anxiety starts young and builds over the years with negative experiences snowballing. Well-meaning educators and others who try to support along the way do not know how to intervene and help. This builds until the mathematics anxiety culminates in panic once the stakes become meaningful to the student. Despite her severe mathematics anxiety, Priscilla was able to overcome it and pass not just precalculus, but also two semesters of Calculus. The origin of Priscilla's mathematics anxiety, resulting from different factors and experiences, is common among students affected by mathematics anxiety in the United States.

Conclusion

The various voices in this chapter illustrate the global extent of mathematics anxiety and also the extent of the concern that is felt worldwide. Mathematics is considered vital to both economic prosperity and to tackling the issues presented by climate change. All the countries represented here see mathematics anxiety is stopping their young people becoming qualified and willing to take on the higher studies that require mathematical understanding that they see they need.

Each study points to the way teachers present mathematical ideas as one of the many causes of mathematics anxiety. They also understand the effect of the way that mathematical knowledge is assessed on the way that teachers teach and on the learners that are examined. Where procedural knowledge and speed of response is required to be considered successful at mathematics, the complete understanding of mathematical ideas, the connectivity between all of mathematics and the ability to use mathematics to solve realistic problems are likely to be neglected by teachers attempting to 'cover' the curriculum. However, understanding, connectivity and problem solving are what employers need in their employees and what the world needs to create novel solutions to the climate crisis.

The next chapter will focus on what these countries are doing to help their learners either to not become anxious in the first place or to learn to cope with their anxiety so that they can study mathematics effectively.

References

Alkan, V. (2011). Etkili matematik öğretiminin gerçekleştirilmesindeki engellerden biri: Kaygı ve nedenleri [One of the barriers to providing effective mathematics teaching: Anxiety and its causes]. *Pamukkale University Journal of Department of Education*, 29(29), 89–107.

Alkan, V. (2018). A systematic review research: 'mathematics anxiety' in Turkey. *International Journal of Assessment Tools in Education*, 5(3), 567–592.

Argentinos por la educacion. (2022). *Learn tests 2021*. https://argentinosporlaeducacion.org/informe/pruebas-aprender-2021-2/

Ashcraft, M. H., and Kirk, E. P. (2001). The relationships among working memory, math anxiety, and performance. *Journal of Experimental Psychology: General*, 130(2), 224–237. 10.1037/0096-3445.130.2.224

Atweh, B., and Goos, M. (2011). The Australian mathematics curriculum: A move forward or back to the future? *Australian Journal of Education*, 55(3), 183–278. 10.1177/0004 94411105500304

Ávila-Toscano, J. H., Rojas-Sandoval, Y., and Tovar-Ortega, T. (2020). Perfil del dominio afectivo en futuros maestros de matemáticas. *Revista de Psicología y Educación*, 15(2), 225–236. 10.23923/rpye2020.02.197

Baloğlu, M. (2001). Matematik korkusunu yenmek [Coping with math anxiety]. *Kuram ve Uygulamada Eğitim Bilimleri Dergisi*, 1(1), 59–76.

Baloğlu, M. (2004). The relationship between different ways of coping and mathematics anxiety. *Eurasian Journal of Educational Research*, 16(2), 95–101.

Baloğlu, M. (2010). An investigation of the validity and reliability of the adapted mathematics anxiety rating scale-short version (MARS-SV) among Turkish students. *European Journal of Psychology of Education*, 25(4), 507–518.

Barrera, N. P., Castaño, L. J., Reinoso, L. M., Ruiz, I. S., and Villarreal, J. E. (2015). La contextualización de la enseñanza de las matemáticas en el desarrollo de los niveles de motivación. *Revista Colombiana de Matemática Educativa*, 1(1), 324–329. http://www.ojs.asocolme.org/index.php/RECME/article/view/67

Baylan, H. N. (2020). *2000–2017 yılları arasında matematik kaygısı ile ilgili Türkiye'de yapılan çalışmaların bazı değişkenlere göre incelenmesi [Investigation in reference to some variables of studies conducted in Turkey between the years 2007–2017 which related to anxiety of mathematics]*, Unpublished MS Thesis, Turkey: Selçuk University.

Blazer C. (2011). *Strategies for reducing math anxiety* [Information capsule] 1102 Available from: https://eric.ed.gov/?id=ED536509

Boaler, J. (2016). *Mathematical mindsets: Unleashing students' potential through creative math, inspiring messages and innovative teaching*. San Francisco: Jossey Bass.

Brown, A., Westenskow, A., and Moyer-Packenham, P. (2011). Elementary pre-service teachers: Can they experience mathematics teaching anxiety without having mathematics anxiety? *Issues in the Undergraduate Mathematics Preparation of School Teachers*, 5.

Brown, N. W. (2010). *Psychoeducational groups* (3rd Edition). New York: Routledge.

Burchinal, M., Howes, C., Pianta, R., Bryant, D., Early, D., Clifford, R., and Barbarin, O. (2008). Predicting child outcomes at the end of kindergarten from the quality of pre-kindergarten teacher–child interactions and instruction. *Applied Developmental Science*, 12(3), 140–153. 10.1080/10888690802199418

Carmo, J., and Henklain, M. (2022). Ansiedade à matemática: Uma leitura analítico-comportamental [Matehmatica anxiety: A behavioral-analytical interpretation]. In Aline Beckman C. Menezes (Org.), *Ensinar e Aprender: Desafios para a educação do séc. XXI*. (pp. 113–131). ABPMC.

Carmo, J., Gris, G., and Palombarini, L. (2019). Mathematics anxiety: Definition, prevention, reversal strategies and school setting inclusion. In D. Kollosche et. al. (Eds.), *Inclusive mathematics education. State of the art research from Brazil and Germany* (pp. 403–418). Switzerland: Springer.

Cengiz, N. (2017). *Teknoloji destekli matematiğin öğrencilerin başarıları ve matematik kaygıları üzerindeki etkisi [The effects of technology-supported mathematics on students' success and math anxiety]*. Unpublished MS Thesis, Turkey: Gaziantep University.

Çetin, Ş. Durmaz, B., and Girit, D. (2018). Matematiksel yılmazlık/dayanıklılık ölçeğini Türkçeye uyarlama çalışması [Adaptation of mathematical resilience scale into Turkish culture]. In Ş. Çınkır (Ed.), *Fifth international Eurasian educational research congress proceedings book* (pp. 875–881). Antalya, Türkiye: Anı Publishing.Yayıncılık.

Conway, P., and Sloane, F. (2005). *International trends in post-primary mathematics education: Perspectives on learning, teaching and assessment*. Dublin: Research report commissioned by the National Council for Curriculum and Assessment (NCCA).

Cosgrove, J., Perkins, R., Shiel, G., Fish, R., and McGuinness, L. (2012). *Teaching and learning in project maths: Insights from teachers who participated in PISA 2012*. Dublin: Educational Research Centre.

Demir, G. (2017). *Gerçekçi matematik eğitimi yaklaşımının meslek lisesi öğrencilerinin matematik kaygısı, matematik öz yeterlik düzeyi, algı ve başarıları üzerindeki etkisi [The effect of realistic mathematics education approach on mathematical anxiety, mathematical self-efficacy, perceptions and achievement of vocational high school students]*. Unpublished Ms Thesis, Aydın, Turkey: Adnan Menderes University.

DES (2011). *Literacy and numeracy for learning and life*. Dublin: Department of Education and Skills (DES). Sourced at https://www.education.ie/en/Publications/Policy-Reports/lit_num_strategy_full.pdf on 15/04/2023.

DES (2017). *STEM education policy statement 2017–2026*, Dublin: Department of Education and Skills.

Dowker, A., Sarkar, A., and Looi, C.Y. (2016). Mathematics anxiety: What have we learned in 60 years? *Frontiers in Psychology*, 7, 1–16. 10.3389/fpsyg.2016.00508

Durán-Sánchez, H. V. (2022). *Guía de aprendizaje para mejorar la ansiedad hacia las matemáticas en el grado sexto y séptimo del colegio Gimnasio Moderno Americano de Melgar-Tolima [Tesis de pregrado, Universidad de Cundinamarca]*. Repositorio institucional de la UDEC. https://repositorio.ucundinamarca.edu.co/handle/20.500.12558/4694

Durmaz, M. (2012). *Ortaöğretim öğrencilerinin (10. sınıf) temel psikolojik ihtiyaçlarının karşılanmışlık düzeyleri, motivasyon ve matematik kaygısı arasındaki ilişkilerin belirlenmesi [Identifying the relationships among the degrees of basic psychological needs satisfaction, motivation and mathematics anxiety of high school students (10th grade)]*. Unpublished MS Thesis, Bolu, Türkiye: Abant İzzet Baysal University.

Dweck, C. (2000). *Self-theories: Their role in motivation, personality, and development.* Psychology Press.

Fernández-Cézar, R., Hernández-Suárez, C. A., Prada-Núñez, R., and Ramírez-Leal, P. (2020). Beliefs and anxiety towards mathematics: A comparative study between teachers of Colombia and Spain. *Bolema Boletim de Educação Matemática*, 34(68), 1174–1205. 10.1590/1980-4415v34n68a16

Finlayson, M. (2014). Addressing math anxiety in the classroom. *Improving Schools*, 17(1), 99–115. 10.1177/1365480214521457

Franco-Buriticá, E., León-Mantero, C., Maz-Machado, A., and Casas-Rosal, J. C. (2019). Un análisis de las actitudes hacia las matemáticas en estudiantes para maestro de educación media en Colombia. En J. M. Marbán, M. Arce, A. Maroto, J. M. Muñoz-Escolano yÁ. Alsina (Eds.), *Investigación en Educación Matemática XXIII* (p. 617). Valladolid: SEIEM.

Ganley, C., Conlon, R., McGraw, A., Barroso, C., and Geer, E. (2021). The effect of brief anxiety interventions on reported anxiety and math test performance. *Journal of Numerical Cognition*, 7(1), 4–19. 10.5964/jnc.6065

Gill, O., and O'Donoghue, J. (2008). *A theoretical characterisation of service mathematics.* 11th International Congress on Mathematics Education Mexico. Sourced http://tsg.icme11.org/document/get/319 on 20/05/2013

Gökbulut, Y., and Şahin, S. (2022). 2010–2021 yılları arasında Türkiye'de yapılan matematik kaygısı ile ilgili lisansüstü tezlerin bazı değişkenlere göre incelenmesi [Examination of graduate thesis on mathematics anxiety done in Turkey between 2010–2021 according to some variables]. *Turkish Scientific Research Journal*, 7(1), 147–159.

González, F. (2018). *Impactos de corto plazo del programa extracurricular de refuerzo escolar "Con Las Manos" en un colegio de Bogotá* (The Short-Term Effects of the Extracurricular Program Con Las Manos in a School in Bogotá, Colombia). Documento CEDE No. 23, Disponible en SSRN: 10.2139/ssrn.3175523

Granados, R., and Pinillos, O. (2008). *Actitudes hacia las matemáticas: Un estudio con estudiantes de enfermería [presentación de paper]*. 9° Encuentro Colombiano de Matemática Educativa, Valledupar, Colombia. https://core.ac.uk/reader/12341411

Gunderson, E., Park, D., Maloney, E., Beilock, S., and Levine S. (2018). Reciprocal relations among motivational frameworks, math anxiety, and math achievement in early elementary school. *Journal of Cognition and Development*, 19(1), 21–46. 10.1080/15248372.2017.1421538

Hattie, J. (2008). *Visible learning*. Routledge.

Hattie, J., and Donoghue, G. (2016). Learning strategies: A synthesis and conceptual Model. *NPJ Science Learn*, 1, 16013. 10.1038/npjscilearn.2016.13

Helvacı, B. T. (2010). *Bilgisayar destekli öğretimin ilköğretim 6. sınıf öğrencilerinin matematik dersi "çokgenler" konusundaki akademik başarılarına ve tutumlarına etkisi.[The effect of computer assisted teaching on academic achievement of the sixth-grade student's in math lessons polygons units and their attitudes]*. Unpublished MS Thesis, Ankara Turkey: Gazi University.

Howie, S. (2003). Language and other background factors affecting secondary pupils' performance in mathematics in South Africa. *African Journal of Research in Mathematics, Science and Technology Education*, 7(1), 1–20, 10.1080/10288457.2003.10740545

Hunt, D., and Lawson, D. (1996). Trends in mathematical competency of a-level students on entry to university. *Teaching Mathematics and its Applications*, 15(4), 167–173. 10.1093/teamat/15.4.167

INDEC (2022). *The national institute of statistics and censuses.* https://www.indec.gob.ar/

INEP (2021). Instituto Nacional de Estudos e Pesquisas Educacionais Anísio Teixeira (Inep). *Sinopse Estatística da Pesquisa Resposta Educacional à Pandemia de COVID-19 no Brasil - Educação Básica*. Brasília, DF. https://download.inep.gov.br/dados_abertos/sinopses_estatisticas/sinopses_estatisticas_pesquisa_covid19_censo_escolar_2020.zip

Johnston-Wilder, S., Baker, J., McCracken, A., and Msimanga, A. (2020). A toolkit for teachers and learners, parents, carers and support staff: Improving mathematical safeguarding and building resilience to increase effectiveness of teaching and learning mathematics. *Creative Education*, 11, 1418–1441. 10.4236/ce.2020.118104

Johnston-Wilder, S., and Lee, C. (2019). How can we address mathematics anxiety more effectively as a community? In *15th International Conference of The Mathematics Education for the Future Project Theory and Practice: An Interface or A Great Divide?* Maynooth, Ireland, 4–9 Aug 2019.

Kakoma, L., and Sun Zuma, G. (2023). Instructional intervention to address mathematics anxiety in Sub-Sahara Africa: Systematic review (1980–2020). *Journal of African Education Review*. 1080/18146627.20232201660

Kartalci, S., Acar, G., Zihar, M., and Işık, C. (2021). 9 ve 10. Sınıf öğrencilerinini matematiğin doğası hakkındaki felsefik düşünceleri ile matematiksel yılmazlıklarının incelenmesi [Investigation of 9th and 10th grade students' philosophical thoughts on the nature of mathematics and mathematical resilience]. *Uşak University, Journal of Educational Research*, 7(1), 119–141.

Knight, C. (2021). The impact of the Dyslexia label on academic outlook and aspirations: An analysis using propensity score matching, *Educational Psychology*, 94(40) 1110–1126.

Kwayumba, D., Piper, B., Onyango, A., and Oyugi, M. (2017). *Longitudinal midterm report for the Tayari early childhood development and education programme prepared for the children's investment fund foundation.* Research Triangle Park, NC: RTI International.

LEE (2022). *Brechas en resultados de Pruebas Saber 11: Colombia antes y durante la pandemia por covid-19.* Informe del Laboratorio de Economía de la Educación -LEE No. 46, Disponible en: https://lee.javeriana.edu.co/-/lee-informe-46

Luneta, K., and Sunzuma G. (2022). Instructional interventions to address mathematics anxiety in Sub-Saharan Africa: A systematic review (1980–2020). *Africa Education Review*, 19(1), 103–119. 10.1080/18146627.2023.2201660

Luttenberger, S., Wimmer, S., and Paechter, M. (2018). Spotlight on math anxiety. *Psychol Res Behav Manag*, 8(11), 311–322. 10.2147/PRBM.S141421

Lynch, K., Hill, H. C., Gonzalez, K. E., and Pollard, C. (2019). Strengthening the research base that informs STEM instructional improvement efforts: A meta-analysis. *Educational Evaluation and Policy Analysis*, 41(3), 260–293. 10.3102/0162373719849044

Lyons, M., Lynch, K., Close, S., Sheerin, E., and Boland, P. (2003). *Inside classrooms - The teaching and learning of mathematics in the social context.* Dublin: Institute of Public Administration.

Mac an Bhaird, C., Fitzmaurice, O., Ní Fhloinn, E., and O'Sullivan, C. (2013). Student non-engagement with mathematics learning supports. *Teaching Mathematics and Its Applications*, 32, 191–205. 10.1093/teamat/hrt018

MEB, (2021). *Liselere Geçiş Sistemi (LGS) Merkezi Sınavla Yerleşen Öğrencilerin Performansı (Report No: 17).* Ministry of Education of Turkey. https://cdn.eba.gov.tr/icerik/2021/07/rapor/No_17-LGS_2021-merkezi_yerlestirme_211730.pdf

Mercado, D., Oquendo, K., Ávila-Toscano, J., and Vargas, L. (2018). Ansiedad estadística en estudiantes de licenciatura en matemáticas: papel de las estrategias metacognitivas y la ansiedad ante los exámenes. En S. Valbuena, L. Vargas, & J. Berrío (Eds.), *Encuentro de Investigación en Educación Matemática* (pp. 89–94). Puerto Colombia, Colombia: Universidad del Atlántico.

Mkhize, M. V. (2019). Mathematics anxiety among pre-service accounting teachers. *South African Journal of Education*, 39(3), 1–14.

Ministerio de Educación (2022). *Resultados Aprender.* https://www.argentina.gob.ar/sites/default/files/2022/06/resultados_aprender_2021.pdf

Moran, G., Perkins, R., Cosgrove, J., and Shiel, G. (2013). *Mathematics in transition year: Insights of teachers from PISA 2012.* Dublin: Educational Research Centre.

Mullis, I., Martin, M., Gonzales, E., and Chrostowski, S. (2004). *TIMSS 2003 international mathematics report.* International Association for the Evaluation of Educational Achievement. https://files.eric.ed.gov/fulltext/ED494650.pdf

Mullis, I., Martin, M., Foy, P., and Hooper, M. (2016). *TIMSS 2015 international results in mathematics.* International Association for the Evaluation of Educational Achievement. https://timssandpirls.bc.edu/timss2015/international-results/wp-content/uploads/filebase/full%20pdfs/T15-International-Results-in-Mathematics.pdf

Mullis, I., Martin, M., Foy, P. Kelly, D., and Fishbein, B. (2020). TIMSS 2019 International Results in Mathematics and Science, International Association for the Evaluation of Educational Achievement. TIMSS-2019-International-Results-in-Mathematics-and-Science.pdf (timss2019.org)

Mulryan-Kyne, C. (2010). Teaching large classes at college and university level: Challenges and opportunities. *Teaching in Higher Education*, 15(2), 175–185. 10.1080/13562511003620001

Mutodi, P., and Ngirande, H. (2014). Exploring mathematics anxiety: Mathematics students' experiences. *Mediterranean Journal of Social Sciences*, 5(1) 283–294.

Nazlıçiçek, N. (2007). *A modeling study to explain mathematics achievement of tenth grade students* (Unpublished doctoral dissertation). İstanbul: Yıldız Teknik University.

NCCA (2016). *Background paper and brief for the development of a new primary mathematics curriculum.* Dublin: National Council for Curriculum and Assessment.

NCCA (2020). *Primary curriculum framework.* Dublin: National Council for Curriculum and Assessment.

NCCA (2023). *Primary curriculum framework for primary and special schools.* Dublin: National Council for Curriculum and Assessment.

Ní Ríordáin, M., and Hannigan, A. (2009). *Out-of-field teaching in post-primary mathematics education: An analysis of the Irish context.* Research Report NCE-MSTL, University of Limerick.

Ní Shúilleabháin, A. (2014). Lesson study and project maths: A professional development intervention for mathematics teachers engaging in a new curriculum. In S. Pope (Ed.), *Proceedings of the 8th British Congress of Mathematics Education* (pp. 255–262), 14–17 April 2014, Nottingham, UK.

O'Donoghue, J. (1999). *An intervention to assist 'at risk' students in service mathematics courses at the University of Limerick.* Teaching Fellowship Report, Ireland: U.L.

OECD. (2004). *Learning for tomorrow's world: First results from PISA 2003.* Paris: OECD publishing.

OECD (2013). *PISA 2012 assessment and analytical framework: Mathematics, reading, science, problem solving and financial literacy.* Paris: OECD publishing. https://doi.org/10.1787/5f07c754-en

OCED (2019). *PISA 2018 results (volume I): What students know and can do.* Paris: PISA, OECD publishing, https://doi.org/10.1787/5f07c754-en

Oldham, E. (2006). A lot done, more to do? Changes in mathematics curriculum and assessment 1986–2006. In D. Corcoran, & S. Breen (Eds.), *Second International Science and Mathematics Conference, Drumcondra, Dublin, September 2006* (pp. 161–174). Dublin: Dublin City University.

O'Meara, N., and Faulkner, F. (2021). Professional development for out-of-field post-primary teachers of mathematics: an analysis of the impact of mathematics specific pedagogy training. *Irish Educational Studies*, 1–20. 10.1080/03323315.2021.1899026

O'Meara, N., Fitzmaurice, O., and Johnson, P. (2017). Old habits die hard: An uphill struggle against rules without reason in mathematics teacher education. *European Journal of Science and Mathematics Education*, 5(1), 91–109. 10.30935/scimath/9500

O'Reilly, M., Dooley, T., Oldham, E., and Shiel, G. (2017). Mathematics education in Ireland. In G. Kaiser (Ed.), *Proceedings of the 13th International Congress on Mathematical Education.* ICME-13 Monographs. Springer, Cham. 10.1007/978-3-319-62597-3_24

OSYM, (2021). *Sayısal veriler.* Retrieved from https://dokuman.osym.gov.tr/pdfdokuman/2021/YKS/sayisal_veriler_28072021.pdf

Perkins, R., Shiel, G., Merriman, B., Cosgrove, J., and Moran, G. (2013). *Learning for life: The achievements of 15-year-olds in Ireland on mathematics, reading literacy and science in PISA 2012.* Dublin: Educational Research Centre.

Piper, B., Ralaingila, W., Akach, L., and King, S. (2016). Improving procedural and conceptual mathematics outcomes: Evidence from randomised controlled trial in Kenya. *Journal of Development Effectiveness*, 83, 404–422.

Piper, B., Merseth, K., and Ngaruiya, S. (2018). Scaling up early childhood development and education in a devolved setting: Policy making, resource allocations, and impacts of the tayari school readiness program in Kenya. *Global Education Review*, 5(2), 47–68.

Piper, B., Sitabkhan, Y., and Nderu, E. (2018). Mathematics from the beginning: Evaluating the Tayari pre-primary program's impact on early mathematic skills. *Global Education Review*, 5(3), 57–81.

Prendergast, M., O'Meara, N., and Treacy, P. (2020). Is there a point? Teachers' perceptions of a policy incentivizing the study of advanced mathematics. *Journal of Curriculum Studies*, 52(6), 752–769. 10.1080/00220272.2020.1790666

Puentes, L. (2016). Motivación, estrategias de aprendizaje autorregulado y ansiedad matemática en estudiantes de pregrado en Arauca, Colombia. *Revista Internacional de Estudios en Educación*, 2, 62–82.

Quiceno, J. M., Remor, E., and Vinaccia, S. (2016). *Fortaleza: Programa de Potenciación de la Resiliencia para promoción y el mantenimiento de la salud [FORTRESS. Empowerment program of resilience for health promotion and maintenance]*. Manual Moderno.

Reali, F., Jiménez-Leal, W., Maldonado-Carreño, C., Devine, A., and Szücs, D. (2016). Examining the link between math anxiety and math performance in Colombian students. *Revista Colombiana de Psicología*, 25(2), 369–379. 10.15446/rcp.v25n2.54532

Ramirez, G., Gunderson, A., Levine, S., and Beilock, S. (2013). Math anxiety, working memory, and math achievement in early elementary school. *Journal of Cognition and Development*, 14(2), 187–202. 10.1080/15248372.2012.664593

Richardson, F. C., and Suinn, R. M. (1972). The mathematics anxiety rating scale: Psychometric data. *Journal of Counseling Psychology*, 19, 551–554.

Ryan, M. (2019). *An investigation into the extent and derivation of mathematics anxiety among mature students in Ireland [Unpublished Ph.D. Dissertation]*. University of Limerick https://ulir.ul.ie/bitstream/handle/10344/8146/Ryan_2019_Investigation.pdf?sequence=4

Ryan, M., Fitzmaurice, O., and Johnson, P. (2023). Investigating mathematics anxiety among mature students in service mathematics courses using the mathematics anxiety scale UK. *International Journal of Mathematical Education in Science and Technology*. 10.1080/0020739X.2023.2179950

Samuel, T., and Warner, J. (2021). "i can math!": Reducing math anxiety and increasing math self-efficacy using a mindfulness and growth mindset-based intervention in first-year students. *Community College Journal of Research and Practice*, 45(3), 205–222. 10.1080/10668926.2019.1666063

Schleicher, A. (2018). *PISA 2018: Insights and interpretations*. Retrieved from https://www.oecd.org/pisa/PISA%202018%20Insights%20and%20Interpretations%20FINAL%20PDF.pdf

Sithole, A., Chiyaka, E., McCarthy, P., Mupinga, D., Bucklein, B., and Kibirige, J. (2017). Student attraction, persistence and retention in STEM programs: Successes and continuing challenges. *Higher Education Studies*, 7(1), 46–59.

Smith, T., and Capuzzi, G. (2019). Using a mindset intervention to reduce anxiety in the statistics classroom. *Psychology Learning and Teaching*, 18(3), 326–336. 10.1177/1475725719836641

Spangenberg, E., and van Putten, S. (2020). Relating elements of mathematics anxiety with the gender of preservice mathematics teachers. *Gender and Behaviour*, 18(2).

Sümen, Ö. (2013). *Geogebra yazılımı ile simetri konusunun öğretiminin matematik başarısı ve kaygısına etkisi [The effect of teaching symmetry subject by Geogebra software to mathematics success and anxiety]*. Unpublished MS Thesis, Samsun, Turkey: Ondokuz Mayıs University.

TIMSS (2019). *Trends in international mathematics and science study* https://timssandpirls.bc.edu/timss2019/

Turkish Council of Higher Education (2023). Thesis center. Retrieved May 02, 2023, from https://tez.yok.gov.tr/UlusalTezMerkezi/giris.jsp

Uwezo (2016). *Are our children learning: Kenya sixth learning assessment report*. UNESCO IIEP Portal https://learningportal.iiep.unesco.org/en/library/are-our-children-learning-2016-uwezo-kenya-sixth-learning-assessment-report

Villamizar, G., Araujo, T., and Trujillo, W. (2020). Relationship between mathematical anxiety and academic performance in mathematics in high school students. *Ciencias Psicológicas*, 14(1), 1–13. 10.22235/cp.v14i1.2174

Yeager, D., Carroll, J., Buontempo, J., Cimpian, A., Woody, S., Crosnoe, R., Muller, C., Murray, J., Mhatre, P., Kersting, N., Hulleman, C., Kudym, M., Murphy, M., Duckworth, A., Walton, G., and Dweck, C. (2022). Teacher mindsets help explain where a growth-mindset intervention does and doesn't work. *Psychological Science*, 33(1), 18–32. 10.1177/09567976211028984

Yenilmez, K., and Özbey, N. (2006). Özel okul ve devlet okulu öğrencilerinin matematik kaygı düzeyleri üzerine bir araştırma [An investigation of the mathematics anxiety levels of public and private school students]. *Uludağ University Journal of Department of Education*, 19(2), 431–448. Retrieved from. https://dergipark.org.tr/tr/pub/uefad/issue/16684/173379

Yorgun, A., and Mert, S. (in press). Investigation of mathematical experiences of high school students in Turkish context. *Journal of Humanistic Mathematics*.

Mathematical resilience global developments

International authors

Introduction

This chapter consists of reports about the way that the authors from the same countries as Chapter 13 have developed ways to introduce mathematical resilience to their learners. They discuss the way that the principles behind mathematical resilience are helping them to work to mitigate the effects of mathematics anxiety for learners in their country. The contributors are the same as in Chapter 13, so once again they come from countries across the globe: Argentina, Brazil, Colombia, Ireland, Kenya, Southern Africa, Turkey and USA. Also, the contributions are in the authors' own voice, a voice that reflects their position on the continuum of those interested in developing mathematical resilience in learners. The continuum ranges from those in academia with a reach that can affect many learners to those working with individual learners. Each has a story to tell that offers valuable insights into how mathematical resilience could address mathematics anxiety to the benefit of societies worldwide.

Brazil

Telma Silveira Pará, João dos Santos Carmo, Karina Lumena de Freitas Alves

Although there are few Brazilian studies of reversing mathematics anxiety, the promising results they have so far yielded indicate that we are on a productive path of investigation.

A first reversal study, reported in the national literature, is by Colombini, Soji and Pergher (2012). These authors conducted a behavioural-analytical study that sought to improve the relationship between a 16-year-old adolescent and mathematics. During the consultations, coping strategies were developed, valuing successes, modelling pro-study behaviours and providing a model to perform mathematics tasks. The authors point to an increase in the frequency of pro-study behaviours, improvement in school grades, and a probable decrease in math-related anxiety.

To reduce mathematics anxiety, Haase et al. (2013) proposed an intervention that is partly focused on pedagogical aspects and partly dedicated to cognitive-behavioural psychotherapy. The results showed a significant reduction in mathematics anxiety and 100% accuracy in mathematics performance.

The proposal of a multidimensional intervention by Haase et al. (2013) to reduce mathematical anxiety has also been tested and expanded by other researchers. A program to help students with mathematical anxiety was developed at the Federal University of São Carlos by the second author of this chapter. This is an intervention project offered to Psychology students from the third to the sixth semester of the Psychology course at the Federal University of São Carlos. The experience has been developed in public and private schools responsible for Elementary and High School education. The proposal aims to help students with high and extreme degrees of mathematics anxiety to deal productively with the situation, through interventions and an alteration of specific contingencies of their daily lives. The intervention program involves the following steps:

1. Identification of high and extreme degrees of mathematics anxiety through the application of a Mathematics Anxiety Scale (named MathAS), developed by Carmo (2008) and psychometrically validated by Mendes (2016). In addition to the scale, an inventory of mathematics study habits and a brainstorming technique was applied. This last technique helped us to identify the conceptions and attitudes that the participant presents about mathematics.

2. From this stage, the attendances are individualised. Each participant is taught the use of some relaxation techniques: diaphragmatic breathing; Jacobson's progressive muscle relaxation; systematic desensitisation.

3. Each participant is taught study habits and social skills in the classroom.

4. At this stage, pre-test instruments are reapplied in order to verify if there was any reduction in mathematics anxiety.

Parallel to the attendance of the participant, the facilitator of the program meets fortnightly with one of the student's parents and with the mathematics teacher to verify how the student is performing in homework and in the classroom. At each interview, the facilitator provides tips and instructions to parents and the teacher about the organisation of the student's study environment and the release of reinforcements. The results have been promising and are reported in Mendes, Carmo and Muniz (2020), and Carmo and Crescenti (2022).

De Souza Domingues et al. (2022) analysed the effects of physical exercises on the brain, in cognitive and academic performance and mental health, from the perspective of Neuroscience, and propose new investigations on the effects of physical exercise on mathematics anxiety. The authors show some positive evidence using qualitative data of the effects of physical exercises on the reduction of mathematics anxiety.

A recent study was performed in a public high school of Rio de Janeiro aimed at addressing mathematics anxiety and building mathematical resilience among students. It was the first action in Brazil that used the Mathematical Resilience Toolkit - Growth Zone Model (Lee and Johnston-Wilder no date); the Hand Model of the Brain (Siegel, 2010) and the Relaxation Response (Benson, 2000). The study also used Participatory Action Research (PAR) as methodology (Pará and Johnston-Wilder, 2023). The levels of mathematics anxiety were measured before and after the intervention using the revised MAS scale (MAS-R), and the data indicated positive evidence in performing this kind of intervention. The study also presents a concept based on the Scottish Report (Homes and Grandison, 2021) that was adapted and that added to the literature on building Mathematical Resilience: anxiety-informed mathematics teaching which considers the necessity of adopting practice that raises awareness of anxiety, recognises mathematics anxiety in learners and seeks to resist causing further harm.

There is still a long way to go for Brazilian researchers dedicated to addressing mathematics anxiety. It will be necessary to build and maintain international bridges with partners of renowned institutions and researchers abroad. This is crucial for the development of research in building mathematical resilience that can make a difference in students' lives in Brazil. We can mention some initiatives: a recent project coordinated by Professor João dos Santos Carmo from the Federal University of São Carlos in collaboration with the University of Warwick in the United Kingdom with the supervision of Associate Professor Sue Johnston-Wilder. The project adopts the knowledge and apprehension of the methodologies used in the UK, verifying "in loco" how they may be applied in Brazil. These methodologies propose two types of interventions: 1 to 1 and group interventions. The 1 to 1 intervention aims to develop mathematical resilience using three tools, the growth zone model, the hand model of the brain and the relaxation response to reduce mathematics anxiety. The group interventions aim to combine the study with the issue of using STEAM education (Science, Technology, Engineering, Arts, and Mathematics) to minimise mathematics anxiety and develop mathematical resilience.

Ireland

Maria Ryan

While international evidence shows that mathematics anxiety exists across the education spectrum as well as in everyday contexts (Buckley et al., 2016; Dowker et al., 2016), there is a paucity of research on mathematics anxiety in the Irish context (Ryan, 2019). Consequently, the prevalence of mathematics anxiety in Ireland has not been quantified, international tests such as PISA provide limited insights into the existence of mathematics anxiety among Irish 15-year-olds (Perkins et al., 2013). Ryan's (2019) doctoral thesis explored mathematics anxiety among undergraduate mature students in Ireland and was the first PhD thesis to exclusively look at mathematics anxiety in the Irish context. Subsequently, O'Hanlon's (2023) mathematics anxiety research explored the environmental antecedents of mathematics anxiety among further education (FE) students.

Awareness and exploration of mathematics anxiety and mathematical resilience in Ireland has been heightened through a variety of events in recent years, including information seminars, workshops, professional development events, and conference presentations, and hosted by a mix of educational organisations, for example:

- Initial Teacher Education providers around Ireland run dedicated mathematics anxiety workshops for student teachers in preparation for the school placement experience. For example, at Mary Immaculate College – School of Post-Primary Education one such workshop focuses on numeracy and the impact of mathematics anxiety (Ryan, 2022).

- Research by Santos at University College Dublin has led to the development of the Teachers' mathematics anxiety classroom management programme under the ARITHMOS project (Santos, 2022).

- Regional Education Centres run in-person and online professional development seminars for primary and secondary teachers, special needs assistants; parents; educational leaders. For example, in 2020 mathematics anxiety awareness webinars were run by seven education centres throughout Ireland (Ryan, 2020).

- The Teaching Council's Researcher-in-Residence Scheme (RiRS) was launched in 2021 and links HE researchers with schools, whereby the former will help the latter with an identified research project relating to a number of areas – including numeracy – as outlined in the Cosán framework (Teaching Council, 2022). One such project includes addressing mathematics anxiety within the Junior Cycle (ages 12–15) in the secondary school context (Teaching Council, 2022).

- Irish Learning Support Association (ILSA) hosted a presentation on mathematics anxiety at their annual conference in 2022.

- The National Adult Literacy Agency (NALA) helps adults to improve their literacy, numeracy and digital skills, and conducts training for tutors of literacy, numeracy and digital skills (NALA, 2023). They also conduct webinars and workshops on mathematics anxiety and dyscalculia, to heighten awareness and give guidance for numeracy tutors and students.

- EpiSTEM 'Numeracy Meets' (in collaboration with NALA) adult numeracy tutor continuing professional development webinar series, including one webinar dedicated to 'overcoming mathematics anxiety' with a focus on the role of the numeracy tutor (EpiSTEM, 2022).

- Munster Technological University (MTU) hosted the 'Raising Awareness Around Dyscalculia' (RAAD) project in 2022.

- Mathematics Resilience Network – Ireland branch was officially launched in June 2023 at MTU.

- Conferences dedicated to mathematics learning difficulties, mathematics anxiety and mathematical resilience including Sustaining Mathematical Accessibility at University College Dublin (March 2023), and Developing Awareness of Mathematics Anxiety Workshop (MRN-Ireland branch) at Munster Technological University, Cork (June 2023).

The launch of the MRN-Ireland branch in June 2023 provided a platform to raise awareness of mathematics anxiety and mathematical resilience within an Irish context as well as providing an opportunity to gauge the extent of research projects with a focus on mathematics anxiety and mathematical resilience. The range of topics spanned the higher, further and primary education contexts, as well as initial teacher education in both the Republic of Ireland and Northern Ireland:

- Higher education
    - Mathematics anxiety and the impact on the adult learner (Ryan, M.);
    - Raising awareness around dyscalculia and mathematics anxiety in higher education (Murphy, R. Casey, D. and Crowley, J.);
    - Mathematics anxiety in undergraduate business students (McCullagh, O.);
    - Barriers to a career choice in STEM: untangling mathematics anxiety and mathematics self-efficacy (Gomides, M.)

- Further education
    - Mathematics anxiety and adult numeracy education in Ireland (Prendergast, M., O'Meara, N., O'Sullivan, K. & Faulkner, F.)

- Primary education
    - Untangling mathematics anxiety, attitudes towards mathematics, home numeracy and mathematics performance (Santos, F.)

- Initial teacher education
    - Cognitive behavioural therapy for pre-service primary school teachers experiencing mathematics anxiety (Carroll, T.);
    - Implementing mastery approaches to reduce mathematics anxiety using pedagogical action research with students in initial teacher education (Northern Ireland) (Parks, G.).

While the findings of the aforementioned research projects provide welcome contributions to the body of knowledge for the Irish context, there is much scope to extend research on mathematics anxiety and mathematical resilience in this jurisdiction, especially within the early years, primary and secondary education sectors, both for pre-service and in-service teachers, as well as within the further education sector. Close collaboration across professional networks and with higher education institutions must be central to continuing to address the various issues that impact mathematics teaching and learning across the education spectrum. Such collaboration will also heighten awareness of mathematics anxiety among educators, students, parents and all others with a vested interest in helping to address mathematics anxiety and encourage mathematical resilience.

In conclusion, while there is collective impetus to spread the word on mathematics anxiety and mathematical resilience in Ireland, there is still a dearth of research in this area and across education sectors, including in initial teacher education and as part of teacher professional development. Research into mathematics anxiety would benefit from an understanding of how prevalent it is in the Irish context across the education system; normative data to that effect would be welcome to allow for the quantification of mathematics anxiety in Ireland. Further, the outcomes of more extensive research on mathematics anxiety and mathematical resilience in the Irish context would contribute immensely to the teaching and learning of mathematics across Ireland's education sectors, and with relevance to institutional, national and international contexts.

Colombia

Japcy Margarita Quiceno and Stefano Vinaccia

Resilience is a construct that has undergone many years of evolution. Therefore, it is not surprising to find a diversification in its definition, even within the same discipline. In the field of education, the term educational resilience can be found, and within this, there is the concept of mathematical resilience. The common thread among all definitions is that humans have the ability to adapt and bounce back during challenging life situations. Regarding mathematical resilience, it is defined as "the positive attitude that helps students overcome obstacles during the process of learning mathematics" (Johnston-Wilder et al., 2015, p. 1). This learning process is mediated by multiple factors, ranging from individual to psychosocial and contextual, and is evidenced in mathematics outcomes by country.

In several areas of knowledge, including mathematics, Colombia has consistently performed significantly below average on both national and international exams, ranking among the countries with the lowest scores (Díaz-Pinzón, 2021; Sanabria-James et al., 2020). The gender gap is marked and women consistently score lower than men, affecting their choice of STEM careers (Botero-Guzmán and Marín-Díaz, 2022).

According to Murcia and Henao (2015), several factors are likely to contribute to poor academic performance in mathematics in Colombia, including: a) independent contents and competencies without integration with other knowledge, b) errors in school promotion procedures, c) educational practices imported from other countries, which, although successful, have not been contextualised to the reality of the country, d) the proliferation and diversity of curricular proposals as The General Education Law grants curricular autonomy to each institution in the elaboration of the Educational Project, e) cognitive difficulties and unfavorable reading and comprehension levels in students, f) absence of family support, and g) inappropriate practices and pedagogical models used by teachers. In addition, teacher remuneration is inadequate and is not commensurate with their responsibilities (Herrera, 2020). Furthermore, prolonged violence, forced displacement, poverty, hunger, and inequality have negatively impacted the student population, hindering healthy and stable school development (Caro and Kárpava, 2020).

However, despite years of efforts by the Colombian government to address deficits in educational quality through teacher training and program promotion, the results are worrying and continue to be a warning sign that something is not working properly, from primary to higher education. In addition, some professionals in the field of education and even health have been working to identify the factors that enable some children to develop and achieve better academic results than others even though they come from the same socioeconomic background, and the same challenging conditions of Colombian reality. In search of an answer, they have stumbled upon the construct of resilience.

Now, in academic circles in Colombia, psychologist Japcy Margarita Quiceno has tackled the topic of mathematical resilience and mathematics anxiety. In 2018, she addressed those topics in the "Elective Seminar IV" course of the master's degree in education at the University of Medellin, Colombia. The students in this master's course were mathematics teachers. Later that year, Professor Quiceno was invited to talk about mathematical resilience and mathematics anxiety at a "special conference" and a "short course" at the XXXII Latin American Meeting of Educational Mathematics hosted by the University of Medellin.

Three years later, in 2021, she was invited as a speaker to a "special conference" in the cycle of conferences on "Elements of Neuromathematics" led by the University of Medellin (Colombia) and the Pontifical Catholic University of Valparaiso (Chile). Her lecture, entitled "Mathematical Resilience: A Positive Protective Factor to Cope with Math Anxiety", aimed to present the advances and developments and the most representative leaders in the subject of mathematical resilience and mathematics anxiety to the academic community in mathematics, going beyond what had been presented in previous events. From this last conference, a review document on the same topic was derived from her work.

During the last conference, Professor Quiceno had her first communication with Sue Johnston-Wilder, one of the authors in this new area of knowledge, who provided relevant information for Professor Quiceno to continue her work. Furthermore, Professor Quiceno has invited Professor Stefano Vinaccia, another expert in resilience and positive psychology, to collaborate on joint projects.

It is worth noting that most studies on resilience in the educational field in Colombia have focused on assessing levels of resilience, distinguishing and identifying resilient students (Alvarán-López et al., 2021; Dueñas et al., 2019; Vergel-Ortega et al., 2021), determining resilient factors associated with academic performance (Velásquez et al., 2022), and even designing instruments for this purpose (Peralta et al., 2006), as well as assessing academic resilience (Bayona-Rodríguez and López-Vera, 2021).

In the area of mathematical resilience, studies are scarce. There are isolated investigations that have sought to analyse the relationship between academic performance in mathematics and school success. For instance, Santiago-Carrillo et al. (2020) evaluated resilience in 1500 students from different university programs who were studying between the first and fourth semester and who were enroled in at least one mathematics course. This study concluded that successful students demonstrate creative/formal thinking skills, application of mathematical ideas, and resilience.

Colombian studies on resilience in educational settings have used general scales, such as Wagnild and Young's Resilience Scale and CD-RISC 10, or specific scales designed for school children, such as the JJ63/JJ43 Scale, Salgado-Lévano's Resilience Inventory for Children, and CRE-U Questionnaire for University Students by Peralta-Díaz et al. However, there is an urgent need for the development of studies that prioritise the design and validation of specific scales for measuring mathematical resilience. This would enable the results of the studies to be more objective as they would be related to the construct and would facilitate the development of research lines in mathematical resilience focused on assessment and intervention. Furthermore, with regards to public policies, proposals could be further refined and contextualised to ensure that the strengthening of each resilient factor (individual, socio-family, and socio-educational) is both permanent and transversal throughout the entire academic and evolutive cycle, so that students can have a positive experience with mathematics, regardless of their socioeconomic status.

Colombian studies emphasise the importance of continuing to measure resilience in academic environments (Velásquez et al., 2022) and promoting it through intervention programs that involve families, teachers, and students (Acevedo and Restrepo, 2012), especially in regions where students have low to medium levels of resilience (Alvarán-López et al., 2021; Sánchez-Arias, 2016) and also to foster hope in the construction of peace in Colombia (Pérez-Ibarra, 2017). Although there are resilience intervention programs for students, teachers, administrators, and graduates of higher education (Camargo-Goyeneche et al., 2014), and for students between 8 and 11 years of age in public schools (Sierra-Barón, 2012), these may not be sufficient to address the social and educational complexity that the country is facing. Furthermore, it is recommended that Colombia focuses on interventions in mathematical resilience, with a particular emphasis on cognitive factors associated with paradigms and irrational beliefs towards numbers. This could help to reduce the gender gap and strengthen the self-confidence of women in their choice of STEM professions.

In Colombia, it is suggested that strengthening collaborative work in the field of mathematical resilience with other disciplines specialised in positive emotional health to design interventions would be beneficial. Emotional mobilisation can have an impact on the entire psychological structure of an individual; thus, professional training is necessary in this area as it could exacerbate health issues and increase demotivation, especially towards numbers, in the participating individuals. In Colombia, there is a program that focuses on empowering resilience through positive targeting at the emotional and cognitive level (see Quiceno et al., 2016).

The literature has consistently shown that positive emotions contribute to balance the affective state in mathematics learning (Villavicencio and Bernardo, 2016), facilitate mathematical problem solving (Greensfeld and Deutsch, 2022), promote mathematics self-regulation strategies and achievement (Ahmed et al., 2013), and enhance self-efficacy (Liu et al., 2018). Although success and failure in learning are mediated by cognitive competence, emotional competencies are buffering and protective variables that help individuals cope with the vicissitudes of learning and promote resilience.

The difficulties that children face in their relationship with mathematics can offer opportunities to develop active coping strategies and enhance their resilience (Darhim and Dahlan, 2019). While mathematics classes can provide valuable environments for fostering resilience, repetitive and decontextualised activities, or excessive demands can result in *"routine fatigue"*, as described by Japcy Quiceno, which can impede learning, motivation, and resilience.

In conclusion, there is a compelling demand for greater advancement and development of mathematical resilience in all dimensions in Colombia. This requires the involvement of the entire academic community, including students, teachers, parents, and school staff, as well as the government. It is essential to instill in children from their earliest years of schooling habits and styles that foster a resilient attitude towards mathematics. This will not only lead to improved academic performance in mathematics but also enable them to apply mathematical concepts to real-life situations and their surrounding contexts.

Turkey

Abdulvahap Yorgun

Interestingly, there are very few studies on how to cope with or treat mathematics anxiety in Turkish literature. A few studies aimed to reduce mathematics anxiety using different teaching techniques (Cengiz, 2017; Demir, 2017; Helvacı, 2010 and Sümen, 2013). However, therapeutic methods used in psychotherapy to cope with anxiety can be used to reduce mathematics anxiety too. The main therapeutic methods may be listed as systematic desensitisation, exposure, expressive writing, guided imagery, ABCD Analysis, EMDR. Individual or group programs including these techniques to help students alleviate their mathematics anxiety may provide effective outcomes.

On the other hand, the Turkish Ministry of National Education announced that a Math Campaign has been started in May 2022. The main goals of the Math Campaign are to make students love mathematics using different teaching techniques and strategies. These strategies include some real-life activities showing the mathematics in daily life, technology-based techniques, etc. Unfortunately, although it is one of the main reasons behind the students' negative attitudes towards mathematics, there is no focus on mathematics anxiety.

Similarly, Alkan (2018) reviewed 59 papers on mathematics anxiety and none of them included empirical methods. Therefore, there is a gap in Turkish literature related to treating mathematics anxiety and the development mathematics anxiety intervention programs will contribute to filling this gap. Such programmes could provide school counsellors with a therapeutic tool to intervene in mathematics anxiety. In this way, they may help learners to decrease their mathematics anxiety, increase their mathematics success and develop positive attitudes towards mathematics, resulting in loving and learning it.

The author of this chapter has developed the Mathematics Anxiety Intervention Programme (MAIP) to help high school students cope with their mathematics anxiety. This is the first pyschoeducational program developed to help students reduce their mathematics anxiety in Turkey. The program contains ten sessions, and the goals of the sessions are developing mathematical resilience, reconceptualising adverse memories related with mathematics, being aware about personal mathematics attitudes and myths and gaining emotion regulation skills. This program is a psycho-educational group and this type of group is an effective and widely used method in school counselling, teaching learners certain life skills such as anger management, stress management, test anxiety intervention etc. (Brown, 2010).

A typical session structure begins with a warming-up exercise, the summarising or remembering the previous session, working on the content of the present session and evaluation and termination of it. In the first session of the MAIP, group rules are established, the group process is introduced and the expectations of participants are evaluated. In the second session, a lecture is given on the neurobiology of traumatic anxiety. The relationship between anxiety, its effect on the brain and body and learning is discussed. In this session, a role-play is performed involving the reaction of various organs during anxiety. In the third session, participants are asked to write their mathematics stories reflecting their relationship with mathematics. This is known as an expressive art technique and facilitates the expression of emotions. In the fourth session, they create a Mathematics Anxiety Graph and determine their lowest and highest anxiety states. In addition, they fill out an attitude inventory to identify the myths they know about mathematics. In the fifth session, the ABCD model is used to analyse adverse memories related to mathematics. In the sixth session the Growth Zone Model is used. In the seventh session, a conference about emotional regulation is held. The members are given some scenarios and they rehearse how they would use the 4 emotion regulation skills in these scenarios. In the eighth session a conference about the four dimensions of mathematical resilience is used. The members share their "Past Positive Experiences with Mathematics" making them aware of their strengths. In the ninth session, the Miracle Question technique is used. The participants imagine that after a miracle they cure their mathematics anxiety totally. Given this scenario how do they behave? What do they do? What differences are perceived? They are then asked to write a personal slogan to give them power and hope. The tenth session is reserved for evaluation, feedback and endings.

The pre-test – post-test control group design was used to investigate the effectiveness of the program and statistical analysis showed that there are significant differences between control and experiment groups mathematics anxiety post-test scores and also between pre and post-test scores of experiment group. Next, there are some scenes from the various sessions (Figure 14.1).

In addition, the concept of mathematical resilience (Johnston-Wilder et al., 2020) provides children with a holistic structure for recognising and coping with mathematics anxiety. With its four-dimensional structure, the mathematical resilience model can be integrated with emotion regulation skills. Meanwhile, two research projects are investigating mathematical resilience in the Turkish context.

Figure 14.1 The Growth Zone Model in action.

One of the includes adaptation of the mathematical resilience scale into Turkish culture and the other focusing on the mathematical resilience level of students (Kartalci et al., 2021; Çetin et al., 2018). In the future, various programs to teach students the four dimensions of mathematical resilience may be developed and implemented at schools.

Kenya

Harrison Njaru Mbogo

Mathematics learning needs not to cause stress and anxiety even though it is not perceived as easy by majority of the children. Enjoyment of learning mathematics relies on the meaningfulness and dynamics of the mathematics concepts being learned. It is, therefore, an essential role of the teachers to guide and direct mathematics learning in such a way that will safeguard children from anything which may arouse mathematics anxiety. Ideally, children should be instructed in a way that will enhance optimal mathematics functioning in the classroom (Johnston-Wilder et al., 2021).

Mathematics anxiety is known to limit the children's career base. Fortunately, research in mathematics (Camilla and Hanna, 2020; Trakulphadetkrai, 2018) reveals that mathematical story picture books can safeguard young children from mathematics anxiety. The very important reason why mathematical story picture books may aid in building sustainable mathematical resilience has to do with the meaningful rich emotional mathematical context of the stories included in the mathematical story picture books.

This international subsection contributes to the mathematical resilience building toolkit geared towards aiding teachers and children in navigating the mathematics growth zone without the danger of experiencing mathematics anxiety through mathematical story picture books. Research is ongoing that surfaces how young children are persistently underachieving in mathematics (OECD, 2013) necessitating the need for a mathematics anxiety safeguarding toolkit in order to develop mathematical resilience. The mathematical resilience toolkit is developed to allow children with mathematics anxiety to acquire an affective adaptive stance towards mathematics (Johnston-Wilder et al., 2020). Mathematically resilient children develop approaches to mathematics learning which aid them to overcome affective mathematics setbacks and barriers that may essentially be part of learning mathematics. A positive affective adaptive stance towards mathematics may be engineered through explicit focus on the culture of mathematics teaching, using approaches that enhance mathematical resilience. As part of this engineering, this study develops a notion of teaching mathematics through mathematical story picture books. These books are aimed at enhancing the development of mathematical resilience that will aid learners to navigate the mathematics growth zone, without the danger of falling into the mathematics anxiety zone.

The work described here is deliberately focused on developing approaches that explicitly develop mathematical resilience in learners experiencing mathematics anxiety through mathematical story picture books. Based on the Johnston-Wilder et al. (2020) mathematical resilience toolkit the books develop a group culture of 'can do' mathematics which is intended to work to counter the prevalent culture of mathematics helplessness and mathematics anxiety in Kenyan schools and in the general population when faced with mathematics.

Many children in mathematics need to develop a more positive affective stance towards learning mathematics. William (2013) describes it as, "When a child is given something challenging to do the first reaction is to say, 'Can I do this or not'". Many children go down one or two pathways, one is the well-being pathway which is to preserve one's sense of self and well-being. If the child thinks they can do the task, then they will do the task because they will get approbation from their teacher. If they think they cannot do the task, then they may switch off and be disengaged on the grounds that it is better to be thought lazy rather than stupid. Children usually decide not to try the mathematics task.

This mathematics behaviour disorder leads to attitudes of self-exclusion from mathematics and a self-depriving focus on an individual's inability to engage with mathematics. On the other hand, other children faced with a challenging task may choose to activate a growth pathway, they may choose to use the situation as an opportunity to improve their ability to get smarter. The important thing is that teachers learn to aid children in managing their emotions so that they can engage in the growth pathway. The growth pathway will enable learners to engage in the task in order to improve their capability-rather than just focusing on preserving their sense of well-being (Figure 14.2).

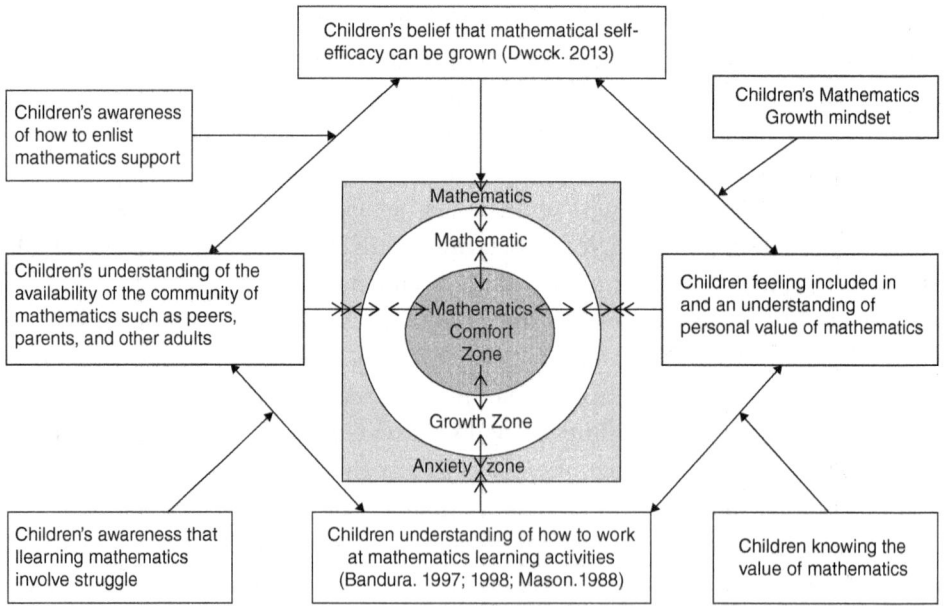

Figure 14.2 Schematic diagram showing how children can be assisted to navigate mathematics growth zone through mathematical story picture books.

Using the Growth Zone Model (Johnston-Wilder et al., 2020), I built an affective intervention through the mathematical story picture books designed to aid children to navigate and stay in the mathematics growth zone. The children need to know in advance that being in mathematics growth zone may trigger productive level of adrenalin and not too much, but just right. The mathematics learning environment needed is one of trust, courage, articulation, collaboration and patience. Ideally, children will find themselves motivated and having opportunities to enter into mathematics growth zone often when being encouraged to take on mathematics tasks while appropriately supported through mathematical story picture books.

Southern Africa

Brighton Mudadigwa, Sakyiwaa Boateng, Nomzamo Xaba, Tawanda Chinengundu, Folake Adelabu and Royda Kampala

It is against the background of prevalent mathematics anxiety that the Southern Africa branch of the Mathematics Resilience Network was formed. The network aims to explore the STEM anxiety levels of undergraduate university students and further develop interventional strategies to build their resilience in mathematics, science, and technology subjects. Our initial project in South Africa endeavours to answer three questions: (a) How prevalent are high anxiety levels among preservice teachers doing STEM subjects? (b) To what extent do high anxiety levels in STEM subjects affect preservice teachers' performance? and (c) To what extent does the intervention plan develop preservice teachers' resilience in STEM subjects?

At the beginning of the project, the study employed Action Research (AR) to develop resilience-informed practices among STEM lecturers and their students (purposeful sampling). Krause and Eilks (2019) highlight that AR thoroughly connects domain-specific STEM education research to curriculum development and teaching practices. An extension of this study will employ a design approach in high schools to coach teachers on their practices to reduce STEM anxiety levels among learners and teachers through an intervention design. The study intends to use interventions, questionnaires, semi-structured interviews, and field notes to collect data. It will be carried out in five institutions of higher learning (four in South Africa and one in Zambia), with over 500 first- to third-year students in mathematics, science, and technology education.

The lecturers initially travelled to Warwick University for training on how to use the mathematics anxiety reducing and resilience building tools. The Hand Model of the Brain (HMB), the Relaxation Response (RR), the Ladder Model, and the Growth Zone Model (GZM) (Johnston-Wilder et al., 2013) are the intervention tools. These tools will be employed to support pre-service teachers in managing their emotions, reducing anxiety, and developing resilience in STEM subjects. The preliminary results indicate that pre-service teachers have anxiety about STEM subjects. The extent of the levels of anxiety will be revealed after the analysis has been completed. The final results will highlight the effectiveness of the intervention program in reducing mathematics anxiety and developing STEM resilience among preservice teachers in Southern Africa.

We assert that working to help those who are willing to study STEM is reasonable for educational and economic reasons in countries with high skill shortages. This project also brings to the fore the importance of collaboration among researchers and higher education institutions to share practices on how to best build STEM resilience.

Argentina

Silvia Renata Figiacone

I have learned, through the years of working with learners who have issues with learning mathematics which seem to amount to dyscalculia and have usually been caused by mathematics anxiety, that there is one real solution these issues: hands on learning (and learning made visible) plus developing mathematical resilience.

As a clinical neuropsychologist, I work every day with children who have been diagnosed with a specific learning disorder. Each individual has a cognitive rehabilitation process planned for them. To do this, we define visible and explicit objectives that should be addressed, and we have an *achievement corner* where anybody can post an achievement which has made them proud (Figure 14.3).

Figure 14.3 "I managed to do divisions. I am happy. Fran."

Building resilience and independent learning are our prime objectives. Everything else aligns behind that intention. Resilience as considered by Lee and Johnston Wilder (2017) goes beyond mathematics and is a concept which describes the phenomenon of how some young people avoid negative consequences and succeed despite significant adversity.

Having a learning disability faces a learner with stress and adversity at school almost every single day. In NeuroEduca, we conduct neuropsychological rehabilitation programmes for children with dyscalculia and other developmental disabilities. For each neuropsychological rehabilitation process, we design a plan that takes into account five pillars and specific objectives (Figure 14.4):

Building rapport is essential to let resilience grow, and to reduce mathematics anxiety. Hence, we focus on building a *positive climate* in every session, using secure attachment (Bowlby, 2005) as a framework. Figiacone (2021) makes clear that it is not possible to learn in a classroom where the emotions that circulate are perceived and constructed as negative. Neither is learning possible if the learner does not value the relationship they have with the person who teaches them. Figiacone (2021) sees emotion and relationship as the keys to learning in any educational space and that these aspects require as much work, planning, and design as the presentation of the content.

Psychoeducation is fundamental to growing as learners and increasing resilience. Learners need to know what is happening to them and that they can take control of their learning. We offer psychoeducation about dyscalculia, learning, mathematics anxiety and of course, mathematical resilience. One of the most powerful strategies we found to psychoeducate children about the process of learning is the Growth Zone Model (GZM) (Lee and Johnston-Wilder, 2018). We

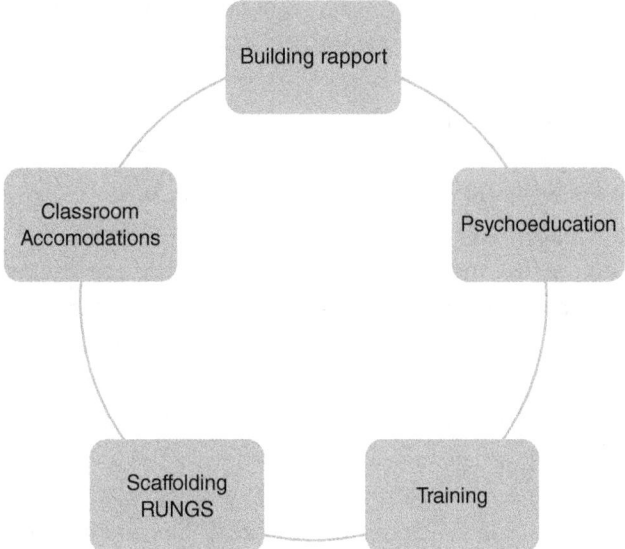

Figure 14.4 The 5 pillars.

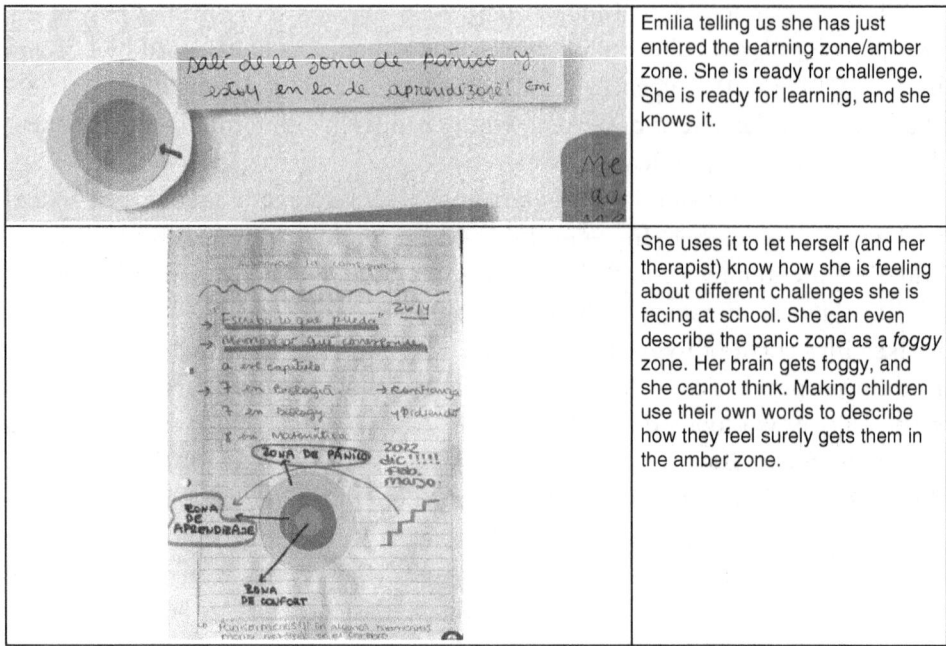

Figure 14.5 Learners' illustrations of their feelings when learning mathematics.

use it in combination with the ZPD (Vygotsky, 1979) to help the children know how they feel and what they need to feel better. Then we help them reflect and post in our achievement corner when they feel they understand (Figure 14.5).

When Emilia told us she has entered the learning zone/amber zone, she knew she was ready for learning. She uses the GZM idea to let herself (and her therapist) know how she is feeling about different challenges she is facing at school. She even describes the panic zone as a *foggy zone*. Her brain gets foggy, and she cannot think. Encouraging children to use their own words to describe how they feel surely gets them in the amber zone.

Training is indispensable for learning. Dehaene (2019) considers that learners must train every day to let their brain consolidate what they have learnt and make powerful memories. Training is essential to reduce anxiety and build resilience. Building resilience is important to increase mental well-being, deal with stress, perform better academically and "enjoy greater productivity" (Ang et al., 2022). Training should consciously avoid the GZM Panic Zone. If a learner is in the panic zone, they will be stressed and if they are stressed, they will not be learning, they may even be fighting, freezing, or escaping. In any case, not learning but surviving.

We use the GZM to help children know explicitly where they are at training sessions. They need to know where they are, and they need to tell us where they are, so we can scaffold their learning effectively. Considering the ladder model, as

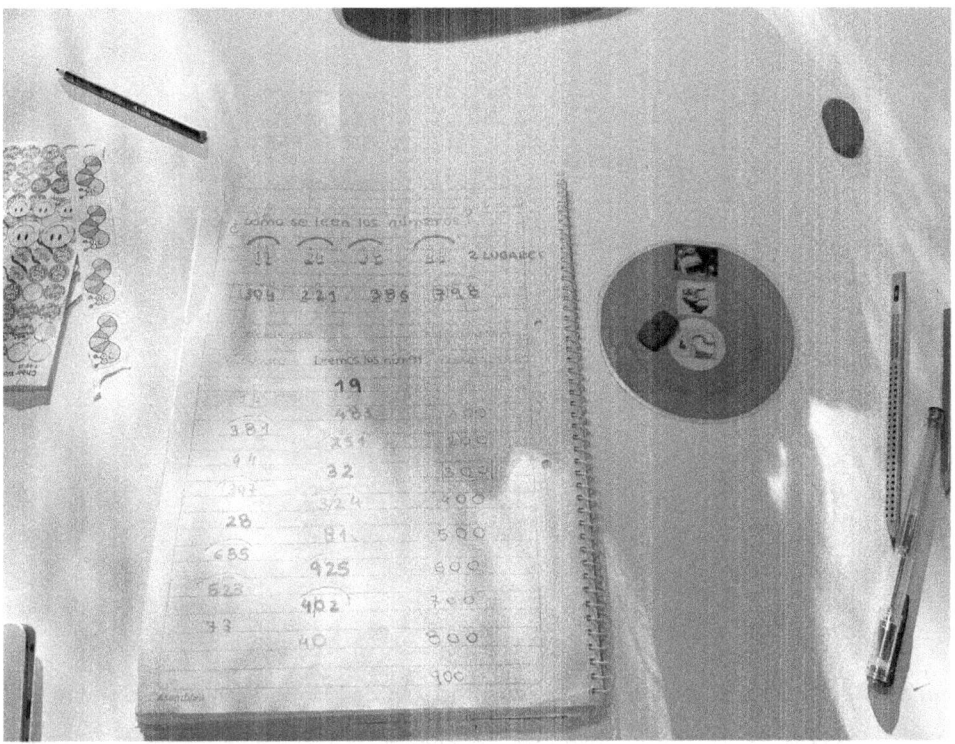

Figure 14.6 Ana using the Growth Zone Model.

shown in Chapter 3, we need to use as many rungs as every child needs, so it is an artisan work in the hands of the clinician.

In Figure 14.6, Ana used the Growth Zone Model to let herself and her neuropsychologist know where she is when training. She can train if she focuses on where she is. She can address any difficulties she sees coming, she can ask for help, and she can struggle in safety. She uses a little gem to move from zone to zone and so her therapists know when to intervene. Since we want her to become an independent learner, we need to let her learn how and when to ask for help. We use many resources for training, most of them taken from the curriculum and school materials of the children to make generalisation easier.

Being resilient has numerous benefits; resilient individuals have better mental well-being and health (Gheshlagh, et al., 2017; Joyce et al., 2018, Leppin et al., 2014), enjoy greater productivity and obtain better academic outcomes then non-resilient people. In regulating mathematics anxiety, we go for psychoeducation, so we create what is called MateTrucos (Mathematics Tricks) to help children visualise mathematics and reduce mathematics anxiety. MateTrucos opens with a brain welcoming you to propose yourself to develop a growth mindset (Boaler, 2019) (Figures 14.7 and 14.8).

With this kind of material, we help children see what literally happens when they are doing mathematics. Along with the Growth Zone Model, these images help the

Figure 14.7 My brain is like a muscle. With effort and dedication I can achieve whatever I propose to myself.

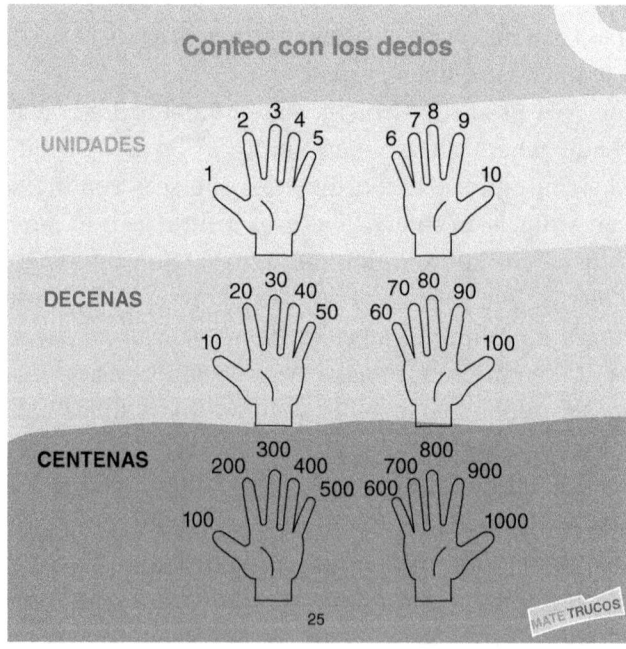

Figure 14.8 With many, many rungs we help children to see that mathematics is more accessible than they think.

1) Clasroom Emotional Climate
2) Planinning using ZPD
3) Building rungs to be consistent dimbers
4) using the GZM to aid metacognition while learining

Figure 14.9 The key to Mathematical Resilience.

children to be metacognitive, aware of the learning going on, and this helps to reduce anxiety and promote resilience.

Classroom Accommodations are indispensable for performing at school if a child has a learning disorder, but they are also useful in promoting resilience at whole class level. Using the Ladder Model, the Growth Zone Model and rungs for learning, we can aid each child to learn better, and we can build a more equitable environment. In teacher training, we promote the use of these kinds of resources at every level, building materials to help teachers promote anxiety regulation and resilience (Figure 14.9).

Hands on learning and building resilience, working with parents, teachers and learners, using visuals such as the Growth Zone Model, the Ladder Model and others, help to strengthen self-confidence which, in turn helps learners to learn in a better fashion.

It is not an easy task but, in my opinion, when thinking about teaching in formal contexts, thinking about five similar pillars to those used in clinical settings could make the teacher a more valued and emotional support for children when they are learning. Think about these five pillars in order to plan and reflect on and about your practice:

- The emotional climate – is it suitable for building resilience?
- Have I made the learning visible?
- Have I designed and planned interventions carefully, taking in consideration the GZM and the Ladder Model?
- Do my plans promote self-regulation, using ideas and images to help learners be in control of their learning?
- Are visual aids everywhere, to foster resilience and autonomy and to diminish anxiety?

Also, reflect about your own learning zone and building your own resilience; we as teachers, of course, are still learners.

USA

Allison Dillard

Is mathematical resilience being developed in the United States? Through my interviews with leaders in mathematics education on the *Allison Loves Math Podcast*, I have observed that the various subcategories of mathematical resilience, especially growth mindset, are gaining traction in the United States. However, this is happening slowly and often in an unorganised way. I firmly believe that the key to enhancing mathematics education in the United States lies in helping educators transition from developing mathematics mindsets in students to developing mathematical resilience in students. Now I will examine how the four subsets of mathematical resilience are developing in the United States.

- **Growth mindset:** Boaler's (2016) Mathematical Mindset books have contributed significantly to the popular growth mindset movement among mathematics educators in the United States. Growth mindset slogans and posters are commonplace in elementary school classrooms and are increasingly adopted by middle and high school mathematics classrooms. Despite the value and widespread acceptance of the ideas around a growth mindset in education, there is a common surface-level promotion of the concept that fails to address the root causes of mathematics anxiety or provide learners with the strategies and experiences they need to succeed in mathematics. I also find it concerning that many of my community college students have never heard of growth mindset and a significant number of them begin the semester believing that they are inherently bad at mathematics.

- **Personal value of mathematics:** Despite the majority of students in the United States reporting a negative attitude towards their mathematics classes, the majority of Americans do consider mathematics to be the most important subject in school. However, as a mathematics educator, there seems to be a disconnect between these statistics and what I observe in the classroom. Once, when I asked my students why we need mathematics, there was resounding silence. Nobody knew. Eventually, one student raised their hand and suggested that we need mathematics to know how to tip our waiter. I was alarmed at the lack of awareness of how mathematics plays a vital role in improving different aspects of our lives. I set out to show my students the many ways that mathematics added value to their lives, but I found that it was difficult to find resources, support, or time.

- **Knowing that mathematics requires struggle:** With the growing popularity of books such as *Productive Math Struggle* by John San Giovanni, an increasing number of teachers in the United States are making the challenging transition away from more traditional and procedural "I do - we do - you do" teaching methods. Teachers recognise the importance of encouraging students to embrace challenges and view mistakes as a natural part of the learning journey. However, creating a classroom environment and crafting lessons that support this can be challenging. As a teacher, once I learned the importance of fostering productive mathematics struggle in class, my primary challenge was finding the necessary resources, time, and know-how to create such lessons and cultivate a safe learning environment for such lessons.

- **Knowing how and when to recruit support:** Teaching students how and when to ask for support can be tricky. On one hand, some students have become overly reliant on assistance. They tend to seek help too frequently, which can hinder their ability to learn how to work through difficult problems. On the other hand, there are also students who avoid asking for help even when they need it, leading to unproductive struggles and even failure. In many schools, there is funding for support services for students, but they are often underused. For example, our community college has a fully funded mathematics tutoring centre with long hours and excellent support staff. However, it often remains underused, and this is not because the students do not need help.

Although the United States is making progress in all four subcategories of mathematical resilience, there is still much work to be done. While I have been fascinated to learn from guests on the *Allison Loves Math Podcast* about their strategies and advice on how to promote mathematical resilience and other similar goals, I have been shocked to find that many students in my community college classrooms have never heard of growth mindset, fail to see the value of mathematics, are reluctant to embrace mathematics struggle, and fail to ask for support when it is so widely available. Despite the progress made, we still need to create a national culture of mathematical resilience. Nonetheless, I am optimistic about what we can achieve in the future.

Conclusion

This chapter has shown the many ways that mathematical resilience is being developed in countries around the world. The common feature of all these reports is the vital need they see for learners of whatever age in their countries to be supported in developing mathematical resilience so that they can succeed in learning mathematics. The need for people who can learn and use mathematics remains high globally.

The interventions reported on in this chapter are varied. There is a clear sense of conviction that the principles and tools recommended for developing mathematical resilience will make a difference for learners in their countries and should be deployed country wide. Many of the ideas reported here are pilots and try outs, seeking out the best way to work with students in their cultural contexts. Some are working with individuals, particularly those whose relationship with mathematics learning is already deeply scarred by anxiety, others are seeking to develop programmes that can support learners and their teachers across their country to build a positive stance towards mathematics and work with resilience towards a future where mathematics thinking and reasoning takes its proper place in building a better future for all.

References

Acevedo, V. E., and Restrepo, L. (2012). De profesores, familias y estudiantes: fortalecimiento de la resiliencia en la escuela. *Revista Latinoamericana de Ciencias Sociales, Niñez y Juventud*, 10(1), 301–319.

Ahmed, W., van der Werf, G., Kuyper, H., and Minnaert, A. (2013). Emotions, self-regulated learning, and achievement in mathematics: A growth curve analysis. *Journal of Educational Psychology*, 105(1), 150–161. 10.1037/a0030160

Alkan, V. (2018). A systematic review research: 'Mathematics anxiety' in Turkey. *International Journal of Assessment Tools in Education*, 5(3), 567–592.

Alvarán-López, S. M., Bedoya-Mejía, S., and Grisales-Romero, H. (2021). Valoración de la resiliencia en escolarizados: línea base para programas de intervención, Antioquia, 2018. *Hacia la Promoción de la Salud*, 26(2), 83–101. 10.17151/hpsal.2021.26.2.7

Ang, W., Chew, H., Dong, J., Yi, H., Mahendren, R., and Lau, Y. (2022). Digital training for building resilience: Systematic review, meta-analysis, and meta-regression. *Stress and Health*, 38(5), 848–869.

Bayona-Rodríguez, H., and López-Vera, D. (2021). Factores asociados a la resiliencia académica: evidencia para Colombia (Factors associated with academic resilience: Evidence for Colombia). Documento CEDE No. 1, Disponible en SSRN: https://ssrn.com/abstract=3773657

Benson, H. (2000). *The relaxation response*. New York: Avon Books.

Boaler, J. (2016). *Mathematical mindsets: Unleashing students' potential through creative math, inspiring messages and innovative teaching*. Jossey-Bass.

Boaler, J., Munson, J., and Williams, C. (2019). *Mindset mathematics: Visualizing and investigating big ideas, grade 7*. John Wiley & Sons.

Botero-Guzmán, D., and Marín-Díaz, A. (2022). Brecha de género en matemáticas en exámenes estandarizados: el caso de Colombia. *Equidad y Desarrollo*, 1(40). 10.19052/eq.vol1.iss40.4

Bowlby, J. (2005). *A secure base: Clinical applications of attachment theory* (Vol. 393). Taylor & Francis.

Brown, N. W. (2010). *Psychoeducational groups* (3rd Edition). New York: Routledge.

Buckley, S., Reid, K., Goos, M., Lipp, O.V. and Thomson, S. (2016). Understanding and addressing mathematics anxiety using perspectives from education, psychology and neuroscience. *Australian Journal of Education*, 60(2), 157–170. 10.1177/0004944116653000

Camargo-Goyeneche, M. C., Macías-Gómez, E., and Quintero-Zapata, M. (2014). La alegría de vivir: metodología didáctica desde la resiliencia. Desempeño con niños preescolares en situación de vulnerabilidad social. *Revista Iberoamericana de Educación*, 66, 159–174.

Camilla, B., and Hanna, P. (2020). Pre-schoolers reasoning aboutNumberson picture books. *Mathematical Thinking and Learning*, 23(3), 195–213.

Carmo, J. (2008). *Escala de Ansiedade à Matemática* (EAM) [Mathematics anxiety rating scale]. Laboratório de Estudos Aplicados à Aprendizagem e Cognição (LEAAC). Universidade Federal de São Carlos.

Carmo, J., and Crescenti, E. (2022). Mathematics anxiety and successful reversion strategies: A Brazilian experience. In L. R. V. Gonzaga, A. M. B. Da Silva, & L. V. Dellazzana-Zanon (Eds.), *Handbook of stress and academic anxiety: Psychological processes and intervention with students and teachers* (pp. 115–126). Springer Nature.

Caro, F. E., and Kárpava, A. (2020). La calidad educativa, un análisis desde la violencia en Colombia. *Revista Espacios*, 41(18), 27–37.

Cengiz, N. (2017). *Teknoloji destekli matematiğin öğrencilerin başarıları ve matematik kaygıları üzerindeki etkisi* [The effects of technology-supported mathematics on students' success and math anxiety]. Unpublished MS Thesis, Turkey: Gaziantep University.

Çetin, Ş., Durmaz, B., and Girit, D. (2018). Matematiksel yılmazlık/dayanıklılık ölçeğini Türkçeye uyarlama çalışması [Adaptation of mathematical resilience scale into Turkish culture]. In Ş. Çınkır (Ed.), *Fifth international Eurasian educational research congress proceedings book* (pp. 875–881), Antalya, Türkiye: Anı Publishing.Yayıncılık.

Colombini, F., Shoji, F., and Pergher, N. (2012). Ansiedade matemática e desenvolvimento de hábitos de estudo: algumas possibilidades de atuação do acompanhante terapêutico [Mathematics anxiety and the development of study habits: Some possibilities for the therapeutic companion to act]. In C. V. V. B. Pessoa, C. E. Costa, & M. F. Benvenuti (Orgs.), *Comportamento em Foco* (pp. 131–142). ABPMC.

Darhim, M. H., and Dahlan, J. A. (2019). Comparison of mathematical resilience among students with problem based learning and guided discovery learning Model. *Journal of Physics: Conference Series*, 895, 012098. 10.1088/1742-6596/895/1/012098

Dehaene, S. (2019). *¿Cómo aprendemos?: Los cuatro pilares con los que la educación puede potenciar los talentos de nuestro cerebro*. Siglo XXI Editores.

Demir, G. (2017). *Gerçekçi matematik eğitimi yaklaşımının meslek lisesi öğrencilerinin matematik kaygısı, matematik öz yeterlik düzeyi, algı ve başarıları üzerindeki etkisi* [The effect of realistic mathematics education approach on mathematical anxiety, mathematical self-efficacy, perceptions and achievement of vocational high school students]. Unpublished MS Thesis, Aydın, Turkey: Adnan Menderes University.

De Souza Domingues, M., Silva, M., Cordeiro, F., de Souza, M., da Rocha, R., Torres, N., and Neto, J. (2022). Exercício físico e ansiedade matemática: perspectivas para educação matemática a partir das neurociências. *Amazônia: Revista de Educação em Ciências e Matemáticas*, 18(40), 2.

Díaz-Pinzón, J. (2021). Análisis de los resultados de la prueba pisa 2018 en matemáticas para américa. *Revista de Investigaciones Universidad del Quindío*, 33(1), 104–114. 10.33975/riuq.vol33n1.463

Dowker, A., Sarkar, A., and Looi, C.Y. (2016). Mathematics anxiety: What have we learned in 60 years? *Frontiers in Psychology*, 7, 1–16. 10.3389/fpsyg.2016.00508

Dueñas, X., Godoy, S., Duarte, J., and López, D. (2019). La resiliencia en el logro educativo de los estudiantes colombianos. *Revista Colombiana de Educación*, 76, 69–90.

EPISTEM (2022). Adult numeracy practitioner continuous professional development. Sourced at: https://epistem.ie/numeracy-meets/ on 15/04/2023. National Centre for STEM Education, University of Limerick.

Figiacone, S. (2021). *Aprender para Enseñar. Enseñar para Aprender. Claves para entender el proceso de aprendizaje y Estrategias para potenciarlo desde la Enseñanza*. NeuroEduca.

Gheshlagh, R. G., Sayehmiri, K., Ebadi, A., Dalvandi, A., Dalvand, S., Maddah, S. B., and Tabrizi, K. N. (2017). The relationship between mental health and resilience: A systematic review and meta analysis. *Iranian Red Crescent Medical Journal*, 19(6), 8.

Greensfeld, H., and Deutsch, Z. (2022). Mathematical challenges and the positive emotions they engender. *Mathematics Education Research Journal*, 34, 15–36. 10.1007/s13394-020-00330-1

Haase, V. G., Lopes-Silva, J., Starling-Alves, I., and Antunes, A. (2013). Com quantos bytes se reduz a ansiedade matemática? A inclusão digital como uma possível ferramenta na promoção do capital mental. In *Educação digital. A tecnologia a favor da inclusão* (pp. 188–202). Porto Alegre: ARTMED.

Helvacı, B. T. (2010). *Bilgisayar destekli öğretimin ilköğretim 6. sınıf öğrencilerinin matematik dersi "çokgenler" konusundaki akademik başarılarına ve tutumlarına etkisi. [The effect of computer assisted teaching on academic achievement of the sixth-grade students' in math lessons polygons units and their attitudes]*. Unpublished MS Thesis, Ankara Turkey: Gazi University.

Herrera, J. (2020). Evaluación de la calidad en la educación básica y media en Colombia. *Cultura, Educación y Sociedad*, 11(2), 125–144. 10.17981/cultedusoc.11.2.2020.08

Homes, A., and Grandison, G. (2021). Trauma-informed practice: A toolkit for Scotland. The Scottish Government. https://www.gov.scot/publications/trauma-informed-practice-toolkit-scotland/

Johnston-Wilder, S., Baker, J. K., McCracken, A., and Msimanga, A. (2020). A toolkit for teachers and learners, parents, carers and support stuff: Improving mathematical safeguarding and building resilience to increase effectiveness of teaching and learning mathematics. *Creative Education*, 11, 1418–1441.

Johnston-Wilder, S., Lee, C., Brindley, J., and Garton, E. (2015). Developing peer coaching for mathematical resilience in post-16 students who are encountering mathematics in other subjects. In *ICER2015 Proceedings* (pp. 6002–6011). IATED Academy. http://wrap.warwick.ac.uk/73249/1/WRAP_ICERI%202015%20paper%201%20submitted.pdf

Johnston-Wilder, S., Lee, C., Garton, E., Goodlad, S., and Brindley, J. (2013). Developing coaches for mathematical resilience. In *ICERI2013 Proceedings*. IATED Academy. https://warwick.ac.uk/fac/soc/ces/research/current/mathematicsresilience/cfmr/iceri_paper_sjw_et_al.pdf

Johnston-Wilder, S., Lee, C. and Mackrell, K. (2021). Addressing mathematics anxiety through developing resilience: Building on self determination theory. *Creative Education*, 12(9), 2098–2115.

Joyce, S., Shand, F., Tighe, J., Laurent, S. J., Bryant, R. A., and Harvey, S. B. (2018). Road to resilience: A systematic review and Meta-analysis of resilience training programmes and interventions. *BMJ Open*, 8(6), e017858.

Kartalcı, S., Acar, G., Zihar, M. and Işık, C. (2021). 9 ve 10. Sınıf öğrencilerinini matematiğin doğası hakkındaki felsefik düşünceleri ile matematiksel yılmazlıklarının incelenmesi [Investigation of 9th and 10th grade students' philosophical thoughts on the nature of mathematics and mathematical resilience]. *Uşak University, Journal of Educational Research*, 7(1), 119–141.

Krause, M., and Eilks, I. (2019). Using action research to innovate teacher education concerning the use of modern ICT in chemistry classes. *Action Research and Innovation in Science Education*, 2(1), 15–21.

Lee, C., and Johnston-Wilder, S. (n.d.) Getting into and staying in the growth zone. NRICH. https://nrich.maths.org/13491

Lee, C. and Johnston-Wilder, S. (2017). The construct of mathematical resilience. In Xolocotzin Eligio, Ulises ed. *Understanding emotions in mathematical thinking and learning* (pp. 269–291). Elsevier.

Lee, C. and Johnston-Wilder, S. (2018). Getting into and staying in the growth zone. NRICH.

Leppin, A., Bora, P., Tilburt, J., Gionfriddo, M., Zeballos-Palacios, C., Dulohery, M. M., ... and Montori, V. (2014). The efficacy of resiliency training programs: A systematic review and meta-analysis of randomized trials. *PLoS One*, 9(10), e111420

Mendes, A. (2016). *Ansiedade à matemática: Evidências de validade de ferramentas de avaliação e intervenção [Mathematics anxiety: Evidence of validity of assessment and intervention tools]*. Tese de doutorado. Programa de Pós-Graduação em Psicologia. Universidade Federal de São Carlos.

Mendes, A., Carmo, J., and Muniz, M. (2020). Aplicação de um programa de auxílio a uma estudante com ansiedade à matemática. [Applying an aid program to a student with math anxiety]. In Miriam C. Utsumi (Org.), *Pesquisas em psicologia da educação matemática: Avanços e atualidades* (pp. 161–181). São Carlos, SP: Pedro e João.

Murcia, M., and Henao, J. (2015). Educación matemática en Colombia, una perspectiva revolucionaria. *Entre Ciencia e Ingeniería*, (18), 23–30.

NALA (2023). *National adult literacy agency.* https://www.nala.ie/

OECD (2013). *PISA 2012 assessment and analytical framework: Mathematics, reading, science, problem solving and financial literacy.* Paris: OECD publishing.

O'Hanlon, A. (2023). *Exploring mathematics anxiety and potential associated environmental antecedents amongst a group of adult learners.* Unpublished Master's dissertation, University of Limerick.

Peralta, S., Ramírez, A., and Castaño, H. (2006). Factores resilientes asociados al rendimiento académico en estudiantes pertenecientes a la universidad de Sucre (Colombia). *Psicología desde el Caribe*, 17, 196–219. https://rcientificas.uninorte.edu.co/index.php/psicologia/article/view/2054/1303

Pérez-Ibarra, A. de J. (2017). Educación en resiliencia. *Revista de la Universidad de La Salle*, (74), 191–207.

Perkins, R., Shiel, G., Merriman, B., Cosgrove, J., and Moran, G. (2013). *Learning for life: The achievements of 15-year-olds in Ireland on mathematics, reading literacy and science in PISA 2012.* Dublin: Educational Research Centre.

Pará, T., and Johnston-Wilder, S. (2023). Addressing mathematics anxiety: A case study in a high school in Brazil. *Creative Education*, 14(2), 377–399.

Quiceno, J., Remor, E., and Vinaccia, S. (2016). *Fortaleza: Programa de Potenciación de la Resiliencia para promoción y el mantenimiento de la salud [FORTRESS. Empowerment program of resilience for health promotion and maintenance].* Manual Moderno.

Ru-De, L., Zhen, R., Ding, Y., Liu, Y., Wang, J., Jiang, R., and Xu, L. (2018). Teacher support and math angagement: Roles of academic self-efficacy and positive emotions. *Educational Psychology*, 38(1), 3–16. DOI: 10.1080/01443410.2017.1359238

Ryan, M. (2019). *An investigation into the extent and derivation of mathematics anxiety among mature students in Ireland* [Unpublished Ph.D. dissertation], University of Limerick https://ulir.ul.ie/bitstream/handle/10344/8146/Ryan_2019_Investigation.pdf?sequence=4

Ryan, M. (2020). *Invitation by education centres to conduct webinar on mathematics anxiety.* Personal correspondence by email.

Ryan, M. (2022). *Invitation to speak to students on mathematics anxiety and numeracy for school placement.* Personal correspondence by email.

Sanabria-James, L., Pérez-Almagro, M., and Riascos-Hinestroza, L. (2020). Pruebas de evaluación Saber y PISA en la Educación Obligatoria de Colombia. *Educatio Siglo XXI*, 38(3), 231–254. 10.6018/educatio.452891

Sánchez-Arias, G. (2016). Programa de intervención terapéutica en resiliencia en institución de educación superior. *Revista de Psicología Universidad de Antioquia*, 8(1), 49–64.

Santiago-Carrillo, M., Gallardo, H., and Vergel, M. (2020). Resiliencia en estudiantes exitosos en matemáticas. *Praxis & Saber*, 11(26), e9973. 10.19053/22160159.v11.n26.2020.9973

Santos, F. (2022). Teachers' maths anxiety classroom management programme. *PsychArchives*. 10.23668/psycharchives.6906

Siegel, D. (2010). *Mindsight: Transform your brain with the new science of kindness.* London: Oneworld Publications.

Sierra-Barón, W. (2012). Promoción de resiliencia en niños de instituciones educativas oficiales de Neiva, Colombia. *Revista Iberoamericana de Psicología: Ciencia y Tecnología*, 5(1), 19–27.

Sümen, Ö. (2013). *Geogebra yazılımı ile simetri konusunun öğretiminin matematik başarısı ve kaygısına etkisi [The effect of teaching symmetry subject by geogebra software to mathematics success and anxiety].* Unpublished Ms Thesis, Samsun, Turkey: Ondokuz Mayıs University.

Teaching Council (2022). *Cosán national framework for teacher's learning.* Sourced at: https://www.teachingcouncil.ie/en/teacher-education/teachers-learning-cpd-/cosan/ on 15/04/2023

Trakulphadetkrai, N. (2018). Story picture books as a mathematics teaching and learning tools. *Primary Mathematics*, 22(2), 3–7.

Velásquez, J., Bedoya, S., Alvarán, S., González, D., and Grisales-Romero, H. (2022). Prevalencia de no resiliencia y factores asociados en escolares de un municipio colombiano, 2019. *Revista Facultad Nacional de Salud Pública*, 40(2), 1–12. 10.17533/udea.rfnsp.e346304

Vergel-Ortega, M., Gómez-Vergel, C., and Gallardo-Pérez, H. de J. (2021). Resiliencia en mujeres universitarias en Norte de Santander, Colombia. *Revista Latinoamericana de Ciencias Sociales, Niñez y Juventud*, 19(3), 1–19. 10.11600/rlcsnj.19.3.4590

Vygotsky, L. (1979). *El desarrollo de los procesos psicológicos superiores.*

Villavicencio, F. T., aBernardo, A. B. I. (2016). Beyond math anxiety: Positive emotions predict mathematics achievement, self-regulation, and self-efficacy. *The Asia-Pacific Education Researcher*, 25, 415–422. 10.1007/s40299-015-0251-4

William, D. (2013). *Emotions and learning education Scotland journal of excellence.* Retrieved from htttp:// w.w.w.journeytoexcellent.org. uk/videos/expertspeakers/emotionsandlearningdylanwilliamap.

PART 5
Looking forward

Mathematical resilience – What needs to change?

Clare Lee and Sue Johnston-Wilder

Introduction

An important outcome of a learning environment that uses the principles of mathematical resilience is that everyone can progress with mathematics in a way that helps them to function and thrive (Sen, 2009) in their lives and careers. Teaching using the principles of mathematical resilience works to remove both the stigma and psychological difficulties which prevent many from progressing. Being allowed and encouraged to develop mathematical resilience is an important step towards personal wellbeing and social justice. However, if mathematical resilience is to be developed in every mathematical learning environment, changes need to be made.

Some changes can be enacted by individuals, through adopting the principles of mathematical resilience and tapping into some of the resources available in this book and through online communities. Other changes can be enacted by teachers; we have already seen how teachers can, for example, build positive learning environments in Chapter 4 and research their own practice in Chapter 9. However, for the goals of everyone able to progress in mathematical learning, growing their personal wellbeing and increasing social justice, to be reached fully, systemic changes will also be needed, and they will involve, for example, tackling widespread inequitable practices and changing norms in education.

Systemic changes

The assessment system

Much of school practice is shaped, conducted and managed in the shadow of formal assessments. This is true of all subjects but examinations in mathematics cast an especially significant shadow because of its high-stakes, gatekeeper nature and common use as a proxy for general academic performance. Many people see an assessment system which is based on timed examinations as fair and objective, see for example Saxton (2023). All learners sit the same examination at the same time and those that do very well are awarded a top grade and those that do not perform well, fail. These grades are then considered to mark certain learners as suitable for certain jobs and not others, which is the gatekeeper effect of mathematics examinations.

There is an obvious flaw in the 'fairness' of these tests: What if you do not feel well on the day? What if your mathematically anxious teacher has "infected" (Dowker et al., 2016) you with mathematics anxiety? What if your school simply could not recruit a good mathematics teacher and you have been taught by substitute teachers for the last two years? A circumstance that is the likely consequence of the current recruitment crisis in England (Stripp, 2023) where secondary schools rarely have a full complement of specialist mathematics teachers. Stripp also reports that this shortage is particularly true for schools in disadvantaged areas. The examinations are not an objective measure of someone's capability to learn and use mathematics when they are not under pressure and have been offered the time and support, they need to learn. Examination grades can only ever represent a person's performance on a particular day.

Dylan Wiliam made the point that:

> The trouble with such 'objective' approaches is that while many things can be measured, there are also many important things that cannot, and the danger is that things that can be measured easily come to be regarded as more important that those that cannot (Wiliam, 2001, p. 57).

Examinations are in part written so that they can be quickly and reliably marked (Wiliam, 2001a). An effective way to increase reliability in the marking of examinations is to ask questions that have single unambiguous answers and methods to get those answers. Questions assessing thinking, reasoning and problem-solving skills are largely absent from GCSE examinations at age 16 in the UK because these skills are hard to measure reliably. The outcome of their absence from the examinations is that these vitally important skills are not part of most learners' school experience, and an inferior version of mathematical learning is often offered, modelled on the expected nature of examinations questions. Ofsted (2023) reports that:

> The teaching of disparate skills to enable learners to pass examinations but not equip them for the next stage of education, work and life, and weaknesses in the teaching of mathematical problem solving, remain areas of weakness across many schools. (Ofsted, 2023. No page)

The GCSE examinations in 2023, were norm referenced in that the pass mark was manipulated to ensure that 70% of the candidates that year were given a pass grade and 30% were not (fft education data lab, 2023). This means that in every GCSE examination in the UK approximately two thirds of the candidates pass and one third will fail, with all the implications that has for their life choices and chances (Jones, 2017). Examination results are 'norm-referenced', as opposed to 'criterion-referenced'. The process is complicated, but in essence GCSE examinations compare learners with each other (norm referencing) rather than with fixed criteria (acknowledging whether they meet certain criteria or not). The National Numeracy Strategy (Jones, 2017) advocates criterion referenced examinations, which are similar to the driving test or the UK citizenship test, where, if you are good enough, you pass, regardless of how well others do. In Scotland, assessments during primary school are completed online, when the teacher considers the learners in their class are ready. No pass or fail grades are given to the learner or the teacher and if the learner shows signs of excessive struggle in one question the following questions are selected to be easier to complete. The stated purpose of these yearly assessments is to identify the child's progress and to support the teachers' professional judgement (Scottish Government, no date).

Changes to the assessment system can be made. Do employers want to know that a certain potential employee was marginally better than one third of the candidates taking the examination? Or do they want to know what mathematics that employee knows and can do? The current assessment system labels one third of people taking a mathematics examination each year as failures, and in doing so it often humiliates them, which inevitably harms their wellbeing. It is time to question whether a national assessment system of the scale that is currently a feature of many countries' education system is needed and if so, what is it needed for? If the political will is to continue with such a system, making a change to a criterion-referenced system would be more informative to learners and stakeholders alike.

Teacher training

There seem to us to be two ways in which more, or maybe just better, training is needed for all teachers who help people learn mathematics. Our research indicates that teachers would be able to help learners to develop mathematical resilience if they understood more about their role in safeguarding the mathematical wellbeing of their learners and if they understood more about how to help learners fill in gaps in their understanding (Lee and Johnston-Wilder, 2013).

All teachers have a role in safeguarding the wellbeing of their learners and in the case of mathematics teaching that requires working to prevent the development of mathematics anxiety by being very cautious when choosing which pedagogies to use. The training we are advocating for both pre-service and in-service teachers is aimed at helping teachers recognise and address mathematics anxiety. The training will help them look out for symptoms of mathematics anxiety in learners and understand how disabling such anxiety can be. The ideas in Chapter 2 will help teachers with tools to gauge how widespread mathematics anxiety is in their learners and the ideas in the rest of the book will help them teach in ways that either prevent anxiety occurring or diminish the anxiety felt by most learners.

There will be some learners that need more help than can be provided in a classroom or by their peers, but most will respond to the use of mathematical resilience principles to structure their learning environment.

Although some countries or nations show understanding of the teachers' role in developing and safeguarding the general wellbeing of their learners and make explicit their expectations in this regard (see for example the Professional Standards for Teaching and Leadership; Welsh Government, 2019), no such document yet recognises specifically the need to understand the development and effects of mathematics anxiety. We know that many learners develop mathematics anxiety in primary school which makes such training vital for people seeking to become primary school teachers. As there are mathematical elements to all subjects in secondary school and beyond, arguably it would be useful for all teachers to receive training that will help them understand and help those who suffer from mathematics anxiety. We know that many teachers themselves suffer from mathematics anxiety (Dowker et al., 2016) so such training must also be prepared to offer support for those who do, along with ways to overcome this disabling condition.

There is already a shortage of mathematics teachers – and there has been for many years (Stripp, 2023). It seems to us that although there are many other factors involved, such as pay and conditions, if fewer learners were taught in ways that disregard their mathematical wellbeing, the pool of those qualified and willing to teach mathematics would grow. Recruiting solely more highly qualified in mathematics teachers may sound logical, as they are likely to have enjoyed mathematics lessons themselves. However, it could make the problem worse, as highly qualified mathematics teachers may not be able to empathise with learners, their struggle to learn and other issues they face. If more in-service teachers are trained to understand how to work to develop mathematical resilience in their learners, there is likely to be, in turn, a greater number of candidates willing and able to take on the onerous but rewarding role of teaching mathematics, building a virtuous circle.

Research that we have undertaken (see for example Lee and Johnston-Wilder, 2013; Lee and Morgan, 2024) indicates that although most mathematics teachers can and do plan and teach well-structured and engaging lessons, fewer teachers know what to do when they see a learner having difficulties. One young woman, speaking about her mathematics teacher, said:

> He could see that I was struggling, and you could see that he just didn't know what to do, he wasn't experienced enough to help me.

Another said:

> I always felt in my school anyway, the teachers were more interested in the people who were naturally able to do it and they were at the front and they got more attention and so the people at the back were, well, I felt like we were just pushed aside.

Both these statements indicate some teachers lack understanding of how to help those that are struggling or, to use the language of the four tools, those that have rungs missing from their ladder to successful understanding. Some people might argue that this is a matter of giving teachers more pedagogical content knowledge, but we think, and the second quote goes some way towards indicating this, that being able to address these issues rests on growing teachers' understanding of their role. When teachers have built the kind of relationships and ethos in the classroom that enables them to ask appropriate questions and allows the learner to express where their difficulties lie, teachers are more likely to be able to help. If teachers then listen actively to the learners' answers, demonstrate appropriate levels of awareness of the way the learner might be feeling, they will be able to offer ways for each learner to succeed.

There is also the widespread assumption that some can do mathematics, and some just cannot; a 'fixed' mindset culture (Boaler, 2015). Thus, those that 'can' receive interest and encouragement from the teacher and those who have gaps in their understanding, that make them seem as though they 'cannot', feel themselves to be pushed aside or moved to group of learners where fewer demands are made on their understanding. Those teachers whose experience, awareness and reflection has enabled them to understand where gaps might be, and who use more thorough formative assessment (Black et al., 2003) to explore their learners' understanding, are able to help their learners fill the gaps and develop secure understanding. Currently, it does not seem there are enough of these expert teachers, but professional development can and should be designed. The situation worsened due to the interruptions Covid-19 necessitated; there will be people discovering 'rungs' that are missing in their understanding due to the Covid-19 pandemic for many years to come.

If we are to see more people with sufficient mathematical resilience to continue to learn and use mathematics to the benefit of society, changes in training for both pre-service and in-service teachers are needed. All teachers must understand how to recognise and mitigate mathematics anxiety and know how to ask questions that help find gaps in learners' knowledge, where those gaps may be and how to support their learners to fill those gaps.

The use of setting in schools

The practice of setting in schools has been shown by research to have many negative influences on those learning mathematics. Amongst them is that being placed in a group according to some measure of your attainment tends to affirm the fixed theory of learning (Yeager and Dweck, 2012) that you either 'can or can't do' mathematics. The practice of setting works against learners adopting a growth theory of learning, which is vital if they are to become mathematically resilient.

'Setting' or 'streaming' are terms used for the variety of approaches that are used to group together learners that have shown similar attainment. These practices are currently used by many schools both primary and secondary, although it is seen more in secondary schools. The aim of both setting and streaming seems to be to allow teachers to focus their teaching more effectively and efficiently by grouping those with similar needs together. However, an extensive review by the UK government funded Educational Endowment Foundation found that the effects of streaming and setting are negligible, and that setting or streaming does not raise attainment for most learners. They say:

> The impact of setting and streaming is 0 months progress, on average, with worse outcomes for low attaining learners. The evidence around setting and streaming is limited (EEF, 2021, Key findings).

That evidence is limited is partly due to the lack of randomised controlled trials in the literature they accessed, which may be because so few schools at secondary level are prepared to abandon setting their learners, making evidence hard to come by. One longitudinal study by Wiliam and Bartholomew (2004) found that it was the set to which learners were allocated that made the difference in learners' outcome, not the school overall. They studied learners with the same scores in their National Key Stage Three (KS3) Test (a government set test taken at age 14 years in England, which has now been abandoned) over four years in six schools. They found that those who had an equivalent KS3 score but were placed in a top set outscored those placed in the next set down by half a grade in GCSE examinations on average. This lowering of the GCSE grade continued throughout the range of sets.

Wiliam and Bartholomew (2004) also found that those with parents who had lower socio-economic status scores were over-represented in the lower sets compared to others with the same KS3 test score but higher socio-economic status. EEF (2021) support this and say that there is evidence that teacher expectations of disadvantaged learners may increase their chances of being placed in lower sets or streams. In practice setting (and the potential to move between sets) is often further confused and conflated with other elements observed by teachers in schools. Learners can be placed in lower sets because of negative behaviour factors in class. This both highlights the illusion of the need for attainment-based setting and leads to many disadvantages. Learners in lower sets and streams are more likely to be taught by less experienced and qualified teachers. It seems that setting or streaming may construct negative self-fulling prophecies for disadvantaged learners.

The EEF (2021) study also notes that grouping learners based on attainment may have longer term negative effects on the attitudes and engagement of lower attaining learners as it may discourage the adoption of a growth mindset (Yeager and Dweck, 2012), that is the belief that their attainment can be improved through effort. Boaler and Foster (2021) also noted that many learners who were placed in higher attaining groups suffered a great deal of anxiety as they did not believe that they belonged there. This is another example of setting endorsing the fixed theory of learning.

The evidence is there that using setting or streaming disadvantages all but the most highly achieving learners and that it makes no difference to higher attaining learners' progress. Most mathematics teachers that we talk to say that they agree that they can see these negative consequences in their learners but have little idea about how to teach a range of attainments in one group. It seems to them to require so much planning and finding or producing resources that they simply would not have the time. We sympathise; it is time consuming and uncomfortable to adopt new ways of teaching, but the harm setting causes to all learners seems to us to demand that these changes are made. Teachers should not have to do this alone and must seek to collaborate in working out how to help all their learners keep up with the mathematical ideas they are teaching so that they do not have to catch up later. Learners staying broadly together as they progress with mathematics is part of the core principles of mathematics mastery (Drury, 2018) which is advocated by the UK government; therefore, there should be support available in changing the way learners are taught when setting is no longer used. One solution is to give learners an element of choice as to which learning group they wish to join, and another is to identify learning groups based on current interests: for example: humanities or physical sciences.

The view of good mathematics teaching promoted by those in authority.

Many ideas advocated in reviews by Ofsted, a government led accountability body set up by the English government, are part of an environment that enables its learners to develop mathematical resilience. We agree that a clear focus on teaching mathematics where teachers model new vocabulary, regularly check learners' understanding and swiftly pick up misconceptions is laudable and that teachers should help learners 'keep up' rather than 'catch up' amongst other ideas in Ofsted's 2023 report. However, the lack of interest in these reports in learners' mathematical wellbeing should change. For example, a recent group of advisers to the UK government acknowledged that many learners who are required to retake their mathematics GCSE several times because of initial failure experience "*alienation and disengagement*" (House of Commons Education Committee, 2023) but did not mention either mathematics anxiety or the need to address this, despite research (e.g., Johnston-Wilder et al., 2014) showing that anxiety was the likely cause of the learners' problems.

If those charged with creating national curricula and the accompanying rhetoric, and those who hold teachers accountable, took seriously the research that traditional ways of teaching mathematics are often implicated in causing anxiety and harm to mathematical wellbeing then changes in the pedagogies used in the classroom would follow, to the benefit of learners of mathematics and society at large.

The way in which mathematics is presented in the media

It is common to hear, people who are talking about anything to do with numbers in the media, say 'sorry I can't do the maths'. Sunak (English Prime Minister at the time of writing) acknowledged this prevalent way of talking about mathematics when he said there is a widespread "cultural sense that it's OK to be bad at maths" (Guardian, 2023). National Numeracy, an independent charity which aims to help raise low levels of numeracy among both adults and children, has been working to challenge negative attitudes in the media as well as to influence public policy and offer practical ways to help adults and children improve their numeracy. National Numeracy team members have managed to get some coverage and make the case that no one would talk about poor literacy skills in the same jokey tones that are used to discuss poor numeracy skills. Their annual national numeracy day is an opportunity to advertise that everyone can do and enjoy doing mathematics, with a little perseverance and support. However, they have been reluctant explicitly to address mathematics anxiety and that prior harm may have been caused by certain ways that mathematics has been taught.

> Sue told a psychologist J she worked with mathematics teachers. J's immediate response, unconsciously, was to move her hands protectively up over her face and chest. When Sue asked her, 'what happened to you?' J's response was 'I hadn't thought of it like that.'

In school you might ask for some time in a teachers' meeting to discuss everyone's role in building mathematical resilience. Your colleagues can understand that the way they talk about mathematics is important and that, if they are mathematically anxious, they can address their anxiety, or the least they could do is not 'infect' (Dowker et al., 2016) others. They can help promote a growth mindset and awareness of mathematics anxiety, which will help your learners feel able to grow mathematically.

There is also a need for visible role models who happily work with mathematics at different levels and in different contexts. Clare should not have been, but was, surprised when her hairdresser started talking about how much mathematics she used in her work and how lucky she had been to have a teacher who helped her love mathematics. As mathematics teachers and researchers, we tend to only reluctantly mention our work in these situations as we have become accustomed to a negative reaction. We suggest that this negative reaction is rooted in prior harm and adverse prior experience with mathematics, something that needs to be brought out into the open.

Furthermore, in the media in England, you might see Rachel Riley, Bobby Seagull or Hannah Fry, who are all excellent at mathematics, but the wide range of people who use mathematics as a normal and natural part of their working day remain invisible. Currently, mathematics seems to be either the invisible chameleon hidden in the background or the exceptional talent, with little in between. Using terms other than mathematician might help. When we see a surveyor, a dentist or a hairdresser at work, we see someone using mathematical thinking, not a mathematician.

What must change in mathematics lessons?

Changes in mathematics lessons can be made without the systematic changes that have been advocated in the first part of this chapter. Just because a teacher is employed in a system that seems to advocate elitism (Nardi and Steward, 2003) and works against social justice does not prevent their classroom being one where wellbeing is considered, and everyone is included, in learning mathematics effectively.

Presenting mathematics in a connected and creative way

Askew et al. (1997) wrote an influential report suggesting that connectionist teachers helped their learners make better progress in mathematics than those who taught mathematics as a collection of facts and standard methods. Connectionist teachers themselves understood that there is a rich network of connections between different mathematical ideas and were able to help their learners to select and use efficient and effective strategies. They used a variety of representations to help their learners to connect different areas of mathematics. Such teachers engaged their learners in discussions of each concept in order to reveal their learners' thinking and address misconceptions. Askew et al. (1997) also established that highly effective teachers believed that almost all learners are able to become numerate and used teaching approaches that ensured all learners were being challenged and stretched. There are similarities between connectionist teaching and the principles of mathematics mastery (Drury, 2018) both of which we believe can help all learners develop mathematical resilience. However, many teachers will need to change some of their ways of thinking, for example, to put aside fixed mindset thinking, before the principles of teaching mathematics for mastery can be put into place in all mathematics learning environments.

Teaching mathematics in a more creative way is part of helping learners understand and believe that practically everyone can make progress in mathematics. There are books which can help, such as Peter Mattock's (2023) Conceptual Maths, which shows how to teach mathematical concepts in ways that give time for learners to develop their understanding, rather than just teaching 'how to do' mathematics and help learners to connect ideas and making progress in creative ways. There is Barton (2018) which makes a research-based plea for teachers to be creative in the way that they teach and Boaler (2015) who offers creative ways to help learners address their mathematical mindset and really understand that with perseverance and support they can make progress with mathematics, and they can enjoy it.

Considering the wellbeing of learners

The role of mathematics teachers in promoting the mathematical wellbeing of learners must be at the forefront of their thinking. This is partly a systemic change in mandating that it is every teacher's role to ensure their learners' wellbeing. However, it is something mathematics teachers must choose to bring into their lessons as, for many of their learners, mathematics is an anxiety-inducing subject in a way many other subjects are not. Time spent ensuring that every learner understands ideas is time saved later. Where learners are taught by teachers who are genuinely interested in them and how well they understand and can use the mathematics being learned, learners make good progress (Hattie, 2023) and build resilience (Yeager and Dweck, 2012) which allows them to build on their mathematical learning throughout their lives.

Conclusion

There is no one change in this chapter that is likely to change the perception of mathematics in wider society overnight. Changing the narrative around mathematics must start somewhere and the formal learning environments where mathematics is taught and learned seems to be a good place to start. Helping more people understand that mathematics anxiety is real and disabling is a first step in starting to ensure that anxiety does not hold learners back and instead, we create "a population of mathematically competent individuals, who might otherwise go undiscovered" (Lyons and Beilock, 2012, p2102).

This book is addressed to teachers, who may understandably say that they are unable to make systemic changes and are required to act under curricula and school structures mandated by others. To a large extent this is true, but in the end, it will be teachers who make the difference, even if society has to wait for them to educate more people to be mathematically resilient, who then move into positions of systemic authority. Understanding where that system is not designed with the best interests of the wellbeing of learners in the twenty-first century at heart, will enable teachers, parents or other interested parties, which could be said to be everyone, to add power to the voice for change. That understanding may enable more people to stand up and say what needs to change whenever and wherever they can.

References

Askew, M., Rhodes, V., Brown, M., Wiliam, D., and Johnson, D. (1997). *Effective teachers of numeracy: Final report.* ISBN 1871984637

Barton, C. (2018). *How I wish I'd taught maths.* John Catt Educational Ltd ISBN-13 978-1911382492

Black, P., Harrison, C., Lee, C., Marshall, B., and William, D. (2003). *Assessment for learning - Putting it into practice.* Maidenhead, U.K.: Open University Press.

Boaler, J. (2015). *Mathematical mindsets.* Jossey-Bass.

Boaler, J., and Foster, D. (2021). *Raising expectations and achievement: The impact of two wide scale de-tracking mathematics reforms.* Youcubed, https://www.youcubed.org/wp-content/uploads/2018/09/Raising-Expectations-2021-JB.pdf

Dowker, A, Sarkar, A, and Looi, C. (2016). Mathematics anxiety: What have we learned in 60 years? *Front Psychol.*, 25(7), 508. 10.3389/fpsyg.2016.00508

Drury, H. (2018). *How to teach mathematics for mastery.* Oxford: OUP.

EEF (2021). *Setting and streaming, education evidence, teaching and learning toolkit.* https://educationendowmentfoundation.org.uk/education-evidence/teaching-learning-toolkit/setting-and-streaming

fft education data lab (2023). + https://ffteducationdatalab.org.uk/2023/08/gcse-results-2023-the-main-trends-in-grades-and-entries/

Guardian (2023). 'Anti-maths mindset' costs UK a huge sum, Rishi Sunak claims https://www.theguardian.com/education/2023/apr/17/anti-maths-mindset-costs-uk-a-huge-sum-rishi-sunak-claims

Hattie, J. (2023). *Visible learning – The sequel.* London: Routledge.

House of Commons Education Committee (2023). *The future of post-16 qualifications, Parliament UK.* https://publications.parliament.uk/pa/cm5803/cmselect/cmeduc/55/report.html

Johnston-Wilder, S., Lee, C., Garton, E., and Brindley, J. (2014). Developing coaches for mathematical resilience: Level 2. In *ICERI2014 Proceedings* (pp. 4457–4465), IATED Academy.

Jones, W. (2017). *A third of kids are written off as failures. It doesn't have to be this way, National Numeracy Strategy* https://www.nationalnumeracy.org.uk/news/third-kids-are-written-failures-it-doesnt-have-be-way.

Lee, C., and Johnston-Wilder, S. (2013). Learning mathematics - Letting the pupils have their say. *Educational Studies in Mathematics*, 83(2), 163–180.

Lee, C., and Morgan, J. (2024). Remembering learning mathematics – We can run but we can't hide. *Teacher Development* [in press].

Lyons, I., and Beilock, S. (2012). Mathematics anxiety: Separating the math from the anxiety. *Cerebral Cortex*, 22(9), September 2012, 2102–2110, 10.1093/cercor/bhr289

Mattock, P. (2023). *Conceptual maths.* Crown House Publishing. ISBN-13 978-1785835995

Nardi, E., and Steward, S. (2003). Is mathematics T.I.R.E.D? A profile of quiet disaffection in the secondary mathematics classroom. *British Educational Research Journal*, 29(3), 345–367. 10.1080/01411920301852

Ofsted (2023). Coordinating mathematical success: The mathematics subject report, Ofsted. https://www.gov.uk/government/publications/subject-report-series-maths/coordinating-mathematical-success-the-mathematics-subject-report

Saxton, J. (2023). Grading exams and assessments in summer 2023 and autumn 2022 Gov.uk https://www.gov.uk/government/speeches/grading-exams-and-assessments-in-summer-2023-and-autumn-2022

Scottish Government (no date). National standardised assessments for Scotland https://standardisedassessment.gov.scot/

Sen, A. (2009). *The idea of justice.* Allen Lane & Harvard University Press 978-1-84614-147-8

Stripp, C. (2023). Lack of specialist maths teachers 'massive injustice', says flagship scheme boss, Schools Week Lack of specialist maths teachers 'massive injustice' (schoolsweek.co.uk)

Welsh Government (2019). Professional standards for teaching and leadership, https://hwb.gov.wales/professional-development/professional-standards

Wiliam, D. (2001). What is wrong with our educational assessments and what can be done about it? *Education Review*, 15, 57–62.

Wiliam, D. (2001a). Reliability, validity, and all that jazz. *Education 3–13*, 29(3), 17–21. 10.1080/03004270185200311

Wiliam, D., and Bartholomew, H. (2004). It's not which school but which set you're in that matters: The influence of ability grouping practices on student progress in mathematics. *British Educational Research Journal*, 30(2), 279–293.

Yeager, D., and Dweck, C. (2012). Mindsets that promote resilience: When students believe that personal characteristics can be developed. *Educational Psychologist*, 47(4), 302–314. 10.1080/00461520.2012.722805

Resilience-building problem-solving tasks
The future

Gaye Williams

Introduction

In this chapter, I will use my experience of teaching and researching in Australia to examine how high-quality learning situations (HQLS) can be established in mathematical learning environments. I found when I started teaching that many of my learners had barriers to learning mathematics even though they performed well in my science classes. I wanted to build learners' positive feelings towards mathematics learning (Williams, 2002b), which I was able to achieve as a teacher. I first came across the ideas of optimistic learners (Seligman, 1995) when I moved from teaching into research. I recognised an optimistic orientation to the world in some of my previous learners who had been inclined to explore unfamiliar ideas, and an absence of this inclination in non-optimistic (including pessimistic) learners.

Through experimentation, I found that accessible unfamiliar problem-solving tasks undertaken in group settings engaged disengaged learners and further engaged other learners. Through my research into developing a problem-solving culture for learning mathematics, I found that both mathematical meanings and insights could be developed by learners and groups of learners without mathematical input from 'expert others' during the exploratory process, *if* the task employed, and the learning environment, had appropriate characteristics.

In this chapter, I distil ideas from my research that help to make features of HQLSs transparent and show enactments of them that contribute to a resilience-building mathematics learning culture. I also demonstrate the need for resilient teachers to facilitate collaborative ways of working. These ways of working are generally very different to ways that mathematics teaching is frequently undertaken. I report on features of professional development found to enable appropriate teacher change, and finish by looking forward to a more resilient future.

Resilience

Resilience has been defined as: "academic and emotional and social competence despite adversity and stress" (Nettles et al, 2000, p. 47), "the mechanisms and processes that lead some individuals to thrive despite adverse life circumstances" (Galambos and Leadbeater, 2000, p. 291) and as an optimistic orientation to the world characterised by a positive explanatory style related to perceptions of successes and failures (Seligman, 1995).

'Academic resilience' (Hernandez-Martinez and Williams, 2013), 'mathematical resilience' (Lee and Johnston-Wilder, 2017), and 'characteristics of optimistic problem solvers' (Williams, 2014; Seligman, 1995) are resilience constructs. All involve dynamic interactions between psychological characteristics of individuals and elements of contexts that result in changes within the individual and context: "individuals are shaped by their context … but also act and in turn shape their context" (Hernandez-Martinez and Williams, 2013, p. 48).

Mathematical resilience (MR) is a learner characteristic that supports the process of maintaining mathematical well-being when faced with personal or social threat (Johnston-Wilder et al., 2021). Strategies employed by tutors to reduce mathematics anxiety (MA) and build MR include supporting learners' capability to manage stressful situations, and providing tools that raise learners' awareness of emotions and emotion regulation (e.g., Baker, 2021).

Optimistic mathematical problem solving (Williams, 2014) is optimism specific to mathematical problem solving, a subset of Seligman's (1995) optimism construct. It is an *exploratory* style involving 'stepping into unknown territory' to autonomously and spontaneously explore mathematical ideas that were not previously apparent. Optimistic (resilient) learners are prepared to explore intellectual challenges almost out of reach, that require the development of mathematical ideas (new to the learner) to overcome them. Optimistic problem solvers do not talk about failure, but rather finding out more. Failures and successes have thus been redefined as 'not yet knowing' and 'finding out more', respectively (Williams, 2002a). Optimistic problem solvers perceive not knowing as temporary, so expend personal effort to find out more by looking into what is not yet working to see what can be varied to increase the likelihood of finding out more. They differentiate between what is and is not within their control, and perceive being able to problem solve as permanent, and take this on as a characteristic of self.

MR and optimistic (resilient) problem-solving constructs both relate to Csikszentmihalyi and Csikszentmihalyi's (1992) affective states, but differently. MR is associated with safeguarding from worry and anxiety that can result when a challenge is too high, or the mathematics presented is beyond what the learner can understand. Mathematically resilient learning environments help learners maintain a productive stance which includes elements of control and arousal. Differently, environments which increase opportunities for optimistic problem-solving limit the likelihood of worry and anxiety because these environments are designed to enable groups to set their own challenge and decide on what mathematics they want to use to help overcome it. Optimistic mathematical problem solving occurs when conditions for flow exist (Csikszentmihalyi and Csikszentmihalyi, 1992). Flow is a state of high positive affect that occurs during creative activity when learners spontaneously pursue a self-set challenge that is almost out of reach and develop new mathematical ideas, that are almost beyond what they are capable of achieving, to overcome this challenge. During flow learners lose all sense of self, time, and the world around.

For the remainder of this chapter, HQLS refers to 'experience associated with the creative development of new mathematical knowledge in a flow state'. As optimistic problem-solving activity has been found necessary for creative mathematical problem solving, and engaging in multiple flow activities over time has been shown to build the characteristics of an optimistic problem solver (resilience) (Seligman, 1995), identifying features of learning contexts that contribute to creating flow conditions (Williams, 2014) is crucial for increasing the prevalence of such learning cultures.

Creative mathematical problem solving: The task

> *An emerging body of literature ... conceives of mathematics learning as inherently social (as well as cognitive) activity, and an essentially constructive activity instead of an absorptive one (Schoenfeld, 1992, p. 340).*

Awareness of the importance of classroom talk as well as internal mathematical thinking in developing mathematical understandings has increased markedly over the past 40 years, as has research into ways to achieve this (e.g., Wood and Yackel, 1990; Wood et al., 2006; Lee, 2006; Littleton and Mercer, 2013; Hino and Funahashi, 2014). Like Schoenfeld (1992), I take task to mean not only the task as set but the task as enacted, context, interactions, and resources employed:

> *Task in the full sense includes the activity which results from learners embarking on a task, including how they alter the task in order to make sense of it, the ways in which the teacher directs and redirects learner attention to aspects arising, and how learners are encouraged to reflect or otherwise learn from the experience of engaging in the activity initiated by the task.* (p. 207).

My Engaged to Learn Approach (E2L) to implementing solving of unfamiliar problems forms part of tasks under the definition employed. Several key features of this approach are now discussed including group composition, task structure and features of the approach that encourage mathematical struggle within an emotionally secure environment.

Structure of E2L

I developed E2L as a teacher and refined it as a researcher and provider of professional learning (see for example, Williams, 2002b, 2020, 2023; Herbert and Williams, 2021). In experimenting to find ways to encourage learners to discuss mathematical ideas, I found that I needed to let learners struggle when they did not know how to proceed because, if I hinted or told, learners tended to stop and wait for me to hint or tell again. Not hinting or telling is a crucial element of the E2L approach.

The basic structure of E2L includes the teacher composing groups and groups allocating their own roles (e.g., mediator / encourager, reader / timer, recorder, and reporter). The overarching problem-solving task is introduced by the teacher. Learners undertake 3–4 cycles of the following activities during the task with the teacher posing questions during each activity: (a) group brainstorming 10–15 minutes including brief interactions with teacher; b) groups focus their reports and 'prime' their selected reporter while the recorder captures discussion and displays; c) the teacher, informed by group discussions and displays, decides on order of reporting; d) reporters presents to class (1–2 min) using the board or a worksheet to aide communication; e) class members ask questions of reporters to clarify or identify flaws; f) the teacher values an aspect of each report, poses questions likely to elicit comparisons across reports that groups might consider in subsequent group work, and points to ways in which reports contribute to class development of ideas. The activity cycle is repeated.

Group composition

Commonly employed group compositions include attainment grouping (performance grouping), mixed attainment grouping (sometimes selected using attainment criteria) (VanTassel-Baska, 2006), random grouping, and friendship grouping. Group composition structures that contribute to (or could inhibit) resilience-building activity in HQLSs (Williams, 2011b; Ward-Penny and Thomas, Chapter 4) are now discussed.

HQLSs frequently include collaborative problem solving where learners build new ideas together (e.g., Francisco, 2013), rather than co-operative activity where learners each work to reach their own goal supported variously by other group members. Friendship groups do not "ensure good collaboration" (Peck and Barnes, 1999, p. 298) for several reasons including social dynamics which may interrupt collaborative construction if a group member feels pressure to follow the *social* leader (Williams, 2009), frequent off-task talk is likely (Peck and Barnes, 1999), and there may be less variation in ideas than in some other groups. Variability 'differences of opinion' in geography groupings (Buchanek, 2016), and 'different areas of expertise' in spontaneously formed engineering research and design groups (Sawyer, 2007) stimulated HQLSs (flow) interactions, likened by Sawyer to jazz syncopation where unexpected, pleasurable contributions may be inserted by unanticipated instruments. As research has shown that boys are more likely to take over equipment and dominate conversation (e.g., Keast, 1998) gender balance or more girls than boys in a group are recommended. Random (Liljedahl, 2014) groups are not guaranteed to fit productive grouping types.

The relative optimism of group members can promote or inhibit opportunities for group HQLSs (Williams, 2014). Groups composed of only optimistic learners frequently manoeuvre HQLSs but, groups of non-optimistic learners rarely progress beyond what they already know. Even in groups with mainly optimistic learners, creative activity does not always occur. For example, a non-optimistic high performing learner can inhibit HQLS opportunities by keeping the talk within what they know, and not allowing other group members to explore.

'Same pace of thinking groups' (Williams et al., 2012) (not 'same level of performance groups') think at a similar pace about new ideas and can thus 'sustain flow' together. Goldfinch (see Williams et al., 2012; Goldfinch and Williams, 2016) was surprised at differences she found in group membership between performance groups and E2L 'same pace of thinking groups' in her class. She now encourages teachers to group learners differently depending on the intended purpose: 'same pace of thinking groups' for exploring new ideas, and mixed levels of performance for peer tutoring.

Group members inclined to encourage others can catalyse co-construction by facilitating productive interactions in which learners, who initially do not know a way to proceed, experiment and discuss until they begin to find ways (Cohen, 1994). These productive learner interactions include exploratory talk (Mercer, 1995; 2008) which can be facilitated by encouraging: a) contributions from all group members; b) active listening; c) building on what has come before; d) asking questions, including about explanations and justifications; and e) treating the ideas and opinions of others with respect. Through these activities an atmosphere of trust, sense of shared purpose and shared decision making, opportunities for cognitive challenge and emotional security occur. Peck and Barnes (1999) and Williams (2009) recommended the inclusion of at least one encouraging group member in each group.

Accessible and authentic tasks

'Accessible tasks' allow all learners opportunity to participate, giving multiple entry points, accessible through varying levels of mathematical sophistication, that can be communicated through multiple representations. Authentic tasks can further increase the ease of participation by "[s]ituating mathematics tasks in students' cultural contexts [which] empowers them to participate through considering mathematics as part of their own identities and lives" (Gibbs and Hunter, 2018, p. 333).

The blue smarties promise (BSP Task)

The BSP Task (Williams, 2011b) is an authentic and accessible task that includes a range of activities that illustrate HQLS features. Year 5 classes worked in groups of four to design a blue Smartie[1] promise to entice blue Smartie lovers to buy. The only criterion was that the promise could not to be broken. Lesson structure followed the general E2L form: cycles of group work, reporting, learners questioning reporter, and brief discussion, before the next cycle began. The first two cycles involved predicting, counting, and sharing findings with the class. In Cycle 1, each group had one box of Smarties, predicted, counted, compared predictions with their count, developed a sentence about their comparison, and primed their reporter who added the group's sentence to the board and explained the group's thinking. In Cycle 2, each learner had a box to open, the group predicted what the whole class results would look like, counted, added data to a table on the board (Figure 16.1), found three interesting points about the table, and primed their second reporter who reported their three points to the class.

Reports on what was interesting included 'most boxes had 5, 6, 8 blue Smarties', the broad range, and various reasons for the data spread. Learners experienced many surprises including when: a) findings did not match predictions; b) different numbers of each colour found in box; c) fifteen blue Smarties were found in one box (Cycle 2), eliciting gasps of amazement; and d) learner Lenny found the data table (Figure 16.1): "... really *really* surprising ... even the *four* (pause) because that is *half* (pause) what I thought it would be". Lenny spontaneously began to think about the average as a start to sense making, and shifted focus from the BSP to focus on average "*Yeah* I didn't really put that much into our ... promise because I was [soft laugh] *trying to figure out the average*".

In Cycle 3, groups struggled with their Blue Smarties Promises so the teacher extended the number of cycles from 3 to 5. In Cycle 3 groups had used only

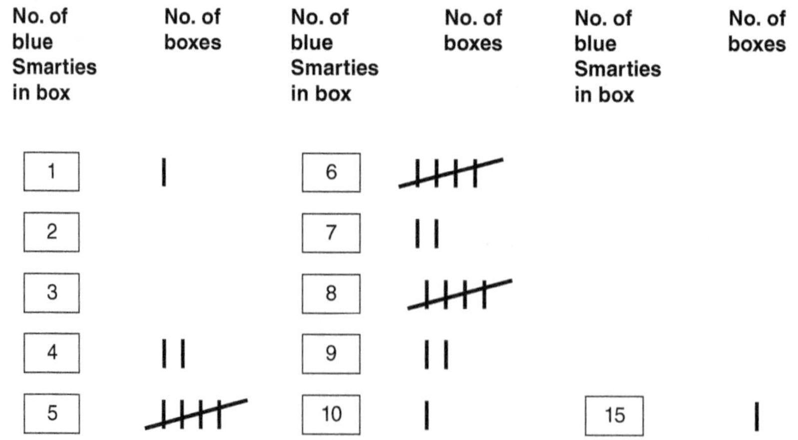

No. of blue Smarties in box	No. of boxes	No. of blue Smarties in box	No. of boxes	No. of blue Smarties in box	No. of boxes
1	I	6	⊪⊬		
2		7	II		
3		8	⊪⊬		
4	II	9	II		
5	⊪⊬	10	I	15	I

Figure 16.1 Display on the board, recording number of boxes with each number of blue Smarties.

'absolute' terms and found their promises could be broken. For reporting, the teacher placed the groups in an order in which ideas developed from 'no promise is possible', to promises produced but not checked to "You will get five blue Smarties in a box", to some groups still trying: "We tried three promises but found all could be broken- we are still trying". The legitimacy of promises was discussed by learners. After class discussion, the teacher posed long 'wondering' questions with suggestions (not obligations) to consider them:

I wonder ... can all of those reported ideas be possible or can only some, or even none? When you go back to your groups you might like to think about this. You might also want to try varying the way you write your promises to see if that gives any useful ideas ... I wonder are there types of wording you could use to make different types of promises?

In Cycle 4, groups began to use a little language of chance, but the promises were mostly not enticing to blue Smarties lovers "There are at least four blue Smarties in each box" or "You are very likely to get four or more blue Smarties in a box" or "You might be lucky and get 15 blue Smarties in your box" (which attracted exclamations). The legitimacy of the Cycle 4 promises was discussed, then the teacher wondered whether there was anything different about some of the wording this time and class members identified 'language of chance words'. Informed by reports and discussion in Cycle 4, learners became more engaged and their thinking more creative. Reports demonstrated greater understandings of chance in enticing wording for blue Smarties lovers: "These boxes are likely to contain six or more blue Smarties, you could be lucky", "If you buy three boxes, you will be surprised at the number of blue Smarties in these boxes" and "Tell the Smarties Company that they will need to change what they want in their promise". At times there was laughter and / or gasps of appreciation.

Emotional security

A caring environment where learners feel secure in contributing new ideas is crucial for HQLSs. The BSP Task as implemented through E2L focuses on the whole child not only their mathematics learning (see for example, Inoue et al., 2019). There is a need for human kindness (Baker and Johnston-Wilder, 2019), establishing humane classroom norms, and connecting learning with learners' personal experiences (Dalton and Watson, 1997). E2L group rules also contribute to an emotionally secure environment: (Williams, 2011b): 'give everyone a turn to talk', 'explain a different way if some do not understand', and 'if you do not understand give your group detail about what you do / do not understand to help them reexplain', 'don't laugh or just say "no you are wrong" instead consider whether this idea could be changed slightly or give a quiet justification for why part of it will not work'. Raise group awareness that with this type of learning, creative ideas may come from learners you had not expected to have such ideas and laughing at them could stopped them contributing so your group could miss out on some good ideas.

Two class rules designed to keep reporters emotionally safe are: Class members can ask the reporter questions to gain a better understanding of what was presented but cannot ask questions beyond the content reported or contradict ideas while the reporter is at the board. Contradictions, if justified, can occur during later reports, discussion at the end of the reporting session, or in reporting in the next cycle. These constraints provided an environment that reduced judgements of mathematical performances thus reducing the likelihood of MA developing (dos Santos Carmo et al., 2019). Priming the reporter is another element of E2L that contributes to emotional security as demonstrated by this statement of a learner who generally worried about what others thought of her: *"[I like the group work in E2L because] if you [reporter] get the wrong answer ... the whole group takes the fall [colloquial for 'share the blame']"*.

Productive struggle

'Telling' by a peer (or teacher) can produce negative emotions when a learner wants to undertake productive struggle. Teachers enabled productive struggle by not hinting or telling but rather supporting such struggle in the BSP Task through their comments prior to group priming, post reporting, and in their enabling of learner autonomy in exploring and communicating their analyses of the table (Figure 16.1).

Lenny did not struggle in the first year of the research, he just 'gave up' when he did not know. By the end of the second year he had changed *"I [now] sit there and I really really think about it ..."*. When asked about how this change had happened, he stated: *"I don't really know ... it just made me think more It was probably actually doing the tasks ..."*. He was surprised in the BSP Task when his calculation of average gave just over two and demonstrated resilience as he puzzled about this. He was uncertain about the procedure he had used (add together tallies and divided by the number of boxes with tallies) and drew on elements of several class activities (a report, class exclamation, and his own thinking) to justify that his calculation was incorrect. He employed these as follows: a) the report "I knew ... if there was eight six and five ... more of them are over five so how is it [around] two?", b) class exclamation, the class responses amplifying fifteen, and c) Lenny's own thinking, his changed ideas on how to read the table and calculate average and his changed ideas on how to read a table "[my calculation] counted the fifteen as one"). Lenny linked big ideas from concrete, verbal, kinetic, tabular, and numerical, representations to develop big ideas (Williams, 2014). Ward-Penny and Thomas (Chapter 4) also draw attention to the simultaneity of teacher actions that develop deep mathematical understanding for learners and actions that are resilience-building for learners.

Building big ideas

Mathematical ideas developed through the BSP Task include big ideas like awareness of variation and randomness, and language of chance. The more nuanced promises developed over time increased the complexity of thinking about probability and ways to apply ideas. There was also an increased understanding of interpreting tables. Other tasks that capture very different big mathematical ideas during HQLSs include the Cane Toad Task (Galbraith et al., 2010), the three-task sequence linking geometric constructions, similarity and congruence, and proof (Williams, 2011a), and the excited exploring of divisibility with a calculator (Kieran and Guzman, 2003).

Requires resilient teachers

Complex teacher actions within E2L require teacher resilience as they try and try again to find ways to elicit and sustain creative learner thinking. These actions are illustrated through teacher decision making in the BSP Task including how to:

- order reports to build ideas
- value each report
- decide whether to question
- sustain interactions when teacher action did not have desired effect
- respond to learner posed question with question without hinting or telling.

Year 1 teacher Earl illustrated resilience in his on-the-run activity of exerting personal effort—looking into the situation to find what he could vary to find out more:

> So, in terms of questions on the fly, just being really conscious of not driving it- It's something that I've put a lot of thought into- it's not come overnight" I've had to chip away at it [the questioning] ... (Williams, 2016, p. 94).

> Quite often it's a similar question that I need to ask or want to ask, but it's just peel back a few layers so that- I'm not explicitly dropping the clues in. (Williams, 2016, p. 94)

Earl had increased his questioning skills and taken this on as a characteristic of self *"When I try a new task now, I can generally think on my feet about what to do"* (Williams, 2016, p. 91).

As quotes from Earl illustrate, a metacognitive overlay to teacher actions is required during E2L. Types of internal teacher questions during and after the lesson include:

- Which groups are / are not interacting productively and why?
- How can this inform future group composition?
- What mathematical ideas are emerging?
- What can I do to help these learners extend their ideas further?

By focusing teachers on such aspects of class activity a metacognitive style can be developed and teacher resilience can build as they find out more. Gu and Day (2007) also identified teacher resilience as necessary for undertaking sophisticated actions involved in effective teaching.

E2L professional learning

Although this section focuses on E2L professional learning programmes (PLPs), it provides insights relevant to other professional learning programmes intended to simultaneously build teacher resilience and pedagogical expertise.

I draw from two whole elementary school PLPs in which I was the external leader, to illustrate several key PLP features. These include, an external leader possessing teaching expertise, knowledge of relevant research, a demonstrated ability to bring about teacher change, and is not a member of the school community. Part of their role is to enhance the expertise of lead teachers so they can lead the PLP between PLP blocks and after completion of the PLP (see for example, Clarke et al., 2014). These PLPs differed from school to school because the lead teachers and team members decided what they wanted to learn more about. In each school, the principal was dedicated to professional learning for her staff so taught classes and used precious emergency teacher funds to release teachers. They also reorganised the school day to facilitate the programmes.

School 1, with five to ten teachers during the time of the PLP, was a small rural school that was initially part of a PLP for a school cluster. The cluster PLP commenced with myself as external leader describing the process and purposes of E2L and what it can achieve. I then modelled the approach in classes across the cluster with teacher observers travelling to take part. A debrief session followed.

The School 1 programme had three PLP blocks (6 days) a year. Several weeks prior to each block, teachers identified a mathematical topic for the task, I provided a skeleton task and with the lead teachers further designed the task. Together with the principal we decided which classes to implement the tasks in. Teachers could accept or decline the invitation to take part and also decide whether they would implement the task with my support or I would implement with teacher support. Lead teachers familiarised class teachers with the task prior to the PLP block. At the start of each block, an after-school session for all teachers was organised. The participating class teachers briefly introduced their task. Teacher year level teams (early, middle, and senior years) explored what learners might do mathematically, possible teacher actions, and anything else that interested them. Teams gave brief reports to staff then the class teacher discussed how they intended to implement the task (modifying their plans in the light of feedback if they wanted to). The class teacher could refine the task further as they saw fit, before implementation. Post task video stimulated discussions were undertaken between the class teacher, lead teachers, observers and myself. The class teacher reflected first, the external leader listened, ensured observers focused on student responses not teacher judgements, and contributed by drawing attention to ideas suggested, and sometimes posing questions.

Subsequently, the class teacher and I worked together to design a workshop for the whole staff (working in year level teams) that follows the E2L structure. It was led by the class teacher (supported by myself) who selected and presented 3–5 lesson excerpts, asking each time for multiple responses to questions like 'What might you do next and why?' and 'What do you see?' After the groups had reported their ideas the class teacher explained why the excerpt was selected, which could be an insight, a positive situation, or a situation that they want to think further about. These excerpts were supported with video and / or student work. After a PLP block, teacher teams implemented the task in another class (with at least one teacher observer). They debriefed before providing feedback to the teaching team, who made decisions about what to emphasise and what to vary about the task, composing groups, 'potential mathematical trajectories', and possible teacher questions that might elicit further learner thinking. Another teacher then implemented the task and the cycle continued. Potential mathematical trajectories are "*prediction[s] of how the students' [learners'] thinking and understanding will evolve in the context of the learning activities*" (Simon, 1995, p. 136).

School 2, an inner Melbourne school with over 25 teachers had worked with me for several years before this PLP commenced. I worked individually with seven teachers (one to three years with each) in conjunction with a whole school PLP similar to the two PLPs School 1 took part in. The PLP reported herein was subsequent to these other two PLPs.

The School 2 PLP reported was undertaken because staff wanted to learn more about designing E2L tasks. In an after-school workshop, year level teams examined tasks in Sullivan and Lilburn (2002) to identify possible E2L tasks or tasks that could be modified to build E2L tasks. This book was selected because only some tasks were appropriate for this purpose so teachers would be learning to identify appropriate tasks. Teams selected tasks, examined them, and made decisions on whether they were appropriate then shared their findings with the whole workshop. A discussion followed. With feedback, teams either shifted to another task, began to modify their task, or made preparations to implement it. Between PLP sessions, all teachers in each team, one by one, implemented the task (with at least one teacher observing). Feedback to the team led to modifications where necessary before the next teacher tried the task. Teams developed short presentations for the next PLP workshop, to share what they had found. They developed their own ways to share (including artefacts, group worksheets, photographs, and video excerpts). The sophistication of what they noticed and / or were able to implement increased over time.

Features of PLPs that develop teacher E2L expertise while building teacher resilience mimic several key features of E2L including attention to emotional security, productive struggle, enabling autonomy, and collaborative interactions.

Emotional security is protected in numerous ways including a) the external leader emphasises that observers focus on learners' responses not the teacher, and observers ask themselves questions such as what worked, what might we want to think more about, and what might be varied to increase the likelihood of learning progressing further; b) the class teacher reflects first during post-lesson discussions, and can limit what is discussed or what video is shown (if they want to do so); c) teachers can accept or decline invitations to implement tasks during the PLP; d) potential teacher anxiety about 'getting mathematics wrong' is limited by asking teacher teams what learners might do (so mistakes are the learners' not the teachers'); and e) as teachers' mathematical understandings build during the PLP, this contributes to their feelings of security.

- *Productive pedagogical struggle* is embedded in the E2L approach through the ongoing decision-making and revisions of decisions on-the-run in class. For example, Earl struggled as he developed his questions. Productive struggle also occurred in staff PLP sessions as teachers struggled to make sense of the task, what mathematics to use and how, and find ways to elicit further thinking from students. In addition, Sharon struggled to compose and innovatively implement 'same pace of thinking' groups (Goldfinch and Williams, 2016). Teacher talk in interviews showed that teacher resilience was built through such struggles.

- *Enabling autonomy* Staff demonstrated autonomy in various ways including their selection of mathematics topic, refinement of the task, identifying of E2L features enacted, and how to structure workshops to deepen teacher thinking. An illustration of enabling autonomy occurred when lead teachers in one school asked me to write a list of questions I ask while implementing tasks, so they could ask them too. I suggested they observe, take notes (questions asked, thoughts on why they were asked, and student responses) and report back during debrief sessions. This became a feature of their learning culture.

- *Collaborative interactions* between teachers were structured into the PLP, including a) debriefing from the initial session modelled by the external leader and subsequent debriefing sessions; b) whole school pre-task sessions, and c) team feedback discussions between PLPs where they developed new ideas together.

Incorporating these features into PLPs can build resilience. It will be a long-term process to build the resilience of all mathematics learners and teachers of mathematics. A possible question becomes: Resilience is a general construct which includes teacher resilience and elements of mathematical resilience. Could we increase the rate of resilience building by identifying other resilience-building initiatives in the community and harnessing them to help build the resilience of mathematics learners and teachers?

Resilience-building in communities

The principals in School 1 and 2, possibly unaware, undertook activity that builds resilient schools (Gu and Day, 2007). They built teacher connectedness through enabling many collaborative opportunities, had faith in their lead teacher and teachers' decision making (see Clarke et al., 2014), and provided opportunities for professional growth.

More broadly, there are many resilience building initiatives in the community including a) parenting initiatives (e.g., Seligman, 1995; Russell and Wright, Chapter 10), b) preschool programmes such as Regio Emilia (e.g., Pope Edwards and Gandini, 2015) and Bush Kinders (e.g., Campbell and Speldewinde, 2018); c) health-related initiatives (Resnick, 2000); and d), workplaces (Goleman, 1998) including those in research and design (Sawyer, 2007). Research into bringing together resilience-building community initiatives and amplify their outcomes could result in a greater number of resilient prospective mathematics teachers, teachers, and learners in schools which could reduce the magnitude of resilience building required in PLPs. The question for us becomes: what commonalities are there across these initiatives that could be relevant to building the resilience of teachers and learners of mathematics?

Resilience: The future

Where are we now? We know that learning rules and procedures without meanings is very unlikely to produce deep learning and frequently builds mathematics anxiety. We also know there are many teachers, and many policy makers who are convinced that learning mathematics only occurs through procedural teaching. We need to change this view. This *may* be accomplished by demonstrating that deep mathematical understandings, increased enjoyment of mathematics, and increased resilience can result from learning mathematics through high quality learning situations that markedly decrease the frequency of mathematics anxiety. A multi-pronged initiative could raise awareness sufficiently to convince reluctant stakeholders that the nature of mathematics learning needs to change. Simultaneous implementing of these prongs is necessary because while procedural teaching prevails, depletion of resilience continues to occur, working against resilience-building initiatives. This initiative includes: a) continuing strategies to reduce mathematics anxiety; b) significantly increasing PLPs that builds relevant pedagogical expertise whilst simultaneously building teacher resilience; c) integrating resilience building into mathematics teacher education programs using similar strategies to PLPs; and d) making connections with other resilience building communities to learn from each other. A pilot programme in a community with many resilience building initiatives could shorten the time frame for raising stakeholder awareness. Over time, as the frequency of procedural teaching reduces and the frequency of mathematical anxiety decreases, resources dedicated to reducing mathematics anxiety could be dedicated to PLPs strengthening resilience building pedagogies.

The question for us is, what can we do as teachers and researchers to contribute in multiple ways to developing multipronged resilience-building approaches and learn from each other during the process? Thinking optimistically, we can look into what is not working yet to identify what we could vary to increase the likelihood of finding out more as we move forward into a future containing high quality mathematics learning. Overwhelming as the situation may seem, giving up is not an option, having faith in each other will hasten that progress.

Note

1 Small multi-coloured (7 colours) sweets with chocolate centres packed in boxes of approximately 50. The outside of the box has pictures of these sweets showing many colours.

References

Baker, J. (2021). 'You see it differently once you calm down': Developing an intervention to support learners to address their mathematics anxiety. In *Unpublished thesis submitted in fulfilment of the requirements for the degree of Doctor of Philosophy in Education*. University of Warwick, Education Studies Department. Accessed 4th May, 2023 at http://wrap.warwick.ac.uk/166960/1/WRAP_Theses_Baker_2021.pdf

Baker, J., and Johnston-Wilder, S. (2019). Mathematics: A place of loving kindness and resilience building. *Journal of the Canadian Association for Curriculum Studies*, 17(1), 11–126.

Buchanek, R. (2016). Negotiating differences: Stimulating creative geographical thinking through critical engagement. *Unpublished doctoral dissertation, Deakin University*, Victoria. https://dro.deakin.edu.au/articles/thesis/Negotiating_differences_stimulating_creative_geographical_thinking_through_critical_engagement_/21108967/1. Accessed 24 June 2023.

Campbell, C., and Speldewinde, C. (2018). Bush kinder in Australia: A new learning 'place' and its effect on local policy. *Special Issue: Affect, Embodiment and Interrelationships: Reconceptualising Educational Policy through Encounters with Learning Spaces and Places. Policy Futures in Education*, 17(4).

Clarke, N., Duncan, A., and Williams, G. (2014). Implementing problem solving at Eagle Point Primary School: A whole school approach. In J. Vincent, G. FitzSimons, & J. Steinle (Eds.), *Maths rocks: The MAV 51st annual conference* (pp. 20–28). Melbourne, Victoria, Australia: The Mathematics Association of Victoria.

Cohen, E. G. (1994). Restructuring the classroom: Conditions for productive small groups. *Review of Educational Research*, 64(1), 1–35. https://doi.org/10.3102/00346543064001001

Csikszentmihalyi, M., and Csikszentmihalyi, I. (Eds.) (1992). *Optimal experience: Psychological studies of flow in consciousness*. New York: Cambridge University Press.

Dalton, J., and Watson, M. (1997). *Among friends: Classrooms where caring and learning prevail*. Armadale, Vic: Eleanor Curtain Publishing.

Day, C., and Gu, Q. (2014). *Resilient teachers, resilient schools: Building and sustaining quality in testing times*. UK: Routledge.

dos Santos Carmo, J., Gris, G., and Palombarini, L. (2019). Mathematics anxiety: Definition, prevention, reversal strategies and school setting inclusion. In: KOLLOSCHE D. et. al. (Eds.) *Inclusive mathematics education. State of the art research from Brazil and Germany.* Switzerland: Springer. 403–418.

Edwards, C. P., and Gandini, L. (2015). Teacher research in Reggio Emilia, Italy: Essence of a dynamic, evolving role. *Voices of Practitioners: Teacher Eesearch in Early Childhood Education,* 10(1), 89–103.

Francisco, J. (2013). Learning in collaborative settings: Students building on each other's ideas to promote their mathematical understanding. *Educational Studies in Mathematics,* 82, 417–438.

Galambos, N., and Leadbeater, B. (2000). Trends in adolescent research for the new millennium. *International Journal of Behavioral Development,* 24(3), 289–294.

Galbraith, P., Stillman, G., and Brown, J. (2010). Turning ideas into modeling problems. In R. Lesh, P. Galbraith, C. Haines, & A. Hurford (Eds.), *Modeling students' mathematical modeling competencies: ICTMA 13* (pp. 243–283). New York: Springer.

Gibbs, B., and Hunter, R. (2018). Making mathematics accessible for all. In J. Hunter, P. Perger, & L. Darragh (Eds.), Making waves, opening spaces *(Proceedings of the 41st annual conference of the Mathematics Education Research Group of Australasia)* (pp. 330–336). Auckland: MERGA.

Gu, Q., and Day, C. (2007). Teachers' resilience: A necessary condition for effectiveness. *Teaching and Teacher Education,* 23(8), 1302–1316. https://doi.org/10.1016/j.tate.2006.06.006

Herbert, S., and Williams, G. (2021). Eliciting mathematical reasoning during early primary problem solving. *Mathematics Education Research Journal,* 35, 77–103. 10.1007/s13394-021-00376-9

Hino, K., and Funahashi, Y. (2014). The teacher's role in guiding children's mathematical ideas toward meeting lesson objectives. *ZDM Mathematics Education,* 46(3), 423–436. 10.1007/s11858-014-0592-0

Hernandez-Martinez, P., and Williams, J. (2013). Against the odds: Resilience in mathematics students in transition. *British Educational Research Journal,* 39(1), 45–49.

Inoue, N., Asada, T., Maeda, N., and Nakamura, S. (2019). Deconstructing teacher expertise for inquiry-based teaching: Looking into consensus building pedagogy in Japanese classrooms. *Teaching and Teacher Education,* 77, 366–377. 10.1016/j.tate.2018.10.016

Johnston-Wilder, S., Lee, C., and Mackrell, K. (2021). Addressing mathematics anxiety through developing resilience: Building on self determination theory. *Creative Education,* 12(9), 2098–2115.

Goldfinch, S., and Williams, G. (2016). Grade 3/4 developing big mathematical ideas: Sharon and her little red book. In W. Widjaja, E. Y-K Loong, & L. A. Bragg (Eds.), *Maths explosion 2016. Melbourne* (pp. 107–115) Victoria: Mathematics Association of Victoria.

Goleman, D. (1998). *Working with emotional intelligence.* Bantam.

Keast, S. (1998). Gender issues and single sex maths classes. *Paper presented at the mathematics: Exploring all angles: Thirty-fifth annual conference of the mathematics association of Victoria.* Melbourne.

Kieran, C., and Guzmàn, J. (2003). The spontaneous emergence of elementary number-theoretic concepts and techniques in interaction with computer technology. In P. Neil A. Dougherty, & J. Zilliox (Eds.), *2003 Joint Meeting of the 27th conference of the International Group for the Psychology of Mathematics Education and the Group for the Psychology of Mathematics Education of North America* (Vol. 3, pp. 141–148). Honolulu, Hawaii: PME.

Lee, C. (2006). *Language for learning mathematics: Assessment for learning in practice*. UK: Open University Press.

Lee, C., and Johnston-Wilder, S. (2017). The construct of mathematical resilience. In Ulises Xolocotzin Eligio (Ed.), *Understanding emotions in mathematical thinking and learning* (pp. 269–291). UK: Academic Press.

Liljedahl (2014). The affordances of using visibly random groups in a mathematics classroom. In Y. Li, E. Silver, & S. Li (Eds.), *Transforming mathematics instruction: Multiple approaches and practices*. New York, NY: Springer.

Littleton, K. and Mercer, N. (2013). *Interthinking: Putting talk to work*. Routledge: Abingdon.

Mercer, N. (1995). *The guided construction of knowledge: Talk amongst teachers and learners*. Clevedon, England: Multilingual Matters.

Mercer, N. (2008). Three kinds of talk. Retrieved from https://thinkingtogether.educ.cam.ac.uk/resources/5_examples_of_talk_in_groups.pdf

Nettles, S., Mucherah, W., and Jones, D. (2000). Understanding resilience: The role of social resources. *Journal of Education for Students Placed at Risk*, 5, 47–60.

Peck, R., and Barnes, M. (1999). Teacher and student perspectives on collaborative learning. In N. Scott, D. Tynan, G. Asp H. Chick, J. Dowsey, B. McCrae, J. McIntosh, & K. Stacey (Eds.), *Across the ages* (pp. 96–305). Victoria: Mathematical Association of Victoria.

Resnick, M. D. (2000). Protective factors, resiliency, and healthy youth development. *Adolescent Medicine*, 11, 157–164.

Sawyer, K. (2007). *Group genius: The creative power of collaboration*. NY: Basic Books.

Seligman, M. E. P., with Reivich, K., Jaycox, L., and Gillham, J. (1995). *The optimistic child*. Houghton, Boston USA: Mifflin and Company.

Simon, M. (1995). Reconstructing mathematics pedagogy from a constructivist perspective. *Journal for Research in Mathematics Education*, 26, 114–145.

Schoenfeld, A. (1992). Learning to think mathematically: Problem solving, metacognition, and sense making in mathematics. In D. Grouws (Ed.), *Handbook for research on mathematics teaching and learning*. (pp. 334–370). New York: Macmillan.

Sullivan, P., and Lilburn, P. (2002). *Good questions for math teaching: Why ask them and what to ask [K-6]*. Sausalito, CA: Math Solutions Publications.

VanTassel-Baska, J. (2006). *Comprehensive curriculum for gifted learners* (3rd Edition). Boston: Pearson.

Williams, G. (2002a). Associations between mathematically insightful collaborative behaviour and positive affect. In Anne Cockburn, & Elena Nardi (Eds.), *Proceedings of the 26th Annual Conference of the International Group for the Psychology of Mathematics Education* (Vol. 4, pp. 401–408). Norwich: University of East Anglia.

Williams, G. (2002b). Have faith in students' ability to think mathematically. In C. Vale, J. Roumeliotis, & J. Horwood (Eds.), *Valuing mathematics in society* (pp. 114–126). Melbourne, Victoria: Mathematical Association of Victoria.

Williams, G. (2009). Why some groups work and some do not. In *Mathematics of Prime Importance, Proceedings of the 2009 MAV Conference* (pp. 271–278). Melbourne, Victoria: Mathematical Association of Victoria.

Williams, G. (2011a). Building optimism in prospective mathematics teachers: Psychological characteristics enabling flexible pedagogy. In O. Zaslavsky, & P. Sullivan (Eds.), *Constructing knowledge for teaching secondary mathematics, mathematics teacher education* (Vol 6, pp. 307–323). NY: Springer.

Williams, G. (2011b). Relationships between cognitive, social, and optimistic mathematical problem-solving activity. In B. Ubuz (Ed.), *Proceedings of the 35th Conference of the International Group for the Psychology of Mathematics Education* (Vol. 4, pp. 345–352). Ankara, Turkey: PME.

Williams, G. (2014). Optimistic problem-solving activity: Enacting confidence, persistence, and perseverance. *ZDM Mathematics Education*, 46(3), 407–422. 10.1007/s11858-014-0586-y.

Williams, G. (2016). Inclining to explore mathematically and pedagogically: Students and teachers possessing the same characteristics. In W. Widjaja, E. Y-K Loong, & L. A. Bragg (Eds.), *Maths explosion 201*. (pp. 87–97). Melbourne, Victoria: Mathematics Association of Victoria.

Williams, G. (2020). Attaining mathematical insight during a flow state: Was there scaffolding? *Hiroshima Journal of Mathematics Education*, 13, 31–56.

Williams, G. (2023). Collaborative approaches to ehancing theoretical perspectives and pedagogical insights. *Theory into Practice*, 62(1), 26–38. 10.1080/00405841.2022.2135906

Williams, G., Harrington, J., and Goldfinch, S. (2012). Problem-solving: 'Same pace of thinking' groups. In Jill Cheeseman (Ed.), *It's my maths: Personalised mathematics learning* (pp. 297–304). Melbourne, Victoria: Mathematics Association of Victoria.

Wood, T., and Yackel, E. (1990). The development of collaborative dialogue within small group interactions. In L. Steffe, & T. Wood (Eds.), *Transforming children's mathematics education: International perspectives* (pp. 244–252). Hillsdale, NJ: Lawrence Erlbaum.

Wood, T., Williams, G., and Mc Neal, B. (2006). Children's mathematical thinking in different classroom cultures. *Journal for Research in Mathematics Education*, 37(3), 222–252.

17 Continuing to work for mathematical resilience

Sue Johnston-Wilder and Clare Lee

Introduction

This book is aimed at teachers, but teachers alone cannot change the whole education spectrum to allow all learners to develop mathematical resilience. If teachers are to be allowed and encouraged to work in ways known to develop mathematical resilience, as detailed in this book, then they will need support from the wider educational community, including other adults in schools, parents, teacher-trainers, researchers and governments. Chapter 15 looked at the systemic and pedagogic changes that will be needed if all learners, and the facilitators of that learning, are to believe that it is possible to learn mathematics without feelings of anxiety and fear. Here we call on everyone involved to act.

We will first look back at the history of working to develop mathematical resilience and address the anxiety that learning mathematics can cause. We will then briefly consider an ongoing argument about the genesis of mathematics anxiety. This debate does not refute either the existence or effects of mathematics anxiety, but it can detract from a clear focus on acting to prevent or mitigate the effects of this disabling condition. We will then emphasise again the necessity of considering the feelings, emotions and wellbeing of learners as they work to learn mathematics. We finish with a summary of what everyone involved can do to further the development of mathematics resilience in all learners.

Resilience is an important attribute, not just when learning mathematics but across all of life's experiences, as for most people life is not 'all hearts and roses'. Everyone will experience barriers that need to be overcome, and thus everyone will need the skills and resilience to ensure their own psychological safety and wellbeing in the face of uncertainty and insecurity. As we have made clear, we do not see resilience as something that a person either has or does not have, nor do we see it as something that a person can be told that they should display more of when encountering a challenging environment or problem. If fact, we think that the more someone is told to be resilient the less they are likely to display the attributes of resilience. Resilience is something that people develop within a supportive, inclusive and encouraging environment where learners are challenged and, crucially, are supported in successfully meeting those challenges.

A resilient individual is able to think when others are panicking, as they know how to calm down and enable themselves to think. A resilient person will persevere in finding solutions to problems, collecting evidence and ideas by listening to others, accessing books and the internet and trying things out until their struggle is rewarded with success. A resilient individual will accept the support of others and will readily give support, as they know that seeking support is not a sign of weakness but a sign of being determined to achieve the best possible outcome. They also know that in giving support they will learn alongside others. All these characteristics of the resilient individual are learned through interaction in an appropriate environment. Southwick et al. (2023) describe key factors that contribute to resilience. Each of these factors relates to ideas that have been discussed in this book: optimism, facing your fears, your beliefs and mindset, having social support, having role models, taking care of mind and body, accepting challenge, thinking flexibly and seeking meaning and purpose.

Teaching and learning mathematics affords opportunities to develop resilience in learners in the context of mathematics that we know from our research is readily applied to other aspects of life (Johnston-Wilder et al., 2021; Lee and Morgan, 2023). Some people develop resilience by serendipity, perhaps having someone in their life who shows them by example how to act resiliently. However, even those who show great resilience in other areas of their lives (Lee and Morgan, 2023) are sometimes unable to bring that resilience to help them overcome the challenges that learning mathematics presents. It seems that they need to be taught to learn mathematics with resilience.

Much of the population can recall adverse prior experiences that have happened to them whilst they were learning mathematics in primary or secondary school (Dowker, 2016; Lee and Morgan, 2023). They use these remembered experiences to explain the feelings that they endure when asked to engage with mathematics. It may be that individuals remember a collection of individually less traumatic experiences which collectively are seen by those individuals to result in the mathematics anxiety and avoidance they currently experience. These individuals do not remember mathematical learning environments as nurturing, supportive spaces but rather as spaces where the threat of shame and humiliation seems very real. They tell of, for example, being picked out by the teacher to answer a question and stuttering or giving the wrong answer because they felt so nervous, they could not think. Ashcraft and Krause (2007) show that anxiety is often the result of teaching mathematics in ways that can result in learners feeling fear and humiliation. Thus, it seems to us that two things should happen:

1. Teachers must give due consideration to the way that they support their learners' wellbeing as they learn mathematics. Teachers must develop a challenging but nurturing and inclusive environment, where everyone is expected to need support at times and to give support at others. An environment that develops mathematical resilience is one where the focus is on supporting everyone's learning and understanding of mathematical ideas and not on who is the quickest, or who can get the most ticks on their paper. Learning mathematics can often be a struggle, which is probably true of doing anything that has personal value to learners. Teachers who are working with their learners to develop their mathematical resilience do not pretend that mathematics could be easy "if only they would try to remember this way of doing it!"; instead, they ask their learners to engage in the struggle to understand and support them in doing so.

2. Even when a generation of learners begins to learn mathematics with facilitators who care for their wellbeing, so that they develop the resilience to overcome the barriers that mathematics will always present, mathematics anxiety will continue to be present in the rest of the population. Most people, even learners in primary school, will need to be taught explicitly how to address and mitigate the effects of mathematics anxiety. Mathematics anxiety is catching (Beilock et al., 2010); you can catch it from your teacher, your parents and from others around you. Therefore, mathematics anxiety will remain prevalent even if all teachers of mathematics work to develop resilience in their learners. Mathematics anxiety must be part of the conversation, because most people want to hide their anxiety, perhaps because it is often shame that has engendered that anxiety. They seem to think, "If nobody is talking about it, it must be me" (Brown, 2021). Therefore, the idea that mathematics anxiety is real and disabling, but that it can be overcome, must remain part of educational discourse and the educational landscape.

> *Everyone needs to hear this vital message (de Calvo, personal communication):*
>
> *The mums were so relieved to hear that maths anxiety is "real", and that they weren't the only ones who felt physically sick before maths lessons and made to feel stupid. They loved the approach of safeguarding and hadn't previously thought about healing the maths trauma they had carried for decades.*

Looking back

Before looking forward, we first look back and see the need for teaching in ways that develop mathematical resilience has been recognised for a long time, as has the idea that widespread attitudes and ideas about teaching mathematics may cause anxiety. In 1982, the influential Cockroft report was published (Cockcroft, 1982); this seminal document remains available on the internet. It was commissioned in 1978 by the UK government because they saw that *"few subjects in the school curriculum are as important to the future of the nation as mathematics"* (p. iii), and because of the *"hesitant grasp many adults have of even quite simple mathematical skills"*.

The Cockcroft committee was invited to consider the implications of the mathematics required in further and higher education, employment and adult life generally, and the prime focus of the brief was on the content of the mathematics curriculum and on pedagogy. Interestingly, the authors report becoming aware of the extent and impact of mathematics anxiety:

> The extent to which the need to undertake even an apparently simple and straightforward piece of mathematics could induce feelings of anxiety, helplessness, fear and even guilt in some of those interviewed was, perhaps, the most striking feature of the study (Section 20).

The report also alludes implicitly to what could be described as mathematics harm, the lowering of expectations of high paid work due to an inability to face the needful mathematics.

> ... the research study into the mathematical needs of adult life revealed the extent to which the need to make use of mathematics could induce feelings of anxiety and helplessness in some people. It also revealed that many people had far from happy recollections of their study of mathematics at school (Section 198).

An example was given of trying to learn school algebra, which was reported to be:

> ... a source of considerable confusion and negative attitudes among pupils. In some cases this was because the work had been found difficult to understand; in other cases it was felt that exercises in algebraic manipulation and topics such as sets and matrices had had little point (Section 201).

It is troubling to look at the date of this report, and to realise if it had been compiled today, forty years later, the researchers would probably find a similar situation. It is possible to teach algebra without generating confusion and negative attitudes. When it is taught in ways that support learners to understand what they are doing and why they might need these ideas, they can approach these ideas positively; for example, Lugalia (2015) reports learners running to algebra lessons when she introduced Grid Algebra as part of their learning. Section 202 goes onto list the resilience-averse actions of some teachers:

Adverse comments on the teaching of mathematics often concentrated on an alleged inability on the part of some teachers to explain clearly, on a tendency to ignore some of those in the class, on an unwillingness to answer questions and on moving through the course too quickly. There was criticism also ... of teachers who had been unable to state the purpose of the work which was being done - 'do it to pass your exams' (Section 202).

Experiences of being left behind, being excluded, questions not being valued, being rushed forward rather than being given time to understand and being given no insight into the value of mathematics, all relate to practices that have resulted in many people wishing to avoid mathematics and worse, feeling anxious and often fearful of engaging with it. Although such practices are still common, there are also currently many, many teachers who are teaching mathematics in ways that build resilience and allow their learners to experience mathematical wellbeing and success. We mention this because it can seem that the generation of mathematics anxiety and avoidance is inevitable when learning because of the nature of mathematics. This is not so.

The Cockroft report (1982) suggested ways that teaching mathematics could be improved, ways that positive attitudes to mathematics could be developed and the learning of mathematics increased. According to the Cockcroft Report, mathematics teaching at all levels should include opportunities for:

- exposition by the teacher;
- discussion between teacher and pupils and between pupils themselves;
- appropriate practical work;
- consolidation and practice of fundamental skills and routines;
- problem solving, including the application of mathematics to everyday situations;
- investigational work (Section 243).

If a combination of these elements were used today in environments where mathematics is learned, then the learners' opportunities to develop mathematical resilience would be increased, along with the likelihood of detecting any learners progressing towards developing anxiety. A quick look through the previous chapters in this book will show that these ideas are incorporated in the ideas discussed.

We will finish this section by asking you to read and reflect on section 810 of the Report, which summarises the importance of every child developing an understanding of mathematical ideas and the confidence to use them.

> ... every [child] needs to develop, while at school, an understanding of mathematics and confidence in its use. In our view this can only come about

as the result of good mathematics teaching by teachers who have been trained for their work and who receive continuing in-service support. It must therefore be the task of all who share this belief to support and encourage the implementation of the changes which we believe to be necessary and to make it clear that, as part of the education which our children receive, mathematics counts (Section 810).

They are still arguing

As you read in Chapter 2, the first discussion of the problem of mathematics anxiety is attributed to Dreger and Aiken in 1957. They termed the idea "number anxiety", but the problem was quickly understood to encompass all of mathematics, not just numerical work. So, mathematics anxiety has been recognised as an issue for more than sixty years. However, there are still some who want to attribute the occurrence of anxiety to someone's poor performance in mathematics rather than to focus on how to teach in ways that prevent it disabling so many people. This is basically the assumption made in the EEF-funded project that aimed to address mathematics anxiety by working to improve primary pupils' working memory (EEF, 2019). There was some improvement seen in the learners' attainment in the post-test they used; however, they made no attempt to measure any changes in the learners' anxiety levels. Carey et al. (2015) explores whether poor mathematics performance elicits mathematics anxiety, or whether mathematics anxiety reduces future mathematical performance, and concludes that the evidence is in conflict. They conclude that what evidence there is may indicate a bi-directional relationship between learners' levels of mathematics anxiety and their mathematics performance; they suggest that mathematics anxiety and performance influence one another in a vicious cycle.

The one-way assumption that underachievement causes the anxiety has been refuted by, amongst others, Ashcraft and Ridley (2005), who argue that mathematics anxiety interferes with the cognitive processes necessary for mathematical problem-solving. Beilock and Carr (2005) discuss the idea of otherwise competent individuals 'choking under pressure', and Young, Wu and Menon (2012) found the negative effects experienced by those with mathematics anxiety were specific to mathematics and unrelated to general anxiety, intelligence, working memory, or reading ability. All these researchers were convinced by their studies that it was the mathematics anxiety that caused underachievement, not underachievement that caused the anxiety.

Dowker et al. (2016) are prepared to say, as a result of their review of published research in the field of mathematics anxiety, that a substantial number of children and adults have mathematics anxiety. They see mathematics anxiety as likely to "*severely disrupt mathematical learning and performance, both by causing avoidance of mathematical activities and by overloading and disrupting working memory during mathematical tasks*" (Dowker et al., 2016, p. 1). They also feel that there needs to be more research into the role, in the development of mathematics anxiety, of pressure from parents and teachers to achieve well in school, and they question whether, and at what point, an increasing emphasis on mathematical achievement may become counterproductive and decrease performance, because of the anxiety induced.

It seems to us, from our research, that where the anxiety is recognised, and addressed, the learners are set free to learn mathematics. Learners will always need support, especially if they have already experienced the disabling effects of anxiety, but the more they are helped to develop resilience the more they can manage their anxiety and prevent it stopping them from learning the mathematics they want to learn. To echo Ben in Chapter 9, mathematically resilient learners struggle with mathematics but not against themselves.

Working with feelings

The evidence presented in this book, and in publications about the construct of mathematical resilience dating from 2010 onwards, demonstrates that the approach outlined in this book is effective and profound for those learners and facilitators of learning who are prepared to try it.

For too long those seriously affected by mathematics anxiety have been asked, or have even asked themselves, "what is wrong with you?". They have been led to believe that they are at fault because they should be able to study mathematics and gain qualifications. However, maybe a better question would be to ask, "what happened to you?" as this begins to explain that any fault does not lie with the learner, it is more likely that they have suffered some adverse prior event or events and have not yet been able to find the kinds of support that they need if they are to learn mathematics.

The support some learners need may lie in applying psychological safety approaches, rather than in more or better explanations of mathematical ideas. Learners can be helped to recognise when they may be beginning to panic, and for example, using a break or some deep breathing, they can become able to return themselves to a place where they can learn. An anxious learner might have learnt that feeling that they have to struggle means they 'can't do it', because learning mathematics is easy for the few who 'can do it'; such a learner needs support to overcome this preconception. The language of the Growth Zone Model was developed to help here. Learners must see struggle as a sign that they are learning. If they are not feeling somewhat challenged, although not too challenged, and that they could and probably will make mistakes, then they are not learning. Struggle does not imply they 'can't do it', it implies that they 'can't do it YET', but with perseverance from them and support from others, they can and will reach the understanding they are seeking.

The Growth Zone Model has proved invaluable in helping learners explain to themselves, and others, how they are feeling as they learn mathematics. It gives learners opportunities to learn how to master their emotions when they experience getting stuck, and to develop an optimistic approach to mathematics. Seligman, as is explained in Chapter 16, defined an optimistic approach as seeing failure as temporary, specific to the situation, and external. The optimistic learner does not see failure as being about them or their characteristics (Seligman, 2006), and is therefore able and willing to apply flexible strategies to solve problems.

The notion of stepping out of the comfort zone, and into the growth zone, involves facing fear and entering what Brown (2021) terms a brave space, where the learner will be supported in taking risks that will be of benefit to themselves. The use of the Growth Zone Model involves both learners and facilitators of learning becoming able to pay attention to the body signals that tell the learners which zone they are in. For example, everyone gets stuck sometimes, and can feel frustrated or even debilitated. However, if the learner notices those feelings, and is able to remind themselves that being stuck is an honourable and useful state, because that is when it is possible to learn about mathematics, about mathematical thinking, and about oneself (Mason, 2015), then being stuck can be seen as stimulating.

When some teachers of mathematics see their learners beginning to display anxiety, they can become over-protective and try to make tasks easy for the learners (Wigley, 1992), in ways which avoids learners struggling or experiencing any challenge. Teachers can see their job as helping their learners avoid any discomfort when learning (Stigler and Hiebert, 2009), which means the positive feelings of successfully overcoming an obstacle are also removed. Sometimes, helping learners to develop their mathematical resilience requires teachers to 'be less helpful' and allow the learners time and space to meet the challenges for themselves. However, it can be equally damaging when teachers are too ready to issue mathematical challenges, without first helping learners understand that they could find themself in the red zone and helping them consider what options they have to remove themselves from that unpleasant space and start learning again.

As mathematics can be difficult and challenging to understand and use, the notion of entering a brave space where learners can meet the challenge of learning mathematics is important. Brave spaces (Brown, 2021) are places where people are invited to say what they think respectfully, without fear of humiliation or similar psychological threats. Once ideas are expressed and made public, they can be discussed and questioned constructively. For some teachers, brave spaces have been facilitated by creating spaces for rough work in the classroom using paper tablecloths or personal whiteboards. Other teachers have focused on creating a brave space in which learners can make, and can learn from, mistakes. Many entrepreneurs report that making mistakes is a very powerful way to learn, and Moser (2011) has reported that when making mistakes the connections within people's brains are strengthened and the power of the brain grows. This is not to say that all learners learn from all mistakes; learning takes place in situations where the feedback a learner receives from their teacher, from their peers or provides for themselves by comparing ideas with others, promotes learning (Black et al., 2003). This might include trying something out using Logo or geometry software or testing out ideas with peers or on a whiteboard.

Throughout this book, the various authors have considered how learners' emotions and feelings promote or compromise their learning. As has been stressed, there are principles that can be seen in action in all resilience-developing environments. In such environments, the teacher or facilitator of learning is above all interested in the mathematical wellbeing of their students. The teacher listens to what the learners say and what they do not say, what they have understood and what they, as yet, do not know. They include everyone and are interested in how each and every learner is progressing. They promote a brave space where learners can accept a challenge, and can accept support in meeting that challenge, where risks can be taken, and mistakes made and learnt from, and where all learners are actively enabled to learn mathematics.

Moving forward

What can we expect to be achieved over the next 10 years? We wish to work towards all learners and supporters of mathematical learning being aware that mathematics anxiety exists, it is real, and it is disabling. We would also like all of them to know that learning mathematics does not need to be accompanied by growing levels of anxiety. Where teachers understand how to work for mathematical resilience and fully take on that role, anxiety will not be generated, and any existing anxiety can be mitigated. However, despite the confirmation of more than sixty years of research, there will be some who choose to continue to teach without regard for their learners' wellbeing. Those who choose to ignore these messages will continue to cause harm to their learners, in the form of mathematics anxiety, and harm to society, as their learners will continue to avoid engaging in mathematics.

> Sarah was a maths student who needed to resit her GCSE examination. We brought out the Smarties to make pie charts to represent the colour distribution in each box. Partway through the activity, Sarah became very excited and called out "now I get it".

Mathematical thinking should be thought of as being intrinsic to every human being. The core message of Boaler (2015) and Dweck (2000) is true; everyone can do mathematics. However, unless the prior adverse experiences of some learners are addressed, many learners will not feel safe enough to engage with mathematical learning. Where all learners of mathematics know they are capable of learning, if they are able to access the support they need to keep themselves safe enough to emerge from their comfort zone, and creep into their growth zone where they feel able to take risks and learn from mistakes and growth-promoting feedback, then the conversation will change – from a culture of 'I can't do maths' to 'I can and will learn maths!'.

Actions to progress mathematical resilience

If the principles of mathematical resilience are to pervade all environments where mathematics is learned, then everyone involved must play their part:

Teachers can – use the many suggestions in this book to transform the learning in their classrooms and truly take on the responsibility for the mathematical well-being of their learners.

Support staff and adults working in schools can – first check, and if necessary, address, the mathematics anxiety that may have been generated through their own adverse prior experiences. They can join the teachers in promoting a language and environment that truly reflects a belief in the growth mindset, that everyone can learn more mathematics. Also, as mathematics anxiety is known to be catching (Beilock et al., 2010), making sure these adults do not pass on mathematics anxiety to the learners is important.

Learners can – be helped to develop mathematical resilience using the ideas in this book. Mathematically resilient learners can safeguard their own mathematical well-being, and they know for certain that they are not the problem. Developing this way of thinking will help them to be able to learn in school and beyond.

Parents can – understand that their main role is to keep their children safe and not contribute to their child's mathematics anxiety. There is help to do this in Chapter 10, and Rosemary Russell's book, "How to help your child do maths even if you don't", is written for all parents and contains help to encourage and support children.

Teacher-training providers can – make sure that all teachers are aware of the prevalence of mathematics anxiety and are introduced to the principles of, and tools to, develop mathematical resilience. Again, many trainee teachers are likely to have existing mathematics anxiety; addressing this anxiety whilst training would mean that each trainee leaves with an understanding that they can learn mathematics, and that they can help their learners to learn mathematics without anxiety.

Government can - recognise that 90% of working people need to work with data, but 30% of the population (or more) are mathematics anxious and will avoid anything to do with mathematics if they can. It is incumbent upon government to take this matter seriously. Taking steps to allow and to actively encourage all teachers to promote mathematical resilience is likely to mean a more numerate and much less anxious population.

There is already an international mathematical resilience network, which has members in more than ten countries across the globe and is currently growing organically. Its members are able and willing to work with teachers in any sector of education, learners, parents, teacher trainers and governments in their countries, to promote the widespread development of mathematical resilience. They have already organised practitioner/researcher conferences, support for teachers undertaking action-research, proposals for policy briefs and help for teachers in understanding how to help learners stop being fearful of getting stuck. The network also works, together with other organisations, to spread understanding of mathematics anxiety and mathematical resilience as widely as possible. In England, the network is aligned with the Maths Anxiety Trust and the Dyscalculia Network; members have provided talks and podcasts for the NCETM and have worked with ETF and NCETM to provide on-line training sessions for teachers in schools and further education. Please join us in this vital work.

Conclusion

As we draw this book to a close, we invite you to reflect on your own practice and that of your colleagues. Perhaps you could consider how you respond when learners make mistakes. Could you help your learners take the risk and understand how they can use mistakes to grow their learning? Perhaps you could consider how to help your learners take on a growth theory of learning and understand that everyone can progress their learning with the right support? Above all, as a teacher of mathematics we would ask you to consider the wellbeing of your students in all you do.

Resilience can be defined as the capacity of a dynamic, living system to endure or recover from a major disturbance and continue to develop in healthy ways (Masten et al., 2011). Southwick and Charney regard those that contributed to their understanding of resilience in the same way we think of the many people who have contributed both to this book and to our understanding of the need for mathematical resilience. We include all, from the youngest learner to the teachers and collaborators when we echo:

> They have become our role models just as we hope they will become yours. They have taught us to look for light at the end of the tunnel; to view adversity as an opportunity for growth and wisdom; to live by our own highest moral and ethical standards; to foster strong personal relationships in which we both give and receive support; to rigorously train our physical, emotional, cognitive, and spiritual selves; and to assume responsibility for our own growth and resilience. They have shown us how to seek the very best in ourselves, and they have taught us that we are each far stronger and more resilient than we think. (Southwick and Charney, 2023, p. viii)

Dear Reader, the rest of this chapter and the future of mathematical resilience is yours to write.

References

Ashcraft, M., and Krause, J. (2007). Working memory, math performance, and math anxiety. *Psychonomic Bulletin & Review*, 14, 243–248. 10.3758/BF03194059

Ashcraft, M., and Ridley, K. (2005). Math anxiety and its cognitive consequences: A tutorial review. *Current Directions in Psychological Science*, 14(5), 181–185.

Beilock, S., and Carr, T. H. (2005). When high-powered people fail: Working memory and "choking under pressure" in math. *Psychological Science*, 16(2), 101–105.

Beilock, S., Gunderson, E., Ramirez, G., and Levine, S. (2010). Female teachers' math anxiety affects girls' math achievement. *Psychological and Cognitive Science*, 107(5), 1860–1863. 10.1073/pnas.0910967107

Black, P., Harrison, C., Lee, C., Marshall, B., and William, D. (2003). *Assessment for learning – Putting it into practice*. Maidenhead, UK: Open University Press.

Boaler, J. (2015). *Mathematical mindsets*. Hoboken NJ, USA: Jossey-Bass, ISBN-13 978-0470894521

Brown, B. (2021). *Atlas of the heart: Mapping meaningful connection and the language of human experience*. London: Vermillion.

Carey, E., Hill, F., Devine, A., and Szücs, D. (2015). The chicken or the egg? The direction of the relationship between mathematics anxiety and mathematics performance. *Frontiers in Psychology*, 6, 1987. 10.3389/fpsyg.2015.01987

Cockcroft, W. (1982). *Mathematics counts*. London: Her Majesty's Stationery Office https://education-uk.org/documents/cockcroft/cockcroft1982.html

Dowker, A., Sakar, A., and Looi, C. (2016). Mathematics anxiety: What have we learned in 60 years? Review article. *Front. Psychol.*, 7. 10.3389/fpsyg.2016.00508

Dreger, R., and Aiken, L. (1957). The identification of number anxiety in a college population. *Journal of Educational Psychology*, 48(6), 344–351. 10.1037/h0045894

Dweck, C. (2000). *Self-theories: Their role in motivation, personality, and development*. Philadelphia, USA: Psychology Press.

EEF (2019). Press release: New EEF trial results: +3 months' boost for primary pupils' maths results. https://educationendowmentfoundation.org.uk/news/new-eef-trial-3-months-boost-maths-results-from-improving-working-memory

Johnston-Wilder, S., Lee, C., and Mackrell, K. (2021). Addressing mathematics anxiety through developing resilience: Building on self-determination theory. *Creative Education*, 12(9), 2098–2115.

Johnston-Wilder, S., Pardoe, S., Almehrz, H., Evans, B., Marsh, J., and Richards, S. (2016). Developing teaching for mathematical resilience in further education. In 9th International Conference of Education, Research and Innovation, ICERI 2016, Seville (SPAIN), 14-16 November 2016.

Lee, C., and Morgan, J. (2023). Remembering learning mathematics – We can run but we can't hide. *Teacher Development* [Early Access].

Lugalia, M. (2015). PhD thesis, Pupils learning algebra with ICT in Key Stage 3 mathematics classrooms University of Warwick

Mason J. (2015). On being stuck on a mathematical problem: What does it mean to have something come to mind? *LUMAT*, 3(1), 101–121.

Masten, A. (2011). Resilience in children threatened by extreme adversity: Frameworks for research, practice, and translational synergy. *Development and Psychopathology*, 23, 493–506. 10.1017/S0954579411000198

Moser J., Schroder H., Heeter C., Moran T., and Lee Y. (2011). Mind your errors: Evidence for a neural mechanism linking growth mind-set to adaptive post-error adjustments. *Psychological Science*, 22(12), 1484–1489. 10.1177/0956797611419520

Seligman, M. (2006). *Learned optimism: How to change your mind and your life*. New York USA: Vintage Books.
Southwick, S. M., Charney, D. S., and DePierro, J. M. (2023). *Resilience: The science of mastering life's greatest challenges* (3rd Edition). Cambridge: Cambridge University Press.
Stigler, J., and Hiebert, J. (2009). *The teaching gap*. NY: Free Press.
Wigley, A. (1992). Models for teaching mathematics. *Mathematics Teaching*, MT141.
Young, C. B., Wu, S. S., and Menon, V. (2012). The neurodevelopmental basis of math anxiety. *Psychological Science*, 23(5), 492–501. 10.1177/0956797611429134

Index

academic resilience 227, 262
accessible tasks 55, 265, 266
action research 5, 26, 34, 36, 127, 128, 143
active listening 52, 53, 79, 80, 83, 265
adult learners 103, 105, 166–169, 171–177
adults learning 174, 176
affect 14, 67, 72, 73, 92, 93, 117, 118, 135, 150, 199, 205, 210, 231, 233
affordances 172
Aiken, L. 26
Ali, M. 14
Alkan, V. 203, 229
amber zone 43, 85, 86, 236
Anderson, R. 151
andragogy 167, 175–177
Andrew, A. 149
anxiety 2–4, 16, 21, 23, 29–31, 46, 128, 130, 152, 199, 201, 207, 251, 283, 284, 286; feelings of 3, 30, 36, 42, 45, 46, 281; zone 104, 106–108, 111, 131, 151–153
Apostolidu, M. 34–35
Argentina 193, 207, 208, 220, 234
Ashcraft, M. H. 194, 279, 283
Askew, M. 257
assessment 16, 72, 73, 135, 188, 227, 251; system 189, 249–251
attitudes 21, 22, 25, 29, 31, 67, 115, 135, 160, 163, 199, 201, 203, 221, 224
Australia 4, 5, 261
authentic tasks 265

autonomy 15–17, 104, 105, 107, 110, 111, 119, 121, 123, 124, 175, 176, 272
Ávila-Toscano, J. H. 199
awareness 103, 104, 106, 107, 116, 123, 124, 222–225, 253, 262, 263, 273

Bandura, A. 173
Barnes, M. 265
Bartholomew, H. 254
Barton, C. 257
Baylan, H. N. 203
Beilock, S. 283
Benson, H. 47
Betz, N. 169
Biesta, G. 181
Boaler, J. 151, 154, 240, 254, 257, 286
bonus points initiative 197
brave spaces 16, 118, 285, 286
Brazil 4, 128, 193–195, 220, 222
Brown, A. 210
Brown, B. 285
Bruner, J. 41
Buckingham, D. 184
building rapport 79–81, 235
building resilience 34, 67, 72, 235, 236, 239, 272, 273
Burton, L. 100

Carey, E. 28, 169, 283
Carmo, J. 61, 194, 221

Carr, T. H. 283
Cates, G. 170
challenge 2, 3, 11, 42, 45, 46, 52, 53, 64, 104, 108, 109, 111, 112, 118, 175, 236, 263, 279, 285
challenging environments 40, 53, 279
Children's Mathematics Anxiety Scale-UK (CMAS-UK) 29, 29, 30, 35
Chisholm, C. 138
classroom accommodations 239
climate change 2, 22, 212
coaching 5, 75–78, 80–87, 89–90; approach 75–77, 79–81, 83–85, 87–90; skills 75, 79, 83, 89, 90; style questions 80; training 88
Cockroft report 281, 282
collaborative learning 121, 176
collaborative tasks 100
Colombia 193, 198–201, 220, 225–228
Colombini, F. 220
comfort zone 41–43, 70, 71, 104–106, 108, 109, 111, 151–154
communicating ideas, parents 149–164
connectionist teachers 257
Crescenti, E. 221
Csikszentmihalyi, I. 262
Csikszentmihalyi, M. 262

Day, C. 270
Deci, E. L. 15, 43
Dehaene, S. 236
De Souza Domingues, M. 221
Dieckmann, J. 151
direct intervention resources 31
Dowker, A. 22, 169, 283
Dreger, R. 26
Drury, H. 174
Durán-Sánchez, H. V. 199
Dweck, C. S. 12, 22, 132, 162, 170, 173, 286

E2L professional learning programmes (PLPs) 270
economic prosperity 181, 212
education 116, 117, 166–169, 187, 188, 196–199, 204, 226, 288
educational design-based research 135

educational system 5, 14, 202
educators 33, 127, 135, 143, 173, 212, 225
Egan, G. 80, 83, 90
Egan model 83, 84, 87, 88
Eilks, I. 233
emotional climate 239
emotional disengagement 61
emotional security 265, 267, 268, 272
emotions 41, 43, 44, 46, 47, 103, 104, 106, 107, 115, 116, 120, 124, 136, 235

Faure, E. 186, 187
FE colleges 167, 168, 176, 177
feedback 68, 69, 72, 73, 79, 81, 82, 89, 90, 154, 271, 272, 285, 286
feelings 41–43, 45–47, 80, 81, 89, 106, 107, 118, 135, 136, 141, 142, 153, 156, 157, 284
Felux, C. 117
Fernández-Cézar, R. 199
Figiacone, S. 235
Finkel, A. 136
formative assessment 52, 72, 81, 188, 253
Foster, D. 151, 254
Francisco, J. 100

GCSE examinations 82, 88, 250, 254, 286
global developments 220, 221, 223, 225, 227, 229, 233, 241
global problem 193, 197, 199, 201, 203, 205, 207, 209, 211
Goleman, D. 116
Goodall, J. 161
green zone 41, 43, 51, 54, 84, 85, 89, 136, 151
Gris, G. 61, 194
group composition 263, 264, 270
group work 65, 92–95, 100, 101, 266, 268
growth mindset 10–13, 17, 43, 128, 130, 162, 171, 173, 210, 211, 240, 241
growth theory 12, 55, 104, 123, 253, 288
growth zone 42, 43, 55, 71, 104–109, 119, 151–154
growth zone model (GZM) 41–44, 51, 71, 103, 104, 106, 107, 119, 152, 153, 160, 162, 222, 233, 235, 236, 239, 284
Gu, Q. 270

Haase, V. G. 221
hand model 34, 41, 44–47, 85, 128, 222, 233; of brain 44–47
Harris, A. 161
Hattie, J. 13, 209
helplessness 46, 281
Henao, J. 225
Hiebert, J. 19
homework 106, 111, 141, 143, 204, 221
Howie, S. 206
Hunt, T. 25, 28, 32
Hyman, I. 14

identity 79, 180, 184, 185, 265
inattentiveness 31
inclusive environment 11, 52, 98, 280
individual learners 26, 30, 41, 75, 103–105, 107, 109, 207, 220
intellectual abilities 12
International Coach Federation 75
interventions 12, 13, 128, 130, 131, 140, 170–172, 200–202, 221, 222, 227; design 131, 140, 233
intrinsic motivation 43, 200
Ireland 4, 193, 196–198, 220, 222–225
Irish Learning Support Association (ILSA) 223

Johnston-Wilder, S. 34, 35, 109, 152, 156, 161, 162, 169, 170, 232, 235

Kenya 193, 204, 205, 220, 231
Kidman, G. 170
Kirk, E. P. 194
Knowles, M. 167
Krause, J. 279
Krause, M. 233

ladder model 41, 49, 50, 53, 132, 233, 236, 239
LaMar, T. 154
Lau, N. 34
Lave, J. 184
learners 13, 15, 16, 31, 48, 54, 77, 81, 84, 93, 105, 107, 109, 118, 130–132, 135, 253, 255; autonomy 103, 112, 268; groups of 25, 27, 28, 30, 88, 93, 95, 97, 101, 253, 261; reactions 171–172; wellbeing of 258, 278
learning mathematics 1, 3, 19, 34, 40, 41, 48, 61, 75, 131, 173, 174, 177, 182, 199, 231, 261, 273, 278, 279
Lee, C. 94, 152, 156, 162, 170, 235
Leshin, M. 154
lifelong learning 180, 181, 183–189
Lilburn, P. 271
Liljedahl, P. 93
Lotan, R. 97
Lucas, M. 149
Luttenberger, S. 22

Macomber, G. 13
Maloney, E. A. 25, 161
Mason, J. 100
MateTrucos 237
mathematical anxiety 21, 22, 25, 26, 31, 150–152, 166, 169, 195, 196, 198, 201, 203, 207, 209–211, 221, 222, 224, 229, 273, 280, 283; addressing 30, 169, 222, 223; development of 4, 22, 196, 203, 251, 284; effects of 193, 220, 252, 278, 280; experiencing 25, 35, 150, 194, 224, 231; extent of 26, 168, 177, 204; impact of 207, 223, 281; levels of 21, 28, 35, 199, 200, 202, 203, 205, 210, 222, 283; presentation 170; prevalence of 22, 23, 36, 194, 196, 200, 205, 210, 222, 287; reducing 26, 34, 199, 234, 274; scales 26, 27, 30, 31, 169; self-report scales for 26
mathematical ideas 15, 17, 19, 49, 50, 54, 55, 175, 176, 182, 187, 188, 262, 269, 270
mathematical learning 2, 3, 12–18, 176, 177, 182, 183, 249, 286
mathematical resilience grid 136
Mathematical Resilience Scale (MRS) 33, 231
mathematical resilience toolkit 34, 132, 136, 142, 222, 231
mathematical skills 114, 181, 187, 196, 198, 201, 281
mathematical thinking 2, 9, 18, 22, 46, 64, 99, 173, 185, 285, 286
mathematical wellbeing 187, 251, 252, 255, 286

Mathematics Anxiety Rating Scale 26
mathematics classrooms 61, 62, 65, 92, 93, 97, 100, 109, 118, 120, 205, 211
mathematics education 25, 29, 33, 174, 196, 209, 240
mathematics learners 29, 36, 41, 47, 48, 104, 115, 117, 124, 272, 273
mathematics problems 30, 32, 83, 84, 136, 169, 175
Mathematical Resilience Network 224, 233
mathematics teachers 83, 88, 95, 135, 221, 226, 252, 255, 256, 258, 285, 288
McKay, C. 13
Mendes, A. 221
Menon, V. 283
micro-mindfulness 48; techniques 48
mindfulness training 48
mindset barrier 172
Morgan, J. 171
Moser J. 285
motivation 1, 16, 32, 33, 77, 79, 128, 130, 131, 167, 168, 196, 200
Mullis, I. 206
Muniz, M. 221
Murcia, M. 225

Nardi, E. 109
National Numeracy 256
negative emotions 81, 116, 157, 268
Nelson, J. 149
NeuroEduca 208, 235
non-judgemental approach 81
number anxiety 26, 283
Núñez-Peña, M. 30

on-line games 183
optimistic learners 261, 265, 285
optimistic mathematical problem solving 262, 263
optimistic problem solvers 5, 262, 263
Özbey, N. 202

Palombarini, L. 61, 194
panic zone 71, 236
Para, T. 34
parental engagement 149, 160, 161

participatory action research 128, 222
path smoothing 19, 50, 63–65
Peck, R. 265
peer coaching 65, 76
Pergher, N. 220
personal value 240
person-centered strategies 117
Petronzi, D. 32, 35
pre-service teachers 207, 233, 234
problem-solving 175, 176, 200, 211
productive pedagogical struggle 272
productive struggle 63, 65, 268, 272
professional learning: programmes 270
professional training 176, 177, 228
psychoeducation 235, 237
psychological education 131, 132
psychological safety 4, 9, 10, 13, 15, 16, 116, 118, 119, 124, 128, 130, 131
psychological variables 26

quality first teaching (QFT) 67, 68
quiet disaffection 61, 92, 109

Ramirez, G. 194
Ramnarain, U. 167
recruit help 163
red zone 42, 43, 51, 53, 55, 84, 86, 119, 120, 152, 154, 156, 162, 163
relaxation response 41, 47, 48, 124, 128, 134, 140, 222, 233
resilience-building 268, 273
resilience-building problem-solving tasks 261, 263, 265, 267, 269, 271, 273
resilience-focused teacher 72
resilient identity 180, 183–188
resilient learners 1, 2, 43, 50, 55, 68–70, 182, 183, 185, 189
resilient mathematical learning environment 63, 65, 69, 73
resilient teachers 261, 269
resourcing 69, 95
Rhymer, K. 170
Ridley, K. 283
Rogers, C. 76
Russell, R. 161
Ryan, M. 198, 222

Ryan, R. M. 15, 43

Santiago-Carrillo, M. 227
Sava, F. 14
Schoenfeld, A. 173, 174
secondary schools 2, 9, 36, 103, 121, 140, 196, 250, 252, 254, 279
self-coaching 76, 89
self-determination theory 15
self-efficacy 1, 29, 32, 33, 105, 108, 114, 132, 135, 194, 198, 224, 228
self-report measures 30, 31
self-report scales 29, 30, 33, 36, 37
Seligman, M. E. P. 262
Shoji, F. 220
Siegel, D. 44, 47
silent coaching 87
Sims, D. 149
Sithole, A. 205, 206
Smith, M. 167
Snook, P. 14
Snowdy, P. 117
social justice 181, 186, 188, 249, 257
South Africa 4, 193, 204–207, 220, 233, 234
Southwick, S. M. 279
Split, J. L. 13
Stacey, K. 100
Starr, J. 79
Steward, D. 98
Steward, S. 109
Stigler, J. W. 19
Stogsdill, G. 136
Sullivan, P. 271
support 11, 16, 20, 41, 43, 46, 53, 54, 65, 90, 93, 115, 174, 175, 180, 183, 241, 279, 280, 284
supportive environment 1, 21, 53, 117
supportive learning environment 2, 56
support staff 5, 109, 114–119, 121, 123, 125
support workers 116–121, 123–125
Swan, M. 100
systemic changes 249, 258

talking mathematics 97

teacher-led mathematical resilience research 131, 135, 141, 143
teachers 14, 30, 32, 51, 52, 62, 64, 72, 92, 127, 176, 177, 251–253, 264, 270–272, 285; training 176, 226, 239, 251
teaching 3–5, 13, 22, 67, 72, 73, 92, 93, 150, 211, 250, 255, 261
teaching mathematics 3, 4, 64, 175–177, 231, 251, 252, 255, 257, 279, 280, 282
team games 97, 98
tools 40, 41, 47, 48, 51, 55, 56, 75, 76, 118, 128, 131, 132, 135, 140
Turkey 193, 202, 203, 220, 228, 229

UK citizenship test 251
USA 34, 193, 209, 211, 220, 240
Uusimaki, L. 170

value 10, 11, 15, 17–19, 30, 43, 53, 55, 79, 130, 143, 175, 240
valuing mathematics 162

Ward-Penny, R. 94
web-based mathematical resilience interventions 140
wellbeing 186, 187, 251, 252, 257, 258, 278, 280, 286, 288
Wenger, E. 184
Whitmore, J. 136
Wigley, A. 50, 64
Wiliam, D. 250, 254
William, D. 232
working groups 123, 150
working memory 70, 117, 118, 132, 194, 283
Wu, S. S. 283

Yates, G. 13
Yeager, D. 12
Yenilmez, K. 202
Young, C. B. 283

Zambia 193, 205, 233
zone of proximal development (ZPD) 44

For Product Safety Concerns and Information please contact our EU representative GPSR@taylorandfrancis.com
Taylor & Francis Verlag GmbH, Kaufingerstraße 24, 80331 München, Germany

www.ingramcontent.com/pod-product-compliance
Lightning Source LLC
Chambersburg PA
CBHW060257240426
43661CB00060B/2820